AF527892

INDIAN MEDICINAL PLANTS

INDIAN MEDICINAL PLANTS

Dr. M.P. Singh, Ph.D. FMA
Head of the Deptt Forest Sciences
Birsa Agricultural University
Ranchi - 834006 (Jharkhand)

Dr. Soma Dey
Department of Forest Sciences
Birsa Agricultural University
Ranchi-834006 (Jharkhand)

SATISH SERIAL PUBLISHING HOUSE

403, Express Tower, Commercial Complex, Azadpur, Delhi - 110033 (INDIA)
Phone : 011-27672469, 27672852 Fax : 91-11-27672046
E-mail : ssp@ndf.vsnl.net.in, hkjain@satishserial.com
Website : www.satishserial.com

Published by :

SATISH SERIAL PUBLISHING HOUSE
403, Express Tower, Commercial Complex, Azadpur, Delhi - 110033 (INDIA)
Phone : 011-27672469, 27672852 Fax : 91-11-27672046
E-mail : ssp@ndf.vsnl.net.in, hkjain@satishserial.com

ISBN 81-89304-02-X

Reprinted : 2015

Printed at

SALASAR IMAGING SYSTEMS, DELHI-35

CONTENTS

Foreword

In spite of the tremendous progress in the development of modern medicine. Plants continue to be an important source of drugs throughout the world, particularly in the developing countries. The last two decades registered significant in the use of higher plants for health care and treatment of several diseases for which modern medicine was not found effective. Because of this, demand for plant drugs enhance enormously. This has necessitated in the large scale cultivation of number of medicinal plants in the recent year in order to maintain sustainable supply to the pharmaceutical industries. As a results, commercial cultivation of large number of medicinal plants has been taken up in the recent years in several countries.

During the thousand of years of early human existence many natural material were identified for combating human ailments either by instinct or intuition or trial and error. The earliest mention of the medicinal use of plants has been found in "Rig veda" which were written between 4000 and 1600 B.C. in the Atharva veda, II find the more varied use of drugs. It is in the "Ayurveda", which is considered as an "Upaveda", (600 B.C.) which lists a total number of 341 plants and plants products for use in health management. "Susruta samhita" also dealth with plants related to medicine. The relationship between man and tree has existed since time immemorial. In ancient India, national prosperity was measured by the abundance of trees. The usefulness of trees and their protection was emphasized in ancient Indian literature like the Puranas and the "Bhagvad Gita". Trees like Bar, Pipal, Arjun Amla, Ber, Jhand, Gullar, Bil, Tulsi, Coconut, Supari etc. were worshipped as Gods and the ancients believed that trees, besides giving solace to the mind, also helped to develop the thinking power of an individual.

The present book in Indian Medicinal Plants, will fill a great want felt both the Research students and the general practitioner. The various sections that have been made out indicate not only the importance of the subject but the unknown and useful information that lies scattered. The authors who has already earned the best of reputation for the writing of such therapeutic works.

I have great pleasure in recommending this book to the notice of the medicinal profession interested in ancient and modern medical sciences. This book stands as the link between the two stores of knowledge and I am sure it overcomes many of the difficulties in a commendable way.

Montek Singh Ahlubalia
Vice Chairman
Planning Commission Govt. of India
New Delhi

New Delhi

November, 25th 2004

Preface

We have attempted to organize this book in such a way that it will be useful to everyone interested in the world of plants particularly those increasing number of people who are concerned with medicinal plants. Many of these valuable and useful group of plants are threatened with extinction, because of over exploitation and habitat destruction. My endeavour here is to make people aware of the potential of medicinal plants from all angles so that these life saving plants can be protected from total destruction. There is a need to identify the natural plant wealth and commercially viable species in each agro-climatic zones; promote industries to produce phyto-chemicals; standardize the agro-techniques for cultivation and production of these plants particularly in forests, waste lands and orchards as an additional income generating activity; and also strengthen the research and developmental activity for evolving new herbal products and their production technology. In this book more than 900 genera are discussed covering nearly 1696 species. Almost all commercially important medicinal plants cultivated in, different part of the world has found place in this publication. An integrated and comprehensive account of all related aspects of the subject such as nomenclature, description, distribution, cultivation, processing, physical and chemical properties of large selection of plants have been incorporated.

However, we would like to mention here that many of the plants are poisonous in the unprocessed form, and self medication with wild plants is not at advisable. Some of the information furnished here, are yet to be proved scientifically. For treating any ailments, the plant drug should be administered only with the prescription of a registered medical practitioner. We hope that the information furnished in this book will make it easier to identify the medicinal plants and make their growing more through and satisfying.

It is our pleasure to acknowledge the help which we received from many publications for completing the book. Much of the information was gleaned from the libraries of Botanical Survey of India. However, Forest Research Institute, Dehradun, Indian Agricultural Research Institute, Delhi, Central Library, Patna University, Patna and Birsa Agricultural University, Ranchi (Jharkhand State).

Our greatest debt of gratitude is to Professor R.N. Trivedi for encouragement, expert suggestions and for living with this book for a great many years. We are indebted to Dr. B.S. Singh, Mr. Ifthekar Alam, Miss Rashmi Renuka Toppo, Miss Leena Ekka, for their sincere and wilting help in correcting the typed manuscripts.

We thank Mr. H. K. Jain of Satish Serial Publication House, New Delhi. for taking keen interest in making the publication as good as possible.

M. P. Singh
Soma Dey

INTRODUCTION

The prehistoric men and women, as they advanced towards civilization in the lap of time, depended on plants. Like the wild animals living forests human beings too in the days of yore used plants intuitively for food, shelter and even curing their many a maladv and thereby kept their health in perfect state of fitness and lived a long life unlike the human folk of the present-day trouble ridden world. With the progress of civilization from the dim past, even before the present century, various charms of plants played a great role in human life, society and activities. No authentic record, however, of any kind is available of the pre-Vedic period in this country and the period before the early part of the Christian era in other countries.

In Egypt, India and China mines of information are available in the old literature, folklores, mythological stories, epic poems, medicinal treatises, housands of years old manuscripts, copper plates, palm leaves and similar other records, many of which are kept preserved even to the present day. A vast field of knowledge on this subject still remains terra-incognita in the innumerable Tibetan manuscripts and writings on herbals, preserved in the archives or the monasteries scattered all over Sikkim, Bhutan and Tibet.

Application of various recipes of herbs and herbal charms, which are most interesting and in some cases highly effective for curing many diseases, seemed to have been in practice accompanied with chanting of spells as early as 4000 B.C. The earliest reference to the use of medicinal herbs as a cure for diseases and as charms is found in the manuscript of "Eber Papyrus" which dates to about 16th century B.C. The mysticism of the medicinal plants is also mentioned in the Atharva and Rig Vedas. Perhaps the oldest repositories of human knowledge were written between 400 and 1600 B.C., by Charaka and Susruta in the treatment of maladies in human beings and animals. These are monumental work and outstanding contributions towards our knowledge of science and art of healing.

Later, during the Buddhist period, considereble progress was made and medicinal plants were cultivated under the direction of highly qualified specialists.

During the centuries, that have gone by, the materia medica of the indigenous systems of medicine has become extensive and heterogeneous. Out of 2000 items recorded in Indian medicinal literature, less then 200 are of mineral and animal origin; the rest are derived from vegetable sources The vegetable materia medica has been built up in the course of centuries and every region of India has contributed to its development. It is well known that due to the impact of other systems of medicine, particularly due to the rapid progress and spread of modern medicine and surgery, faith in and popularity of the herbal medicines have been gradually declining. But there is much to say in favour of the use of native medicinal herbs or the indigenous system of medicine.

The use of different parts of several medicinal plants to cure specific ailments has been in vogue from ancient times. The indigenous system of medicine namely Ayurvedic, Siddha and Unani have been in existence for several centuries. This system of medicine cater to the needs of nearly 70% of our population residing in the villages. Apart from India, these systems of medicines are prevalent in Korea, China, Singapore, Malasia West Asia and many other countries. Besides the demands made by these systems as their raw material, the demand for medicinal plants made by the modern pharmaceutical industries has also increased manifold. Thus, medicinal plants by way of export constitute a group of industrially important crops which bring appreciable income to the country.

India is a vast country where wide variations in climate, soil, altitude and latitude is available. Nature has bestowed on us a very rich botanical wealth and a large number of diverse types of plants grow wild in different parts of the country. Increased awareness about the potential of this group of interesting and useful plants has encouraged many innovative and progressive growers and entrepreneurs to take up their cultivation as a commercial purposes. Apart from health care the companies provides means to scores of people of livelihood.

Our knowledge of medicinal plants has mostly been inherited traditionally. Use of plants for curing various ailments are not confined to the doctors only but is known to several households as well. There are many interesting and sometimes astonishing things to learn from collectors of medicinal herbs. Spreading and preserving this knowledge on medicinal plants and their uses has become important for human existence. There is a growing tendency all over the world to shift from synthetic to natural based products including medicinal plants. People should also timely now to consider neglected and medicinal plants little known.

An Importance of collecting good herbarium material for taxonomic identification of the collected species must be stressed. Nation demand for conservation of all useful plant species, and also cultivation, maintenance and assessment of germplasm for future use, since among the most vulnerable plant species in India. The medicinal plants are mostly over-exploted.

The plants occupy a very significant place in modern medicine also raw material for some important drugs, although synthetic drugs and antibiotics brought about a revolution in controlling different diseases. But these synthetic drugs are out of reach to millions of people. Those who live in remote areas depend on traditional healers, whom they know and trust. Use of medicinal herbs judicious can even cure deadly diseases that are defied synthetic drugs long.

In India pharmaceutical industry is fairly advanced and sophisticated. The plant based drugs, however, have shortened the life-span of the source of material. There is continuous search for more potent and cheaper raw material to feed the industry. With concerted research and development efforts, many medicinal plants could provide raw material in abundance to the indigenous pharmacies and local herbalists. Strong linkages should be developed between medicinal plant growers, health experts and pharmaceutical industries for developing scientific basis.

PAST HISTORY OF MEDICINAL PLANTS

The thousand of years of early human existence many natural material were identified for combating human ailments either by instinct or intuition or trial and error. The earliest mention of the medicinal use of plants has been found in 'Rig Veda', which were written between 400 and 1600 B.C. In the "Atharva Veda", we find the more varied use of drugs. It is in the "Ayurveda", which is considered as an "Upa Veda", that definite properties of drugs and their uses have been given in great detail. "Chakra Samhita" is another earliest treatise on "Ayurveda" (600 BC) which lists a total of 341 plants and plant products for use in health management. "Susruta Samhita" also dealt with plants related to medicine. Subsequent authors of later treatises have extended the list of Ayurvedic single plant drugs to the six hundred species of plants. Bhiksu Atreya, a well known professor of the University of Taxila; his student named Jivaka, who later became the physician of Bimbisara of Magadha; Dhanvantari and Nagarjuna were the well known persons with an intimate knowledge of the characteristics of medicinal plants. Rouvolfia, which has now attained world-wide popularity, finds mention in ancient Hindu scriptures as well as in the monumental work of Charaka. The plant was mentioned as useful antidote for snake bites and insect stings, and has its curative effect in case of insanity. The great importance of this drug have been realised only recently in western medical therapy.

Characterisation and isolation of Active Principles:

The active principles isolation from medicinal plants and characterisation, can be traced to the beginning of 19th century. From crude drug Ma lluang (Ephedra spp) of China, ephedrine was isolated in 1887 and later introduced as drug in 1925. Likewise, from opium (Papaver somniferum), morphine was isolated in 1804 and introduced as drug in 1818. From Cinchona spp. of Peru, quinine was isolated in 1820 and introduced as drug in 1825. Ipecacuanha (Cepliaelis ipecacuanha) of Brazil, emitine was isolated in 1894 and introduced as drug in 1912.

Large number of drugs from medicinal plants were discovered and introduced in modern pharmacopoeias during 1850-1950. Some of the important crude plant drugs are belladona (Atropa belladona), cascara (Rhamnus purshina), digitalis (Digitalis purpurea),ipecae (Cepliaelis ipecacuanha), opium (Papaver somniferum), rouwolfia (Rauvolfia sepentina) and veratrum (Veratrum viride). More than seventy-five pure compounds derived from higher plants find their place in modern-medicine and some of the traditional active plant principles are atropine (Hyocyamus muticus), coedine (Papaver somniferum), digoxin (Digitalis lanata), digitoxin (Digitalis purpurea), colchicine (Colchicum autumnale), hyoscine (Datura metel), pilocarpine (Pilocarpus jnborandi), quinidine and quinine (Cinchona spp.). From 1950 onwards, there was examination of ethnotherapeutics constitute in search of new drugs from higher plants and marine flora. During this period there was outstanding contribution of fine drugs from higher plants like reserpine, deserpidine and rescinnamine from Rouvolfia serpent inn; and vinblastine and vincristine from Catharanthus roseus.

Medicinal System Transformation

Plants still constitute one of the major raw materials for drugs for treating various ailments of human-being, although there has been significant development in the field of synthetic drug chemistry and antibiotics. During the last two decades considerable changes have taken place in the medicinal system all over the world. Because of the general awareness of the widespread toxicity and harmful after-effects associated with the long use of synthetic drugs and antibiotics, the Western Society prefer the drugs from natural sources than the synthetics. In USA and UK, the plant based drugs are being used substantially. Russia and China have adopted an integrated system of allopathic, traditional and folk system of medicinc.

Widely Known Medicinal Plants for Their Curative Effect of different Diseases :

Adhatodn vasica as an expectorant and antiasthamic; Achyranthes aspera and Boerhavia diffusa as diuretic; Acorns cal.imus as tranquilizer; Aegle marmelos, Plantago ovata and Ailanthus malabarica in chronic diarrhoea and dysentry; Butea frondosa as an anthelmintic for round worms; Bacopa monnieri for memory; Boerhaavia diffusa and Curcuma longa as anti-inflamatory; Cassia fistula as catharitic; Centella asiatica for intelligence, Melia azadirachta as an antiperiodic; Terminelia arjuna as a cardiac tonic: Commiphorn mukul as an anti-rheumatic and nervous tonic; Sida rhombifolu as anabolic; Hydrocarpus weightiana for leprosy; Calicopteris floribunda as anthelmintic and laxative; Holarrhena antydisentrica and Salmalia malabricum in dysentry; Alstonia scholaris which contains the alkaloid ditamine as antiperiodic in malaria.

Plants Used in Medicine for Essential Oils

There are several plants yielding essential oils, which are used in medicine. Some of the most important essential oils of this type are eucalyptus oil from Eucalyptus spp., Japanese mint oil from Mentha arvensis; peppermint oil from Mentha piper ita, clove oil from Eugenia caryphyllata; dill oil from Anethum graveoiens and anni seed oil from Pipinella anisum.

Importance of Cultivation of Medicinal Plants Commercialy

Many of the crude drugs which are sources of medicinal preparations still come from the state of their wild growth. The natural resources how- ;o-ever large are bound to diminish. Time has therefore come to bring these)lants under plough to meet the rising demand of the resultant product. "hese plants as a group have a large potential for primary introduction of new commercial crops. There is an increasing demand for plant based drugs, and pharmaceuticals in the world market. India is among the traditional producers and exporters of several medicinal plants. Lack of basic information on different parameters of crop productivity is a limiting factor in this group of plants. There is, therefore, need for intensive agricultural studies leading to genetic improvement and cultivation methods for expansion of area under medicinal and aromtic plants.

Cultivation of Plants Commercial Useful for Modern Medicine :

More than 125 medicinal plants are used in modern medicine. Plants used in traditional systems of medicine are of over 500 different types. Most of the raw material for traditional medicine of pharmaceutical houses is collected from wild sources. Many of the medicinal plants are cultivated commercially now-a-days for extraction of some important active constituents for use in modern medicine. For steroidal hormones which are synthesized from diosgenin, hecogenin or solasodine different species of Dioscorea, Solanum and Agave are cultivated. Costus speciosus is also a source of diosgenin. Papaver somniferum is a source of different types of alkaloids, out of which morphine, codeine, papaverine and nascopine are used in medicine. Various species of Cinchona also contain large number of alkaloids in the bark, and most important are quinine and quinidine. Different species of Datura, Hyoscyamus, Atropa, Duboisia, Scopolia and Physochlaina contain tropane alkaloids which include hyoscyamine, hyoscine, atropine, and are commonly used in modern medicine. Digitalis lanaia is an important plant which contain digoxin and lanatoside-C glycocides. Rouvolfia serpentina, R. canescens and R. vomitoria contain reserpine, resinamine and deserpidine, and the species R. vomitoria also contain ajmaline. Catharanthus roseus has become important in modern medicine and contains more than 100 alkaloids of which Ajmalicine (ranbasine) is of considerable importance. Leaves of this species contain two anticancer alkaloids viz., vincristine and vinblasline. Camelia sinensis is a caffeine yielding plant, and Enjthroxylum coca yield cocaine. Cephaelis ipecacuanha and C. acuminata yield number of alkaloids of which emetine and psychotrine are most important. A drug is obtained from dried sclerotia of the fungus Claviceps purpurea grown on rye (Secale cereale). This contains more than two dozen alkaloids of which ergofamine, ergometrine and ergotoxine group of alkaloids are used in medicine. Plantago ovata seed is mostly used in medicine because of its mucilage, which is colloidal in nature and serves as a safe bulk-laxative. Glycyrrhiza glabra contain saponin-like glycosides of which glycyrrhizin is most important. From Cassia angustifolia and C. acutifolia the crude drug senna is obtained. Various species of Berberis and Mahonia yield berberine. Carica papaya yield peptolytic enzyme. Ammi majus seeds contain furanocoumarin, called xanthotoxine. Strychnos nux-vomica seeds contain strychnine alkaloids.

Medicinal Plants and Phytochemicals used in export :

India is a leading exporter of the medicinal plants in the world trade. The major export of medicinal plant parts or whole plants from India are poppy husk and seeds, opium crude, psyllium (Plantago ovata} husk and seeds, senna (Cassia angustifolia) leaves and pods, chirayata (Swertia angustifolia), galanthus (Alpinia sp.) rhizomes, Tukmaria and Zedoary (Curcuma zedoaria) roots, periwinkle (Catharanthus roseus) roots and leaves, Glycerrhiza glabra dried rhizomes, ipecae dried rhizomes and roots, Kuth roots, nux-vomica dried ripe seeds, sarasparilla, serpentina roots and some ayurvedic herbs.

Apart from the medicinal plant parts, India also exports large quantity of phyto-chemicals like beta-ionone, papin, solanesol, quinine sulphate, atropine sulphate, quinine hydrochloride, salts and derivatives of quinine, quinine alkaloids, berberine hydrochloride, emetine alkaloids, strychmine alkaloids and salts, salts and other derivatives of mix vomica, alkaloids and brucine, vegetable alkaloids, salt and other derivatives of ergot, opium alkaloids, ephedrine hydrochloride, ergot alkaloids, hormones, neosoralen, macsoralen and trimetoxy soralen. But India's share of phytochemical exports compared to the overall world trade is very meagre.

Importer of medicinal plants inlarge from India are U.S.A., Japan, Germany, U.K., France, Switzerland, South Korea and China.

ROLE OF MEDICINAL PLANTS OF FOREST ORIGIN IN HORO-PATHY

Tribal concept of health, of disease of treatment, of life and death is as varied as their culture and accordingly, the tribal society is guided by traditionally laid down customs and every member of the society is expected to conform to it. The fate of the individual and the community a. large depends on their relationship with unseen forces which intervene human affair. If men o!fend them the mystical power punish by sickness, death, or other natural calamities.

Disease is caused by bad spirits and the spirits are bad eitier because one has not properly propitiated them or because some evil person has manipulated them. Thus, to a tribal mind the real enemies of human health and prosperity are the gods and the dead. The usual theory of disea: e in tribal society is that it is caused by the breach of some taboo or by hostile spirits, the ghosts or the dead. Sickness is the routine punishment for every lapse and crime meted out to the by these spirits. Accordingly, they havs taboos and prayers. The pahan (village priest), in addition to his religious functions acts as the mediator between tie people and the mystical powers. By propitiation and prayers he tries to maintain the good relationship between the people on one hand and mystical powers on the other which intervene human affairs. Thus, magico-religious treatment of disease is resorted to. As a matter of fact, disease to a tribal mind is like another phenomenon of the natural world, is many faceted and essentially incomprehensible in the sense that no single theory will cover II the known facts perfectly.

The santhal has a theory of disease which in his scientific < iewpoint is as natural a cause as infection caused by bacteria. But we can look upon as an agency which is ru tural and yet super-natural and scmetimes also diffused with human agency. According to them, disease would be caused by a "Tejos" which may be large or very microscopic. "Tejo" enter the body t,trough food etc. and caus leprosy, hydrophobia etc. The "Tejo" germs are often believed to be collected by witches for spreading diseases.

The Santhal has a belief that evil men by dint of their mac cal power intrude dirts such as stone or wood or pebbles or lumps of hair in the body which cause unbearable pain. The "Ojha" sucks out from the body these things and they believe that by doing so their pain is cured. Thu, the Santhals of Santhai Parganas have a strong belief in witchery and malevolent bongs.

The Malpaharias, the Saurias and other Paharias of the Rijmatial Hill of the Santhai Parganas in Bihar also believe that one of the cause of disease is the intrusion of bits of stone or wood, etc. by sorcerer and which is cured by sucking out these things from the body.

The Oraon sorcerer sometimes carries with him a rag Lundlc containing what arc known as "Nasans" (destructive and harmful agencies) and include human hair, na pairings, claws, fragments of bone and legs of chickens and other birds and animals and also small quantities of rice grains, mustard seeds and certain other grains. Sorcerer mixes these in the food pouncing appropriate spells which kill the victim.

The Oraons have a belief that there are among them so ne powerful sorcerers or witches who by some appropriate mantras extract the liver of an intended victim witho.it the latter perceiving it.

The Hos of Singhbhum believe that women dying in childoirth transform themselves into malignant spirits and delay in dt/ivery as also other complications. Children suffer from rickets owing to the influence of spirits who are but the disengaged souls of dead children. The Hos also telieve in witch known as "Rakti Bhawani" which shoots arrow at night.

The Korwas believe that fever in women and children are caused by "Churail" which is a female spirit and extremely mischieveous.

The Kharias of Dhalbhum believe in spirit basis of diseases.

The Birhor (Uthlu takes up some arua rice in his hand and ir vokes his "Nasan" spirit and throws the rice in the direction of the enemy (House) and injures him.

The Asur call them "Dahlia" or "Brishaha". They are supposed to exercise evil influence.

The Mundas are great believers in the power of evil-eye, md in cases of repeated sickness in a family or among the cattle of a family, a witch-finder - the Sokha, fvtati or B lagat is appealed to for detecting the witch. Thus, the Mundas have a doctrine concerning sickness and a functioning group of curers. They also attriubute a few diseases to particular deity as for instance, epidemic diseases like cholera, small pox etc., are attrubuted to sitala. With an effort to avert the diseases, they annually worship the deity. The belief of the interference of supernatural agency in diseases, the attriubution of some diseases to certain deities and the belief in evF! eye, spirit possession or sorcery clearly indicate that still the Mundas have clung to their traditional modes of beliefs and practices.

Medical facilities in the tribal areas are still inadequate. While the normal health of the tribal people cannot be said to be bad, after infection their condition often becomes chronic. Moreever, the influence of dreams on health and tribal methods of diagnosis are also to be taken into accourvt. Besides, there are many natural remedies, decoctions of forest herbs or healing lotions known to the tribals. That means the tribal people have pharmacopoeia of their own. At times it amounts to one of the major difficulties in the interior regions in the way of persuading the tribal people to come for medical treatment.

In general the tribal people lead a healthy life, though some of them of course, suffer from manyfold diseases like malaria, goitrc, leprosy, scabies, venereal diseases, bowel complaints, influenza, opthaimia. cholera, smallpox, etc.

Owing to the excessive use of alcoholic beverages, most of the tribal people.suffer from liver troubles as well. For some reason, the people of Palamu are especially affiliated with blindness. Small pox outbreaks almost every year. The vaccinator still creates awe among the tribal people.

Venereal diseases are perhaps more common among the tribal people inhabiting the industrial areas and other pockets, persumabiy due to contact with the people of out side. The venereal diseases are contacted almost always by sexual intercourse. The infection transfe red by other avenue is negligible.

Cure of Some Common Diseases

Regarding the origin of the large number of medicines, practically nothing is known. The remedies resorted to in various diseases are quite startling. The tribals have bit upon some forest herbs in some way or another and found it efficacious. They have then tried the same for complaints other than those it was originally used for, and have in this way, been guided by expereicnce.

But their usage is not so far recoreded in literature or atleast that particular medicinal property which the tribals attrubute to the plants is not known.

Some of the medicinal herbs, commonly used by the tribal people investigated for the purpose are discussed here. The ailment and the manner in which the forest herbs are being used have also been discussed.

TABLE

	Name of the herb (with botanical name and natural order)		Ailment and the manner in which the herb or herbs are being used.
1.	Aentha (HELICTERES ISORA-STERCULIACEAE)		Stomach pain - The root is ground with water and decoction is orally administered.
2.	Baghrandi (JATROPHA CURCAS-EUPHORBIACEAE)	(i)	Burns - Baghrandi latex mixed with salt is used against burns.

		(ii) Loose teeth - Baghrandi latex is also used to fix loose teeth.
		(iii) Laxative - A laxative oil is extracted from the seeds of the Baghrandi, (Dose : 15 drops)
		(iv) Body pain - Young seedlings of Baghrandi (Jatropha Curcas), root of chechri latanga (Achyranthes Aspera)and bark of sonapatti (OROXYLON INDICUM) are ground with water. The decoction so obtained is orally administered.
3.	Banabar (ZIZYHHUS SP. - Rhamnaceae).	(i) Bacillary dysentery and diarrhoea - bark which contains much tannin, is ground with water and the decoction is orally administered.
		(ii) Rejuvination tonic - The root is ground with water and the decoction is mixed with indigenous distilled liquor and used as tonic.
		(iii) The fruit is eaten to purify the blood.
4.	Bhuin Jamoon (PREMNA HERBACEA Verbenaceae)	Rheumatism and swelling. The root paste is massaged over the affected part.
5.	Brahmi (CENTELLA ASIATICA-Umbelliferae)	Leprosy - The entire plant parts are boiled in water and this decoction is given to the leper patient.
6.	Chaili (MORINDA TINCTOR1A-Rubiaceae)	Cholera - The bark decoction is used in cholera.
7.	Charaogora (VITEX PEDUNCULARIS - Verbenaceae)	(i) Fever - Mundas use the bark decoction.
		(ii) Snake bite - Oraons use it for snake bite.
8.	Chitwair.(PLUMBAGOZ EYLANICA-Plumbagineao)	(i) Ordinary fever - The root is ground water and the decoction is administered orally.
		(ii) Stomach pain. The root is boiled in one litre of water till the volume is reduced to half. Then, the root is crushed in it and the decoction is orally administered. It is also said to excite digestion and promote appetite.
		(iii) Septic wound - The root is boiled in mustard oil till it begins to char. It is then filtered and the filtrate is applied on the wound.
		(iv) Hydrocele - Fresh roots of chitwair are tied on both the arms just like a string fcr half an hour daily for 8-15 days. It is said that they hydrocele will be completely cured.
9.	Chhoti Dudhi (Indian Sarasprilla) HEMIDESMUS INDICUS - Asclepiadaceae)	Fever, skin diseases and urinary Complaints. The root decoction of the plant is used as drug. The diuretic action of the drug has been shown experimentally. It is largely used as a blood purifier and in rheumatism.
10.	Dudhia (EUPHORBIA HIRTA - Euphorbiaceae)	Body pain, fever and blood dysentry - The root is powdered and taken withwater. It is useful in removing worms in children and bowel complaints. The latex of the herb promotes formation and flow of milk in women.

11. Galphuli (MOGHANIA CHAPPAR-Papilionaceae) Eye sore-leaf juice mixed with salt is used as eye drops.

12. Guiainchi (PLUMERIA ACUTIFOLIA Papilionaceae)
(i) Rangbad - The Oraons use the flowers.
(ii) Lactation - When mothers have no milk, the leaves are applied by the Mundas crushed and mixed with Karanj oil, (PONGAMIA PINNATA) on the breast.

13. Gursukri or Kukurbicha (GREWIA HIRSUTA Tiliaceae)
(i) Diarrhoea and Dysentery - The Oraons use the fruit and also decoction of leaves.
(ii) Sterillity - Three pieces of each of the roots of Gursukri and Kharhar banda (GARDENIA TURGIDA) are ground with water and decoction is orally administered.
N.B. : The treatment is said to be more effective if administered on Sunday or Tuesday.

14. Gusumpuru (ANDROGRAPHIS ECHIOIDES Acanthaceae) Snakebite - The whole plant dried and powdered is used. One teaspoonful of the powder to be mixed in glass of water and drunk; the same amount to be applied to the bitten portion (Recommended by Fr. Lakra, Dumerpat).

15. Harjarwa (VITIS REPANDA - Ampelidaceae) Bone fracture - If the fresh twig of Harjarwa and thungi are chewed for five days the fracture will rejoin.

16. Ich'a, Dhawai (WOODFORDIA FRUTICOSA-Lythraceae) Dysentery - The root is ground with water and the decoction is orally administered.

17. Kalmeg, Bhuin nijim (ANDROGRAPHIS PANICULATA - Acanthaceae) Fever, worms, dysentery and gastritis - All parts of the plant above the ground are employed in preparing the medicinal decoction which is very bitter. Recent experiments have shown that the plant has antityphoid and antibiotic activity.

18. Karigiri, Somraj (VERNONIA ANTHELMINTIC - Compositae) Fever and deworming - The Mundas use the seeds as a febrifuge and the leaves as a vomitive. In paralysis of the legs applied externally. It is a valuable medicine as antheimintic, i.e. to destroy worms.

19. Kharkhasa, Har Singar (NYCTANTHES ARBORTRISTIS - Oleaceae)
(i) Loose motion - The root is ground with water and the decoction is orally administered. The seeds yields medicinal oil.
(ii) Bronchitis, influenza - A decoction of leaves is used.

20. Koraiya(HOLARRHENA ANTIDYSENTERICA - Apocynaceae)
(i) Dropsy and Gout - The bark and root are ground into paste, which is over the affected part.
(ii) Body pain and Fever - The decoction is prepared by grinding the root and bark with water. This decoction is orally administered.
(iii) Snake bite and poisoning - The subject is made to chew the crushed bark and the cure is effected within three minutes.
For cases of poisoning, eight days will be required for complete cure.

	(iv) Dysentery, diarrhoea and intestinal worms - Seeds decoction is orally administere.
21. Kujuri, Malkangni (CELASTRUS PANICULATUS	(i) Body pain - Kujuri seeds oil is massaged.
	(ii) Abortion - Bark decoction is used as abortificient.
22. Lilbaha, Bilaiphar (IPOMOEA HEDERACEA Convolvulaceae)	(i) Purgative - The Mundas use the fruit as purgative.
	(ii) Adhkapar (pain in one side of the head). The flower are crushed and put on the paining side.
23. Mara Chundi (ELEPHANTOPUS SCABER (Compositae)	Swelling - The root crushed and juice squeezed out and applied on swollen part.
24. Parhi or Pathal Kohnra (PUERARIA TUBEROSA - Papilionaceae)	(i) Fever - Including material - The root is ground with water and the decoction is orally administered.
	(ii) Body, muscular or stomach pain - The root is ground with water and paste is prepared which is applied on the paining parts. In addition, the above decoction is orally administered simultaneously.
	(iii) Abortion - If the decoction is given to pregnant women it will result in abortion.
	(iv) Cholera - If a small piece of parhi is tied in the neck, it is said to cure cholera.
25. Piar(BUCHANANIALANZAN-Anacardiaccae)	Snake bite - The young bark of piar together with new earth (preferably out of ant-hill is ground with water and paste prepared which is massaged on the bitten part from to bottom.
26. Pirijadu (CHLOROPHYTUM ARUNDINACEUM - Liliaceae)	Importency - Tuberous roots are used when a young Munda woman does not remain with her husband, the latter applies crushed plant to any part of the body.
27. Putri (CROTON OBLONGIFOLIUS-Euphorbiaceae)	Eye sore - Fresh oozing out juice is applied in the sore eye.
28. Rakatphar (ARDISIA SOLENA-CEA-Myrsinaceae)	(i) Dropsy - The root is well ground and applied as a thick plaster to reduce the swelling.
	(ii) Stomach - Ache after child birth. The decoction of root is said to be effective in reducing stomach pain after child birth.
29. Ranpawan (SMILAX MACROPHYLLA-Lyiiaceae)	Constipation, blood dysentery and renal trouble - The root is ground with water and the decoction is orally administered. It is used as appetiser as well. Childred who wake at night are given food on its leaves and are given its twigs as tooth stick.
30. Sain, Assan (TERMINALIA SERRATA -	Hard of hearing - The bark is powdered and taken with warm water. The powder is also mixed with mustard oil and applied in the ear.
31. Saiai (BOSWELLIA SERRATA -	(i) Wound - Antispetic lotion is prepared from its bark.

	(ii) Stomach pain - Decoction of the bark chipsis given.
32. Satawar (ASPARAGUS RACEMOSUS Liliaceae)	Heat fever - The root of Satawar and the root of Putri (27) are ground with water and the decoction is orally administered. The Birhores use it in blood urination and the Mundas use it against dysentery.
33. Satgithia (SPERMACOCE HISPIDA -	(i) Ear ache - The leaves are squeezed and Rubiaceae) the juice so obtained is mixed with salt and used as ear drops. (ii) Cuts and wounds - The fresh leaves should be crushed on the palm of the hand and the juice be made to drop on the wound to cured, and leaves freshly crushed should be placed on the wound.

During resources survey of medicinal plants, Pahan, Munda and Manaki of various tribes were contacted regarding use of forest species in tribal medicine. Some of them hestitated to divulge proper information whereas some gavc piecemeal information. Hence the above table has been prepared mostly on information quoted in Gupta, S.P. (1974) : Tribes of Chhotanagpur plateau : An Etho-nutritional and Pharmacological cross-section, B.T.W.R. Ranchi. Some information were also furnished by Shri P.P. Hembrom, B.F.S. (Retd.) but since they are mere private communication and are yet to be published by him and confirmed, only those informations which were common in published literature have been included. Intensive field work in Ethno-botany has been made by bodding (1925, 1927) and Elwin in Central India, Gupta (1963) in Bihar, and Joshi (1971) in kumaon Himalayas. Of the recent resources mention may be made of the intensive work Carried out in Madhya Pradesh, Bihar, Orissa, Andhra Pradesh, West Bengal and North East India by the Botanical survey of India and Central Medicinal Plant Organisation, Lucknow under the able guidence of Dr. S.K. Jain 1963 onwards.

According to Jain & Paul (1982) certain medicinal plant species are common in occurence both in Bihar as well as other states like M.P., West Bengal, Andhr-Pradesh, and North East India. Some plant species which are used medicinally by the tribals of Bihar and tribe s of other states like Madhya Pradesh, West Bengal, Andhra Pradesh & North East India are mentioned below :

Name of the plant species and family	**Medicinal use by the tribals**
1. ACHYRANTHUSASPERAL. (Amaranthaceae) Latjira	(i) Plant made into paste with water and fruits of piper longum and taken to cure effects of bite of mad dog. (ii) Root made into paste is given to stop bleeding after abortion. It is also given to facilitate delivery.

		(iii) Placing some roots of the plant in the hair-do of expectant mothers temporarily child-birth.
2.	AGERATUM CONYZOIDES E. (Compositae)	Juice of the leaves with Neem oil is used as antiseptics in skin diseases and on fresh cuts to stop bleeding.
3.	ANDROGRAPHIS PANICULATA (Burm. F.) Wall (Acanthaceae) Kalmegh	Leaves are pounded and applied on body in itching, juice of the leaves is given in fever and colic pain.
4.	ANTIDESMA DIANDRUM Roth. (Euphorbiaceae) Amti	A mixture of leaf and root pounded with water is given in orally in blood dysentry.
5.	ATYLOSIA SCARABAIOIDES (L) Benin. (Papilionaceae) B.an - Kulthi	Decoction of leaves is given in dysentery. It is mixed with honey and given to women in child birth.
6.	BARLERIACRISTATAL.	(i) Roots paste applied in toothach.
		(ii) Plant paste with Karanj oil applied to the swelling of legs.
		(iii) Plant paste with rendi (Ricinum Communis L.) oil applied on body which pains.
		(iv) Leaves chewed and swallowed in toothache.
7.	BRIDELIA RETUSA Spreng (Euphorbiaceae) Khaja, Kajhi	Pounded bark is mixed with the decoction of the gum of keonjhi (Sterculia urens Roxb.) and the mixture is prescribed orally for 2-3 days after menstruation for complete infertility.
8.	CALOTROPIS GIGANTEA (L.) R. Br. (Asclepiadaceae) Akaon	The root is applied on snake and scorpion bite. Root is pounded in water and the (Maduca indica) flowers is mixed and appliedexternally in leprosy and also taken orally.
9.	CASEARIA ELLIPTICA Willd. Syn. C. tomentosa (Samydaceae) chilla	Root bark is tonic in anaemic conditions. Decoction of the root is remedy for diabetes. Powder is used for hallucinations. Fruits used to stupefy fishes.
10.	CISSAMPELOSPAREIRA L. (Menispermaceae) Akhandi	Decoction of roots is taken internally in stomach pair and indigestion. The bruised leaves are taken with water in burning sensation near heart.
11.	CRYPTOLEPISBUCHANANI R. & S. (Asclepiadaceae) Janatiki-bel, dudh- lar	Latex is mixed in hot water and is applied on knees to cure rheumatism. Extract of the root is prescribed internally to cure fits.
12.	GLYCOSMIS PENTAPHYLLA (Retz) DC. (Retaceae) Bano - nimbu	Paste of the roots is given with water thrice daily in post delivery pains. Pounded roots is given to kill worms in stomach.
13.	HETEROPOGON CONTORTUS L. (Gramineae) Shruval, choranth.	Either the roots of the plant are chewed by the victim of snake bite or they are pounded in water and given to him.
14.	LEUCAS ASPERA (Labiatae) Chota-halkusa.	Root juice is put in nostril in Adha shishi (Half-Head) headache, Leaf juice is put into eyes twice or thrice daily in burning sensation and in redness of eyes.

15. MADHUA INDICA Gmel. (Sapotaceae) Mahua — Small quantity of seed-cake is kept in a piece of cloth little water is added; it is beaten slowly and its decoction is dropped in nostrils of the victim of snake-bite (two drops in each nostril). It is believed that the victim vomits out poison. The smoke of the cake is said to drive out snakes.

16. MICHELIACHAMPACAL (Magnoiiaceae) Champa — Bark is employed as an abortifacient for 2-3 months old pregnancy. As an oral contraceptive the root (about 6-8 mm long) is made into paste with 21 fruits of black pepper and given after menstruation for 3 days.

17. MUCUNAPRURITAHook. (Papilionaceae) Kawanchi — Roots are made into paste and prescribed orally in blood dysentery.

18. SEMECARPUSANACARDiUM Lf. (Anacardiaceae) Bheiura — Bruised nut is used in abortion of 3-5 month old pregnancy. Oil of the nut after slight heating is used externally on swelling of joints, on foul ulcers and on eruption of the skin. Seed oil is used in epilepsy.

19. SIDACORDIFOLIA L. (malavaceae) Kunjyi — Leaves and roots are pounded and applied locally on boils to ripen and open them.

20. SMILAXZEYLANICAL Syn. S. macrophylla (Liliaceae) Chop - chini — Root juice is mixed with powdered "Urad Dal" (Vigna Munga Hepper) and made into cakes and given with warm milk in gout and rheumatism. The patient is not allowed to take salt.

21. SOLANUMTORVUMSW. (Saianaceae) — Root is powdered with tamrind water and given as antidote in poisoning. Tender fruits are fried and eaten with Raghi (Eleusine corcana L.) in night blindness.

22. TAMARIXERICOIDES Rotti. (Tamaricaceae) Thau — Leaves are cooked with rice and given to children to releive cough. Decoction of leaves reported to cure enlarged spleen in children as well as in adults.

23. TECTONAGRANDIS L. (Verbenaceae) Sagwan — The oil obtained from distillation of wood chips is excellent in eczema and ringworm.

24. VISCUM ORIENTALS WILLD. (Doranthaceae) Banda — Juice of the plant is put in the ear to stop pus formation. Ash of the plant is also put in ear to cure pain.

25. XEROMPHIS SPINOSA (Thimb.) — Roots are used for treating gonorrohoea or diseases of urinary tract. The root bark with the pith removed is pounded in water and the paste given in milk and sugar thrice a day.

26. XYLOSMA LONGIFOLIUM Clos. (Flacourtiaceae) Katari — Bark is common household remedy in stomach pains and intestinal worms. About 5 gm. of dry powder is used for stomach pain, while a large dose in early morning is for intestinal worms.

MEDICINAL PLANTS OF FOREST ORIGIN IN USE UNDER AYURVEDIC SYSTEM OF MEDICINE

The oasis of Ayurvedic therapeutics rests on the five well-defined pharmacological principles of Rasa, Guna, Virya. Vipaka and Prabhava. These five principles, as a rule, cover the entire range of diseases - whether of constitutional charcter or micro-biological origin and are founded in Tridosa theory of Ayurvedic Physio-pathology. Vata, Pitta and Kapha (syn. Shlesma) are collectively called Tridotha, that is, three dosas. They really represent an aggregate of life system which, like any motive system, must have two counterparts, viz., an organic body and an energy system, the latter acting upon the former. Kapha stands for the organic body whereas Vata and Pitta represent the energy system. To be specific, Vata stands for the nervous system whereas Pitta represents the total bio-thermal system. Kapha represents the colloido-plastic systen differentaiating into various tissues and organs of which the simplest and primal form is protoplasm. Thus the rec.iity of Tridosa is conceivable in any life system. The corrrect conception of Tridosa theory reveals what is health ar.d what is disease. There also exists a correlation between the physio-pathology of Tridosa system and the penta-facet.

Pharmacological principles of Rasa, Guna, Virya, Vipaka anc Prabhava. Rasa means taste; they are basically six in number, that is, sweet (madhura) sour (amla), saline (lava la), pungent (katu), bitter (tikta) and astringent (kashaya). Guna denotes certain physical properties of matter (drug and food) which effects Tridosa in health and disease respectively. Virya means two intrinsic properties of matter, which effects Tridosa in health and disease respectively. Virya means two intrinsic properties of ma ter, that is, Sita - Virya and Usna - Virya. The former has the attribute of coolnes: and the latter of hotness. Vipaka means that the end products of all digestive transformation are either sweet or sour inspite of the original rralter - having different taste. Prabhava is that pharmacological principle which denotes certain special power of substance either drug or food. Prabhava does its action quite independent of the other four basic principles of pharmacology i.e. Rasa, Guna, Virya and Vipaka. But Ayurveda maintains, to get a sustaining effect in uphoding tht normal balance, the other four principles must act concurrently in the form of drug, regimen and environment too.

The vegetable drugs, most of them of forest origin, in vogu^ under Ayurvedic system of medicine, can be classified into two goups, viz. botanical and therapeutic.

A. BOTANICAL DRUG GROUP :

The group contains plants, both of forest and cultivated origin. Under this group drugs have been classified into following sub-groups with examples of forest species found .nostly in forests of Chhotanagpur and Santhal Parganas. Some species which are cultivated in Bihar have also been metioned. The names are in Sanskrit with botanical names in bracket.

I. Puspa varaa (Flowers having medicinal value):

1. Padma (Nelumbo nucifera); 2. Kumuda (Nyrnpheae sp.) ; 3. Juthika (Jasminum sp.); 4. Champaka(Michelia champaka); 5. Rakula (Mimusops elengi); 6. Kadamba (Anthocophalus indicus); 7. Dhataki (woodfordia fruiicosa), etc.

II. Phaia varga (Fruits having medicinal value)

1. Amra (Mangi'era indica); 2. Jambu (Syzygium cumini); 3. Kap ttha (Feronia limonia) ; 4. Madnuka (Madhuca indica) ; 5. Tintidi (Tamarindus indica).

III. Shakavarga (Leaves / whole herb having medicinal value)

1. Vastuka (Chenopodium album) ; 2. Marisha (Amaranthus blitim); 3. Chanka (Cicor arietinum); 4. Sushinnaka (Marsiiia minuta); 5. Kasamarda (cassia occidental), etc.,

IV. Pnala shaka varga (Non- dessert fruits having medicinal valie)

1. Karavella (Momoridica charantia); 2. patola (Trichosanthes cucumerina); 3. Bimba (Cephalar.dra indica); 4. Koshataki (Luffa sp.); 5. Shcbhanjana (Moringa oleifera) etc.

V. Kanda shaka varga (Underground parts like rhizome, bulbs, tubers, etc. having medicinal value)

1. Shurana (Amorphophnlius camapanulatus) ; 2. Manaka (A ocasia indica) ; 3. Lasuna (Allium sativum); 4. Palandu (Allium cepa) ; 5. Mulaka (Rapharus sativers) ; 6. Grinjana (Caucus sativus) etc.

VI. Haritakyadi varga :

In this sdbotanical grouping so many medicinal plants of different families have been collectively put together, of which "Haritaki" is the prime member, amongs. a total of 92 drugs, e.g.1. Haritaki {Tsrminalia chebula); 2. Amalaki (Embilica officinalis); 3. Bibhitaki (Terminalia belerica): 4. Daruharidra

(Berberis sp.); 5. Bhallataka (Semecarpus anacardium); 6. Bhanga (Cannabis indica); 7. Aragvadha (Cassia fistula); 8. Chitraka (Plumbago zeylanica); etc.

Incidentally, it may be noted here that the fruits of the first three items of the above list when combined make the very famous and popular Ayurvedic recipe "Triphala" which is successfully indicated in a variety of ailments like constipaton, acidity and even in eye trouble.

VII. Guduchyadj varga :

Under this group the number of drugs involved are 124, of which the first member is Guduchi. A few of them are cited here, e.g. 1. Guduchi (Tinospora cordifoiia); 2. Langalika (Gloriosa superba); 3. Karavira (N'erium indicum); 4. Dhustura (Datura metal); 5. Nimba (A/adirachta indica); 6. Nirgundi (Vitex negundo); 7. Kutaja (Holarrhena antidysenterica); 8. Karanja (Pongamia pinnata); 9. Ashwagandha (Withania somnifera); 10. Ananta (Hemidesmus indicus)

VIII. Karpuradi varga (Aromatic plants having medicinal value)

1. Karpura (Cinnamomum camphora); 2. Chandana iSantalum album); 3. Mustaka (Cyperus rotundus); 4. Usheera (Vetiveria zizanioides); etc.

IX. Vatadi varga :

This group has forty throe members of which the irst member is Vata not to be confused with.

1. Vata (Ficus bengalensis); 2. Ashwattha (Ficus religiosa);- 3. Udurnbara (Ficus racemosa); 4. Khadira (Acacia cattechu); 5. Rohitaka (Aphanamixis rohituka); 6. Varuna (Crataeva nurvala); 7. Saptaparna (Alistonia scholaris); 8. Shalmali (Bombax ceiba); etc.

B. THERAPEUTIC INDICES :

I. Anti-vatic drugs (Vataghna)

1. Prasarani (Paedaria foetida), 2. Eranda (Ricim 3 communis) etc.

II. Anli-paittic drugs (pittaghna)

1. Patola patram (Trichosanthes dioica); 2. Durva Cynodon dactylon) etc.

III. Anti-kaphic drugs (Kaphaghna) :

1. Kantikari (Solanum xanthocarpum); 2. Shobani; na (Moringa oleifera); S.Tulsi (Ocimum sanctum) etc.

The other classification of Ayurvedic drugs is by C ia,raka into fifty distinct heads of Maha-Kasaya - an excellent therapeutic exposition.

To Summarise, We may say that the fundamental I asis of Ayurveda is cosmologically related to the theories of "Pancha Tanmatra - Panch Mahabhoota - Tridosa" on the one hand and Rasa-Guna-Virya -Vipaka - Prabhava principle on the other hand-constituting together a grand aura of biophysical and biochemical laws which govern the human life in I ealth and diseases. Plants specie, mostly of forest origin, play a major role in Ayurvedic system of medicine.

ROLE OF FOREST SPECIES UNDER UNANI SYSTEM OF MEDICINE

Unani system of Medicine owes its origin to Greek philosopher physician Hippocrates who flourished in B.C. 460. It was further developed and enriched by Aristotle, Dioscorides, Pliny and Galen. During the end of 5th century A.D.. Nestorian Greeks carried the Greek civilisation to the East. Jundishapur (Persia),'Damascus and Baghdad were the famous centres ol earning of pharmacology and the basic concepts of Unani Tibb upto 900 A.D. Greek medicine was translated in Arabic and the Arabs wrote standard books making alterations and additions in the Matereia Medica of Greeks. Important personages who wrote books on Arab medicine were Rhazes, Avicenna, Ibn-al-Baitar and Shaikh Dawood Antaki. When the Arabic system of medicine was brought to India by the Muslims, this process of enrichment advanced further and a large number of medicine was borrowed by the Unani from the Ayurvedic system of medicine.

The Unani system of medicines is primarily based upon the Pythagorian theory of four proximate qualities or states (hot, cold, moist and dry) and Hippocratic theory of fou· humours, that is, blood (khoon), phlegm (balgam), yellow bile (safra) and black bile (sauda). The Arabs further elaborated these theories and set the seven working principles (Umer-a-Tabia) which determine the constitution, health and diseased, conditions besides the behaviour of human body. These are (a) element (arkan), (b) temperament (mizaj), (c) humours (akhlat), (d) organs (aza) (e) life spirit (arwah), (f) energy (guva) and (g) actions (afal).

(a) Element (arkan) :

Four elements i.e. earth, water, air and lire represent so! ds, liquids, gases and heat respectively. The earth and water make organs whereas air and fire make energy. Wht n these elements intimately mix with one another to form compounds, these qualities act and react upon one ano.t ler through intimate contact of the minutest particle qualitatively and quantitatively, and the positive is modified jy the negative, thereby a uniform body emerges possessing the balance of these qualities.

(b) Temperament (mizaj) :

The quality of the uniform body or compound is called its temperment (mizaj). Nine types of ternperment are recognised : hot, dry, cold, and moist and their four combinatici s, that is hot and moist, hot and dry, cold and moist, cold and dry and the ninth type i.e. equable (motadil) which is non-existent.

(c) Humours (akhlat) :

According to the presence of different temperment va ious types of personalities are described (i) hot personality (mizaj-e-har), (ii) cold personality (mizaj-e-barid), (iii) dry personality (mizaj-e-yabis) and (iv) moist personality (mizaj-e-rath). The four humours regulate the functions of the body, whether in health or disease. Humours are the first product of digestion of food which may be bad or good. As the humours are moist mobile fluids, these are carried by the blood stream to different parts o[r] body. The excess quantity of blood, yellow bile and black bile are stored in the liver, gall-bladder and in the spleen respectively. The pheigm circulates in the whole cf the body as it has no specified place for storage. Besides thesa humours, secondary fluids are also recognised in the Unani system.

(d) Organs (aza) :

Different organs of body are built essentially from coarsei and solid particles of the humours. Organs may be simple consisting of one tissue like bones or compound consisting of various tissues. The essential organs (aza-e-raisa) are four, namely, heart, liver, brain and testes.

(e) Life spirit (arwah) :

Life spirits are the finer and higher constituents of humcjrs. They are of three kinds, namely (i) Vital spirit (Rup-i-haiwani) (ii) animal spirit (Ruh-i-naisrni) and (iii) natural spirit (Ruh-i-tabai).

(f) Energy (guva) :

It performs four different body functions and this is also of three kinds, i.e. vital (nafsania), animal (haivania) and natural (tabai).

(a) Actions (afal) :

The causation of disease is said to be mainly due to deficiency of humours, plethora and segregation of humour. The treatment is aimed at restoration of that condition to normal state.

The Unani system of medicine is classified nto three types of drugs-vegetable, animal and mineral. This system, in its armaments to fight disease, utilises mostly the medicinal herbs which are provided by the nature alround. Among vegetable drugs are included root;, bark, leaves, fruits, latex, and gums of large trues like Ficus benghalensis (bargad), Ficus religiosa (peepal), Azadirachta indica (neem), Acacia sp., Shorea robusta (saal), and Tectona grandis (sagwan) and sometimes heirt wood is also used. Of smaller plants, sometimes the roots, sometimes the leaves and flowers are used. Of sHl smaller plants or herbs the whole plant or the roots or the leaves alone is used. There are more than 100 plant species occuring in Bihar forests which are used in Unani system of medicine. It is evident that the these med'cinal plant species have been borrowed from Ayurvedic system of medicine as they have been in use in Ayurvedic system from time immemorial.

ROLE OF FOREST SPECIES IN HOMOEO-PATHIC PHARMACY

Dr. S. Hahnemann, born in Germany in 1755, discovered that there is a law in medicine called the "Law of Similars". There is a similarity between drugs and diseases, he put forward the new principle of "Similia Similibus" - Let likes be treated as likes. The unique feature of homoeopathy is its small dose. It seems that by a special method of dilution used in homoeopathy, all the energy of the drug is liberated and transferred to the medium of sugar or alcohol. The dose may look ridiculous but we now know what a vast amount of energy lies within the infinitely minute atom awaiting to be released and utilised. He is the first physician who claimed to split the atom of the drug and unleashed a trememdous power in the field of medicine.

Compared to the pharmacy of the other systems of medicines, homoeopathic pharmacy is much simpler and less complicated. The main preparation undo; the homoeopathic pharmacy is the mother tincture. The preparations of the mother tinctures are done by maceration and precolation processes. These processes more or less follow the same line as are done in the allopathic pharmacy. Substances soluble in the liquid menstrua or vehicle are made into solutions or tinctures for their subsequent dilutions.

For the preparation of mother tinctures Hannernann introduced the centesimal scale which is based on the principle that the first potency should contain one hundredth part of the original drug, which is now very conveniently done using percolators and each succedding potency should contain one hundredth part of the one proceeding it.

The mother tincture or solution may be further divided a id sub-divided by adding alcohol or distilled water. These divisions and sub-divisions of a substance are called Attenuations or Potencies. In attenuating or potentizing two scales are used : Decimal or Centesimal. The Decimal sea e is based on the principle that first potency should contain one tenth part of the original drug and each succeai ing part should contain one tenth part of the one preceding it. It is denoted by affixing an "X" after the numerals. The centesimal scale is based on the principle that the first potency should contain one-hundredth part of the original drug and each succeeding potency should contain one hundredth part of the one preceding it. The centesinal Scale is denoted by simply affixing the numerals after the name of the drug.

Drugs used in Homoeopathic Pharmacy are numerous and they fall under the following categories :

(i) Drugs of vegetable origin and the parts used.
(ii) Drugs of animal origin.
(iii) Minerals or chemical compounds or synthetic products.
(iv) Disease - product of animals or plants.
(v) Energy available from physical and natural reactions.

The present paper deals only with the first category i.e., dri gs of vegetable origin and the parts used. The bulk of the homoeopathic raw materials have to be imported. Ever where mother tinctueres are made indigenously, most of the vegetable drugs which are needed for their preparation have to imported from Europe and America. The import restrictions are so severe that the quantities available a n only meet the fringe of the requirements. Indian drugs of vegetables origin, however, have been included in receit years but the quality is not of the same standard as that of the imported one. A considerable number of common. drugs of Indian vegetable origin which are now extensively used by the Homoeopathic practioners in India occur in the forests of Bihar, specially in Chotanagpur and Santhal Parganas. They are mentioned below :

1. Adhatoda vasica Nees
2. Achyranthus aspera Linn
3. Aegle marmelos Correa
4. Andrographics panicuiata Nees.
5. Azadirachta indica A. Juss.
6. Bryophyllum sp.
7. Calotropis gigantea Br.
8. Carica papaya Linn.
9. Cassia sophera Linn.
10. Cledendron indicum (Linn.) ktze.
11. Cynodon dactylon Pers.
12. Desmodium gangeticum DC.
13. Ficus rcligiosa L.
14. Hemidesmus indlcus Br.
15. Holarrhena anti-dysenterica Wall.
16. Centella asiatica (Linn.) Urban.
17. Leucas aspera Spreng.
18. Nyctanthes arbortristis L.
19. Ocimum canum Sims.
20. Ocimum gratissimum Linn.

21. Psoralea corylifolia DC.
22. Rauvolfia serpentina Benth. (almost Ext'nct, now cultivated)
23. Saraca indica Linn.
24. Solanum xanthocarpum Sch. & Wendal.
25. Syzigium jambos (Linn.) Alston.
26. Terminalia chebula Retz.
27. Terminalia arjuna W. & A.
23. Vernonia anthelmintica Willd.
29. Vitex negundo L.
30. Withania somnifera Dun.

ROLE OF MEDICINAL PLANTS IN MEDICAL ASTROLOGY

Astrology is the oldest of ail sciences and there is enough scope for applying it in all walks of life. Medical men in all departments can benefit from a knowledge of astrology.

There are various ways of getting free from diseases. It ma/ be done by treatment by different by different systems of medicine like allopathy, homoepathy, ayurvedic or uncni. Naturopathy,

Yoga - Sadhana, Pranayam etc. are also important Hindu rrethods of treatment and cure. The other vvay to keep the body away from diseases is to wear the gems for the ri ling planet of one's constellation or nakshatra. The constellation or nakshatra in which the moon is posited in one's birth chart is known as his birth-star. The birth-star gires diseases according to its nature. Everyone will not suffer from the same disease. It varies in accordance with one's birth-star. It is possible to identify the diseases a man is liable to suffer during his life time from the nature of his birth-star. There are 12 signs of rashis ana 27 constellations or nakshatras.

Plants and trees have also their constellations or birth-stars. A particular type of plant species belongs to a particular nakshatra. This has been dealt in detail in ancient literature like Triskandha - Puran. Ancient Maharshis have also suggested healing through plants. It is not a system o medicine where a certain part of the plant is administered orally or applied externally in its original from or in fo'm of leaf-powder, paste, decoction or infusion. It is much more important than that. They have prescribed a metric.d which is simple and can be pratised even by a common man. If a man worships the plant of his constellation arrJ the plant of the constellation in which Saturn transits in Gochar, it will definitely help him in maintaining sound health. Worshipping here means watering the f plant daily in the morning, causing no harm to the plant concerned a ;d preventing others from doing so. Incidentally these plants have medicinal value end are extensively used in varioi s systems of medicine .like allopathy, ayurveda, unani, horopathy etc. The details of nakshatras, the rulers, the di;.eases one is likely to suffer according to the nature of the constellation and the plant species concerned are given in Table I.

The diseases indicated by the ruler of the nakshatra may be diminished or aggravated according to the planets posited, if any, in "Trik" houses, that is, the 6th, 8th, and 12th house from Lagna in a horoscope. Each planet is also governed by a particular type of medicinal plant spec es. The adverse effect of a planet posited in "Trik-asthanas" may be neutralised by wearing the roots of the pla it species concerned on the arm or round the neck in a thread of the prescribed colour on a particular day. Inci ientally, the plant species of such particular Dlanets have also high medicinal value in various system of medicins. The details of the planets, the roots of the :oncerned plant species, the colour of the thread to be used, and the auspicious day when it is to be worn are given in Table II.0.

Thus the basic principles of plant Astrology are to fulfil three requirements:

1. Worship the tree of your constellation

2. Worship the tree of the constellation in which Saturn is transiting in "Gochar". For instance, Saturn is in "Moola" nakshatra during 1988 whereas it will transit through "Poorva-shada" during 1989. Thus Sal should be worshipped by all during 1988 and Sita-Ashoka curing 1989.

3. Wear the roots of the plant species of the planets which arc mclefic in your birth-chart causing diseases.

One who follows the above rules will find that there is magic in P-ant Astrology as this will not only help him in i laintaining sound health but also bestow him happiness and prosperity.

"FOOLS OBEY PLANETS WHILE
WISE MEN CONTROL THEM"

Table - I

S.No.	Constellation (Signs with degrees)	Ruler	Diseases likely to suffer	Nakshatra or constellation tree	Botanical Name
1	2	3	4	5	6
1.	Ashwini (Aries 0° to 13° 20')	Ketu	Injury in the head, thrombosis, nose-bleed, fainting, dizziness, neuralgia, insomnia, meningitis, and stomach troubles	Aonla	Embilica officinalis Gaertn.
2.	Bharani (Aries 13° 20° to 25° 40')	Venus	Brain-troubles, headache, meningitis, insomnia, high fevers	Yamal (Twin tree)	One Ficus growing on another, two species grown jointly.
3.	Kritika (Aries 26° 40° to 30° & Taurns)° to 10°	Sun	Fever, meningitis, brain-troubles, throat troubles, nose-bleed, gout, pimples.	Gular	Ficus racemosa Linn.
4.	Rohini (Taurus 10° to 13° 20')	Moon	Sore-throat, cough and cold eye-troubles, ailments of legs,		

			apoplexy, menstrual trouble		
5.	Mrigasira (Taurus 23° 20 to 30° & Gamini 0° to 6°40')	Mars	Throat-troubles, fever, nose-bleed, goitre, arms fracture, disorders of secret parts, gout.	Khair	Acacia catechu wild.
6.	Ardra (Gemini 6°40' to 20°)	Rahu	Throat troubles, mumps, gout, ear-troubles, red eyes urinary troubles.	Pakar	Ficus infectoria Roxb.
7.	Punarvasu (Gemini 20° to 30' & Cancer 0° to 3°20')	Jupiter	Pleu isy, bronchitis, stomach-troubles, dysp-epsia, headache, pains in the body, eye troubles	Baans	Bambusa spp., Dendrocalamus spp.
8.	Pushyami (Cancer 3°20' to 16°40')	Saturn	Consumption, gastric ulcer, respiratory troubles galls ones, jaundice, eczema, pyorrhoea.	Peepal	Ficus religiosa Linn.
9.	Aslesha (Cancer 16°40' to 30°)	Mercury	Stomach, diseases of legs and knees, hysteria, dropsy, jaundice, neurosis.	Nagkesar	Mesua ferrea L.
10.	Magha (Leo 0° to 13°20')	Ketu	Heart troubles, backache, gravel in kidneys, fainting, spinal meningitis	Bargad	Ficus benghalensis Linn.
11.	Poorvaphalguni (Leo 13°20' to 26°40')	Venus	Heart troubles, diseases of legs, blood-pressure gout, sciatica, operation	Palash	Butea monosperma (Lam.) Taub.
12.	Uttarphalguni (Leo 26°40' to 30' Virgo 0° to 10°)	Sun	Intestinal & bowel disorders, sore-throat, goat	Rudraksh	Elaeocarpus ganitrus Roxb.
13.	Hastam (Virgo 10° to 23°20')	Moon	Stomach troubles, eye-troubles, gout, worms, diarrhoea, dysentery, neura gia, jaundice, skin-troubles.	Reetha	Sapindus Spp.
14.	Chitra (Virgo 23°20' to 30° and Libra 0° to 6°40'	Mars	Bowel disorders, worms, fever, wounds from animals, skin troubles.	Bael	Aegle marmelos correa
15.	Swati (Libra 6°40' to 20°)	Rahu	Urinary troub es, eczema skin troubles, leprosy, toothhache.	Arjun	Terminalia arjuna W & A
16.	Visakha (Libra 20° to 30° Scorpio 0° to 3°20'	Jupiter	Renal abscess, skin eruption, diabetes, kidney troubles, headache, fevers	Vikankat, katahi	Flacourtia ramontchii L. Herrit.
17.	Anuradha (Scorpio 3°20' to 16°40')	Saturn	Menstrual troubles, constipation, piles, nasal-catarrh, sore throat, hipbone fracture, stomach troubles.	Vakul (Moulsari)	Mimusops elengi L.
18.	Jyestha (Scorpis 16°40' to 30°)	Mercury	Leucorrhoea, piles, fistula, bowel disorders, pain in arms & shoulders.	Vriksha	Samanea saman (Jacq. merr.)
19.	Moola (Sagittarius 0° to 13°20')	Ketu	Rheumatism, sciatica, lumbago, hip diseases, pulmonary troubles.	Saal	Shorea robusta Linn.

20.	Poorvashadha (Sagit 13°20' to 26°40')	Venus	Sciatica, dial etes, gout lung-disrders, lumbago, hip-knee disorders.	Sita-ashok	Saraca indica, Linn.
21.	Uttarshada (Sagit 26°40' to 30° and Capricom 0° to 10°)	Sun	Sciatica, eye-troubles, disorders of limbs, consum-ption, gout, toothache.	Kathal	Artocarpus hetero-phyllus Lamk.
22.	Shravan (Capricorn 10° to 23°20')	Moon	Skin troubles, gout, stomach troubles, heart troubles, consumption.	Akvan, Madar.	Calotropis sp.
23.	Dhanishtha Capricorn (Capricorn 23°20' to 30° & Aquarius 0° to 6°40')	Mars	Diseases of legbones, cough and cold, heart troubles, fainting, respiratory diseases, indigestion.	Kekar	Acacia arabica Wilid., Acacia sup. Ham.
24.	Satbhisha (Aquarius 6°40' to 20°)	Rahu	Heart-troubile, gout, arthritis, highblood pressure, fracture of hip bones, palpitation, eye-troubles.	Kadamb	Anthoce phalus cadamba Miq.
25.	Poorvabhadra (Aquarius 20° to 30° & Pisces 0° to 3°20')	Jupiter	Disease of private parts, eye-troubles, small-pox, heart troubles low blood pressure, dropsy.	Aam	Mangifero indica, Linn.
26.	Uttarbhadra (Pisces 3°20' to 16°40')	Saturn	Toothache, fever, small-pox, rheuma ism, stomach disorders, comsumption, dropsy.	Neem	Azadirachta Indica A. Juss
27.	Revati (Pisces 16°40' to 30')	Mercury	Stomach disorders, feet-disorders, gout, cramps, ear-diseases dysentery, tooth-trouble.	Mahua	Madhuca indica Gmel.

Table - II

Lords of the "Trik" houses, that is the sixth, eight and twelveth houses are considered malefic. They produce evil results by location, aspect and association. Any planet posited in the 8th house becomes more powerful and malefic than the actual lord of the 8th house. In order to minimise or neutrarlise the evil effects of planets, wearing of roots of certain medicinal plant species are recommended. They are tabulated below : -

Planet	Diseases likely to produce	Roots of the plant to be worn on the arm, waist or round the neck	Colour of the thread to be used	Auspicious day for wearing
1	2	3	4	5
1. Sun (Surya)	Eye-trouble, high fever, danger from fire, fickle mind.	Aegle marmelos correa (Bel)	Rosy	Sanday
2. Moon (Chandra)	Danger from water, water contamination diseases, asthma, dropsy, highly sentimental	Mimusops hexandra Roxb. (Khirni)	White	Monday
3. Mars (Mangal)	frequent accidents, wounds, piles fistula, impotency	Hemidesmus indicus Br. (Anant mool)	Red	Tuesday
4. Mercury	nervous breakdown, sciatica, dyspepsia, intestinal trouble	Argyreia speciosa Sweet. (Vidhara)	Green	Wednesday

5. Jupiter (Guru)	Heart trouble, diabetes, diseases no. easily detected.	Prema herbacea Roxb. (Bharangi) or Musa paradisiaca (Kela) Linn.	Yellow	Thursday
6. Venus (Sakra)	Sexual diseases, Urinary troubles, diabetes.	Tephrosia purpurea per. (Sarphonka)	White	Friday
7. Saturn	Stomach trouble, Eye-trouble, piles, diseases of private parts.	Urtica parviflora (Bichhu) found in Himalayas only. Instead Ficus religiosa Linn. (Peepal), or Mimusops elengi. (Moulsari) or Azadirachta indica A. Juss. (Neem) may be used.	Black	Saturday
8. Dragon's head (Rahu)	Frequent accidents,k danger from water enlargement of glands, diseases of private parts.	Santalum album L. (Chandan)	Blue	Wednesday
9. Dragon's tail (Ketu)	Piles, fistula, wounds, stomach trouble, fear of animals.	Withania somnifera Dun (Asgandh)	Sky	Thursday

There is difference of opinion regarding the Hindi name "Vidhara". Some astrologers/ayurvedacharyas by "Vidhara" mean Santaloides minus Sch. (Connaraceae) found in the Himalayas whereas some mean lpomoea biloba Forsk. or Ipomoea petaloidea Chois. However a large number of them call Argyreia speciosa sweet as "Vidhara". It is evident from the various schools of thought that "Vidhara" is a member of Convol-vulaceae family and more close to the medicinal qualities of Argyreia speciosa Sweet.

INDIGENOUS FOREST SPECIES USED IN FERTILITY CONTROL

India is the second largest populous country in the world accounting for more than 15% of the world population, accomodated in just 2% of world land area. The population explosion witnessed in recent times has brought, in its. wako, a crop of problems, by way of excessive strain on resources, steadily decreasing standard of life and adverse effect on general environment. Hence population control has assumed great significance and attracted the attention of governmental and other agencies. In this context, search for harmless, inexpensive and effective oral agcnis for fertility control in human beings has gained trerrc-ndous importance. Plants have boon used for this purpose since ancient times. The folklore and ethnic medicines include a large number of plants reputed to be used as oral contraceptives and abortifacients. Modern practitioners of medicines have often debated and dismissed these claims as superstitions. But extraction of a member of valuable modern medicines from traditional herbal remedies set them rethinking and to treat the information of the, ancient and tribal people on the use of herbs and various plant meterials for fertility control purpose as data worthy of investigation. Many of the plants have been subjected to scientific investigation and the results lie scattered in various journals.

Work relating to pharmacological screening of plants has been carried out and are still being carried out for their antifertility purposes. Out of the plants investigated so far, only some have shown promising results. Plants exhibiting positive activity may produce this effect in different vays in females, such as by preventing ovulation, fertilisation and implantation, destroying the zygote or causing Darly abortion.

On the other hand, in males the plant may produce anti-fertil ty effect by acting as an anti-androgen, Spermicidal activity of the plant has also been recorded.

Nagrajan, Jain and Aulakh (1982) have prepared a list of 121 number of plants showing positive antifertility activity. They have also prepared a list of 64 number of plants sh wing negative activity but may show some positive activity under different conditions. The list includes a number of plants species which are either cultivated in Bihar or occur naturally in the Bihar Forests. Such species are listed below :-

I. Plants occurring in forests showing positive antifertil iy activity.

	Plants Name	Part usei	Remarks
1.	Arbus precatorius Linn. (papilionaceae) Rati.	Roots, leaves	Experiments carried out on rats, mice and guinea-pigs
2.	Achyranthus aspera Linn. (Amaranthaceae) chirchir, Latgira	Stem bark, Whole plant	- do -
3.	Adhatoda vasica nees. (Acanthaceae) Vasak	Leaves, alkatoid veceine	- do -
4.	Albizzia lebbeck Benth. (Mimosaceae) Siris	Pods & Roots	Spermicidal activity in human and rat semen.
5.	Azadirachta Indica Linn. (Meliaceae) Neem	Bark	Aqueous extract causes immobilization of human and bovine spermatozoa
6.	Bombax ceiba (Malvaceae) Semul	Roots of human	Uterotonic activity on isolated uterus
7.	Butea monosperma (Lam.) Kuintze (Papilionaceae) Palas	Leaves, flowers, seeds	Experiments carried out on rats & mice.
8.	Clerodendron serratum (Linn.) Moon. (Verbenaceae) Bharangi	Plant Excluding root	Spermicidal activity in human and rat semen.
9.	Costus speciosus (Koen.) Sm. (Zingiberaceae)	Rhizomes	Experience carried out on rats. guinea-pigs, dogs and rabbits.
10.	Cuscuta reflexa Roxb. (Convovulaceae) Amar bel, Alaj-jari	Whole Plant	Experiments carried out on guinea-pigs.
11.	Cyprus rotundus Linn. (Cyperaceae) Metha, Rote-sila	Essential oil	Experiments carried out on female rates.
12.	Euphorbia tirucalli Linn. (Euphorbiaceae) Lanka-sij	Latex	Experiments carried out on female guinea-pigs.
13.	Grewia asiatica L. (Tiliaceae) Phaisa	Seed oil, seed	Experiments carried out on rats only.
14.	Jatropha cureas Linn. (Euphorbiaceae) Euphorbiaceae) Rori, Kamla	Fruits and seeds from the capsule of fruits	Experiments carried out on rats,
15.	Mallotus philippensis Muell-Arg.	Glands & hairs guinea-pigs and rats.	Experiments carried out on female
17.	Ougeninia, oojeinensis (Roxb.) Hoche (Papilionaceae) Panjan	Stem-bark	Experiments on rats.
18.	Oxalis cerniculata Linn. (Oxalidaceae) changeri	Crude drug changeri	Experiments on female rats.
19.	Plumeria acutifolia Poir. (Papilionaceae)Gulaich	Bark	On female quinea-pigs
20.	Psoralia corylifolia Linn. (Papilionaceae) Babchi	Seeds	On female rats
21.	Randia dumetorum Poir. (Rubiaceae) Mainphal	Fruits, pulp, seeds	On female rats, mice and guinea-pigs.
22.	Rauvolfia serpentina Benth. ex Kurr (Apocynaceae) Sarpgandha	Peets	On female rats and mice.
23.	Salix tetrasperma Roxb. (Salicaceae)	Stem bark	On female rabbits
24.	Sesbania sesban Merril (Papilionaceae) Jaint	Flowers leaves	On rats and mice.
25.	Solanum xanthocarpum Sch. & Wen. Solanaceae) Rangni - kanta	Whole plant	On male rats.
26.	Samanea saman Meril. (Mimosaceae) Rain tree	Plant excluding roots	Spermicidal activity in human semen.
		Saponin from whole plant excluding roots	Sperimicidal activity in human seman

27. Terminalia arjuna W. & A. (Combretaceae) Arjun	Bark	Experiment conducted on female rats.
28. Terninalia belerica Roxb. (Commbretaceae)-Bahera	Saponin from fruits	Sperimicidal activity in human semen.
29. Vitex negundo Linn. (Verbenaceae) Sinuar	Seeds	Significant anti-ovulatory activity on rabbits.
30. Withania somnifera Dumal (Solanaceae) Ashwagandh	Roots	Experiment conducted on mice and guinea-pigs.
31. Woodfordia fruticosa Kurz. (Lythraceae) Dhawai	Flowers	Abortifacient in mice.

Besides, there are certain indigenous plants, which do not occur naturally in Bihar forests but are widely cultivated, also show positive antifertility activity. They are as follows : -

1. Allium cepa Linn (Lillaceae) Pyaj	Bulb	Experiments conducted on mice and rats.
2. Allium sativum Linn. (Liliaceae) Lahsun.	Bulb	- Do -
3. Ananas comosus (Linn.) (Bromeliaceae) Merril Ananas	Rhizomes, unripe fruits, leaves	- Do -
4. Cicer arietinum Linn. (Papilionaceae) Chana, gram	Seeds, seed oil	Experiments conducted on mice and rats.
5. Daucus carota Linn. (Umbelliferae) carrot, gajra.	Seed oil, seeds	- Do -
6. Hibiscus rosa-sinensis Linn. (Malvaceae) Shoe-flower, Urhul.	Flowers	Experiments conducted on rats, rabbits. Anti-implantation activity in women.
7. Mentha arvensis Linn. (Laliatae) Pudina.	Leaves	Experiments conducted on female rats.
8. Momordica charantia Linn. (Cucurbitaceae) Kerala	Roots, leaves	Experiments conducted on guinea-pigs and rats.
9. Ocimum sanctum Linn. (Laliatae) Tulsi	Leaves	Experiment conducted on rats.
10. Pisum sativum Linn. (Papilionaceae) Matar	Seeds	Contraceptive activity on female rats and mice.
11. Punica grantum Linn. (Punicaceae) Anar	Fruit sk 1	Raduces fertility on rats and guinea-pigs.
12. Sesamum indicum Linn. (Pedalianeae) Til.	Seeds	Estrogenic activity on female rats.
13. Zea mays Linn. (Gramineae) Makai	Seeds	Ecbolic activity on mice and rats.

II. Plants occurring in Bihar Forests Ishowing negative anti-fertility activity

1. Annona squamosa Linn. (Annonaceae) Sarisha	Seeds	Negetive activity in rats and guinea pigs.
2. Argemona mexicana Linn. (Papaverceae)	Stems, leaves	- Do -
3. Asparagus recemosus Wilid (Liliaceae) Satawara	Alcoholic extract	- Do -
4. Calotropis gigantea (Linn.) R. Br. ex Ait. (Asclepiadaceae) Madaar.	Roots, flowers, leaves	- Do -
5. Caltropis Procera (Ait.) R. Br.	Roots, leaves	Negative activity in rats and guinea-pigs.
6. Casuarina equisetifolia Linn. (Casuarinaceae) Jhau	Leaves, bark, and stems.	- Do -
7. Dolichos biflorus Linn. (Moraceae) Jangli kulthi	Leaves	- Do -
8. Ficus benjamina Linn. (Moraceae)	Leaves	- Do -
9. Leucas aspera Spr. (Labiatae) Gumbha	Whole plant	- Do -
10. Michelia champaca Linn. (Magnoliaceae) Champa	Undripe fruit,	- Do -

11.	Pongamia pinnata pierre. (Papilionoceae) Karanj	Roots	- Do -
12.	Semecarpus anacardium Linn. (Anacardiaceae) Bhelwa	Seeds	No anti fertility activity in rats and guinea-pigs.
13.	Urena lobata Linn. (Malveceae) Bhidi janeted	Leaves bark	No significant anti-fertility activity in rats.
14.	Urginea indika Kunth. (Liliaceae) Kandri, jangali pyaj.	Bulb	- Do -

ROLE OF INDIGENOUS PLANTS IN CONTROL OF DIABETES

Diabetes is a wasting diseases due to insulin deficiency in human beings of all age groups. The most important symptoms of this disease recorded are thirst, excretion of sweet urine and loss in weight. No medicine capable of giving radical cure of diabetes has yet been discovered. Insulin therapy has made great stride in the past fivo decades but with certain limitations. Many herbal remedies, individually or in combination with different formulations such as loaf-powder, pastes, decoctions and infusions, pills, etc. have been recommended in various medical treatises. Most of them are not necessarily of tried value but collected from the folklore, tribal and traditional practices. However, in the recent years, the emphasis has been to identify as many plants as possible which could have control of the disease; if not able to completely cure the same.

Pharmacological screening and clinical trials, as reported by subsequent and recent workers, reveal the presence of hypoglycaemic activity in a large number of plants hitherto not reported. Nagrajan, Jain and Aulakh (1982), in the course of literature survey, have found that about 75 Indian plants have been known to possess the hypoglycaemic activity. Out of these, hardly 10 to 15 plants have been tried on human beings and hypoglycaemic activity of other species have been tried on rats, nice, guinea-pigs etc.

The list of piants having hypoglycaemic activity that occur in the forestsu of Bihar is tabulated below :-

	Plants Name	Part usei	Remarks
1.	Aegle marmelos Corr. (Rutaceae) Bel	Leaves and roots	Hypoglycaemic activity both on rats and humans.
2.	Anacardium occidentale Linn. (Anacardiaceae) Kaju	Bark	Tincture or extract, when administered orally, showed hypoglycaemic activity in normal individuals.
3.	Azadirachta indica A. Juss. (Meliaceae) Neem	Leaves	The composite drug "Tribang shila" containing neem leaves has been tried successful on rats and dogs. Trials on human beings are in progress.
4.	Ficus benghalensis Linn. (Moraceae) Bargad	Bark	Aqueous extract of bark shows hypoglycaemic activity in normal and diabetic patients.

5.	Lagerstroemia speciosa pers. (Lythraceae) Jarul ripe	Leaves and fruits	Oral administration lowers blood sugar.
6.	Momordica chirantia Linn. (Cucurbitaceae) Kerala	Fruits	Widely cultivated for vegetable. Fresh juice of fruits or tablets prepared from the powder of fruit lowered the blood sugar levels significantly in the patient.
7.	Ocimum sanctum Linn. (Labiatae) Tulsi	Whole plant	Planted by the Hindus almost in every (Labiatae) Tulsihouse as a sacred plant. Aqueous solution of whole plant lowers the blood sugar levels.
8.	Pterocarpus marsupium Roxb. (Papilionaceae) Bija	Bark	Decoction of bark lowers the blood sugar in diabetic human beings and increases glucose tolerence. No side effect.
		Heartwood	Aqueous extract of heart wood reduces urine sugar but does not effect blood sugar in diabetic patients.
9.	Rauvolfia serpentina Benth. (Apocynaceae) Sarpgandha	Extract	Oral administration of extract caused decrease of blood glucose and urine glucose in diabetic patients. Reserpine and inhibited the physiological response in diabetic patients.
10.	Syzygium cumini (Linn.) Skeels. (Myrtaceae) Jamun	Seeds	The drug was found to be useful in diabetic patients.
11.	Zizyphus mauratiana Lamk. (Rhamnaceae) Ber	Leaves	Alkaloidal extracts from leaves produced significant reduction in blood sugar and similar to tolbutamide.

There are some more plants, which occur in Bihar Forests, are known to possess hypo-glycaemic activity. They have been tried successfully on rats, mice, guinea-pigs but are still under trial on human beings. The names of such plants are given below : -

Sl.No.	Name	Parts used
1.	Acacia nilotica (Mimosaceae) babul	Seeds
2.	Adhatoda vasica Nees. (Acanthaceae) vasak	Leaves
3.	Albizzia chinensis (Osbeck) Merr. (Mimosaceae) kala siris	Seeds
4.	Allium cepa Linn. (Liliaceae) piaj	Bulb (cultivated)
5.	Allium sativum (Liliaceae) Lahsun	Bulb (cultivated)
6.	Bauhinia retusa Roxb. (Caesalpiniaceae) Kanla	Seeds
7.	Cassia fistuala Linn. (Caesalpiniaceae) Amaltas	Seeds
8.	Daucus carota Linn. (Umbelliferae) Gajjar	Tuberous roots (cultivated)
9.	Dolichos lablab Linn. (Papalionaceae) Seem	Green pods (cultivated)
10.	Eucalyptus citriodora Hook. (Myrtaceae)	Leaves
11.	Ficus racemosa Linn. (Moraceae) Gular	Bark
12.	Ficus Religiosa Linn. (Moraceae) Peepal	Bark
13.	Glycine max Merril. (Papilionaceae) soyabean	Seeds (cultivated)
14.	Lochnera rosea (Linn.) Reichb. (Apocynaceae) sada-barhar	Whole plant
15.	Morus alba Linn. (Moraceae) toot	Leaves
16.	Pinus roxburghii Sarg. (Pinaceae) chir	Bark and roots
17.	Ricinus communis Linn. (Euphorbiaceae) Randi	Roots (cultivation)
18.	Tecoma stans (Linn.) H.B. & K (Bignoniaceae)	Stem (an ornamental shrub.)

RARE AND ENDANGERED MEDICINAL PLANT SPECIES NEEDING SPECIAL ATTENTION FOR PROTECTION

More than 150 species of medicinal importance occur in Bihar but some of the plant species which were reported to occur commonly or abundantly in Bihar about half a century ago have at present become rare or very rare due to over exploitation and have fallen in tho category of endangered species. The glaring examples are Rauvolfia serpentina (L.) Benth ex kurz. and Gloricsa superba L. which are incidentally highly important medicinal plant species. Such plant species whose roots, rhizomes, bark etc. are of medicinal importance deserve maximum attention for their protection as indiscriminate exploitation of such species will adversely effect their natural population in the forest. In view of the above situation, it has become necessary to prepare a list of species which are immediate danger of extinction and need protection/rehibilitation. On the basis of resources survey carried out in South Bihar, specially in Chhotanagpur and SanShal Parganas, the following priorities have been identified :-

A. Species in immediate danger of extinction.

B. Species seriously thereatened.

C. Species of rare occurence and of unique b Manical and medicinal interest.

A. SPECIES IN IMMEDIATE DANGER OF EXTINCTION

(a) Rauvolfia serpentina (L) Benth ex kurz. (Apoc /naceae) Sarpga-idha.

It is a pretty glabrous under shrub 30-60 cm high bright green shining opposite and 3-4 nateby whorled oblong or obovate rather membranous leaves 7.5 to 17.5 cm long, and small white flowers with pink tubes in terminal peduncled bright red cymes. Drupes are polished green and finally black 6 mm to 12.5 mm diam.

Its occurence in India was occasional in damp places. It was more frequent in moist areas of Singhbhum but it has almost become scarce now. Its root is of high medicinal value. It is hypnotic, sedative, specific for insanity, reduces blood pressure, remed, in painful affections of the bowels, in decoctions employed in labours to increase uterine contractions. Hence this species exploited for its roots for several decades has almost led to its extinction from Bihar Forests. In order to cope with the high cemand of its roots under various systems of medicine, the Forest Department, started regular cultivation of this species about two decades back. R. serpentina requires rich soil, moist conditions and partial shade for contimum growth. Under adverse, climatic and edaphic conditions, yield of root suffers and plantation becomes uneconomocal. Hence

Rauvolfia canoccens L., a harder variety though lower in alkaloid content, has been encouraged and is being successfully grown in different parts of India. However, Rauvoifia Serpentina needs rehabilitation in the moist areas of Chhotanagpur forest.

(b) Gloriosa Superba L. (Liliaceae) Karihari

This branched herbaceous climber with beautiful flowers, which was in the past commonly found in the hedges and low jungles of many parts of India is getting depleted day by day. The annual aerial shoot reaches a height of 10m. It flowers and fruits during the rainy season. The aerial shoots are annual and dio down completely by the end of the cold season. The tip of the ovate lanceolate leaves are modified into tendrils. Flowers are greenish yellow in the beginning and filially they become scarlet or crimson.

The perennial tuberous rhizome is upto 30 en long and 1.5 cm in diam. About two decades back it was observed to commonly occur in low jungles of Chhotanagpur and Santhai Parganas. It has dv/indled now and its occurrence has become rare.

From time immemorial the tuberous rhizome was being used in Indian medicine. It is reported to be a tonic, stomachic and antheimintic. Around 1940, colchicine was extracted from the tuber of this species for the first time. It is reported to yield 0.3% of colchicine (dry weight). As colchicine is used in plant breeding work and in medicine, this species has gained much more importance in pharmaceutical industry during the last 4 decades resulting in indiscriminate exploitation of its natural population. Due to tho removal of tubers, the whole plant is destroyed and as such wild population is getting depleted fast. If the present rate of demand for rhizome continues, the pressure of exploitation from the interior forest areas diminishes. Hence a proper check needs to be enforced on the exploitation of this species from the natural forest and the industres who are the main users should be urged to raise their own plantation.

B. SPECIES SERIOUSLY THREATENED

Change is the rule of nature and the vegetation of any place i.; under constant modification, one type leading to another. During the gradual process of evolution of flora it is quite likely that certain species get either eliminated from the area or their domain gets highly shrunk, whereby they have to struggle for their existence. However, this is a long term process. Plant species usually become threatened due to man's influence or stress on the natural flora mainly due to his tendency to exploit the natural resources for his material comforts. Those plant species, whose rhizomes, tubers, bulbs, corms, roots, and bark are of medicinal value are under constant threat of extinction if they are heavily exploited from the natural forest. Certain medicinal plant species which thrive well under total or partial shade also become threatened when the top canopy is emoved due to illicit cutting of tree species. Medicinal croopors and climbing shrubs require support for then proper growth. Due to removal of such tree species which provide support to climbers, the medicinal climbers entangle themselves among shrubby species struggling for proper growth and survival.

In order to locate the medicinal species which are likely to be endangered in the near future, it is necessary to find out the fc.est species whose rhizomes tubers, bulbs, corns, roots, bark etc. are extensively used for medicinal purposes.

(i) Species whose rhizomes, tubers bulbs, corm etc are used medicinally

	Species	Distribution
1.	ASPARAGUS RACEMOSUS WILLD. (Liliaceae) Hindi - Saiawara	In Hill districts of Chhotanagpur and other hills areas
2.	COSTUS SPECIOUS Koen. Sm. (Zingiberaceae) Hindi - keu	In moist localities in the forest of Chhotanagpur and other parts of country.
3.	CURCUMA AMADA Roxb. (Zingiberaceae) Hindi - Amhaldi	In dense forests of India
4.	CURCUMA ANGUSTIFOLIA Roxb. (Zingiberaceae) Hindi - Amhaldi	Forest of India many in Himalayan region.
5.	DISCOREA BULBIFERA Linn. (Diosocoreaceae) Or & Hindi - Pita-alu	Forests of Chhotanagpur and Santhal Paragnas.
6.	DISCOREA PENTAPHYLIA Linn. (Discoreaceae) H - Kanta - alt	Higher altitude in firest of North-East Region.
7.	DISCOREA HISPIDA Dennst. Syn. D. Daemona roxb. (Dioscoreaceae) Ghatw - Taina	In the damper forest of Jharkhand
8.	LASIA ACULEATA Lour. Syn. L. spinosa Thue. (Araceae)	Swampy areas of Himalayan Region.
9.	URGINEA INDICA Kunth. (Liliaceae) H - kandri, gangli-piaz	Frequent on fire lines in the forest and grassy Uttar Pradesh.
10.	ZINGIBER CASSUMNAR Roxb. (Zingiberaceae) H - Banda	In damp shady forest of Chhotanagpur. Common in the Betla and Netarhat.

(ii) Species whose roots are of medicinal use : -

	Species	Distribution
1.	HEMIDESMUS INDICUS R. Br. (Aselepiadaceae) H - Anant moc.	Under shade and mostly in moist localities Himalayan Region.
2.	LEEA ROBUSTA Roxb. (Ampelidaceae) K-Horm, S-Haramda	Frequent along nalas or on cool aspects in Uttaranchal.
3.	PUERARIA TUBEROSA DC. (Leguminosae) H-Ban Kumra Kharw. Patal Kohnra	Chiefly on the side of rocky streams in Uttar Kashi.
4.	SIMILAX ZEYLANICA Linn. Syn. Smilax macrophylla (Liliaceae) Or. Rajdautoni	Frequent in damper forests of North-East Region.

5. SIMILAX PROLIFERA Roxb. as above drier localities but prefers the side of streams. — More frequent than S. zeylanica. In

(iii) Species whose roots, stem and leaves are of medicinal value : -

1. CISSAMPELOS PAREIRA L. (Menispermacea) H - Akandi — Frequent in open and rocky places.
2. TINOSPORA CORDIFOLIA (Wild.) Miers. (Menispermaceae) — Not in our area, occasional in Santhal Parganas.

(iv) Species whose roots and seeds are medicinal : -

1. MUCCUNA PRURITA Hook. Syn. M. pruriens Bak (Leguminosae) Hindi - Kawanch — In dry areas, frequent along streams.

(v) Species whose root, bark and fruit are used medicinally : -

1. OROXYLUM INDICUM Vent. (Bignoniaceae) H - Sona. — Occasional along streams and moist localities in Chhotanagpur & Santhal Parganas. A few years back its bark was sold but exploitation of bark was not done by the purchasers properly and hence auction sale of bark of this species from the forest has been completely stopped.

(vi) Species whose all parts are used medicinally : -

1. WITHANIA SOMNIFERA Dunal. (Solanaceae) Hindi - Aasgandh — Occasionally found in waste ground.

It is recommended that exploitation of the above medicinal species should be restricted to tribals only for their own medical use and the drug industries who are the major users should raise their own plantations to cope with their requirements.

C. SPECIES OF RARE OCCURRENCE OF UNIQUE BOTANICAL AND MEDICINAL INTEREST :-

Some species of the Himilayan sub-temperate region also occur. Such species are of rare occurence and of unique botanical and medicinal interest. Incidentally both fall respectively in different sanctuary areas where the natural eco-system has to be protected for preservation of wild life and species of rare occurrence and medicinal importance are also to be preserved. Such species are as follows:-

1. BERBERIS ASIATICA Roxb. (Berberidaceae) Hindi - Sumlu — A pretty shrub with spinose toothed leaves in Parasnath. Roots medicinal.
2. CLEMATIS GOURIANA Roxb. (Ranunculaceae) Ur. - Golarang — A climbing shrub in kundrugutu (Singhbhum), Parasnath & Netarhat, Stem and leaves medicinal.
3. EQUISETUM DEBILE Roxb. (Equisetaceae) Horse - tail fern. — Stems lax scrambling and often attaining 3m. in height. Along shady

streams in Singhbhum. Ranchi, ascending to Netarhat. Whole plant medicinal.

4. RANUNCULUS PENSYLVANICUS Linn. Along streams in the higher hills of palamu (Ranunculaceae) Hindi - Mamira and Netarhat. Plant medicinal.

5. RUBUS MOLUCCANA Linn. (Rosaceae) A shrub with sarmentose prickly branches. In Netarhat. Leaves and fruit medicinal.

6. RUBIA CORDIFOLIA Linn. (Rubiaceae) Hindi - Manjit A herb scrambling over bushes by means of its scabrid aculeate stems, petioles and whorled leaves. In Netarhat and Parasnath.

7. THALICTRUM FOLIOLOSUM DC. (Ranunculaceae) A fern like erect herb. In Netarhat and Parasnath. Root medicinal.

8. VIOLA PATRINIL DC. (Violaceae) A herbaceous plant with lilac flowers. In Netarhat. Whole plant and flowers medicinal.

The above rare species of unique botanical and medical interest need complete protection and their collection from natural habitat in Netarhat and Parasnath needs to be complete banned.

POTENTIALITIES OF MEDICINAL PLANT CULTIVATION IN INDIA

Forests in Bihar have been overexploited to the various demands of the growing population and in this process the distribution and population of several plant species have been reduced to sucn an extent that if their removal is not restricted in time, some of them may face extinction in course of time. Medicinal plant species occuring in the forests of Bihar have also badly suffered due to this. Medicinal species which have been endangered, need conservation and multiplication. The requirement of drug industries has to be met by raising plantations of the concerned species. Tarai forests of Champaran, Gangetic alluvial plains and Chotanagpu plateau have varied climatic and edaphic conditions. Medicinal plant rpecies suiting to these climatic conditions can be successfully grown to cope with the demand of local drug indurtry and also for export to earn foreign exchange. The medicinal species that can be successfully grown in Chhotanagpur arc as follows :-

(a) Indigenous species which have become rare due to over exploitation :-

1. RAUVOLFIASERPENTINA Benth.
   (Apocynaceae) A pretty undershrub.
   Roots are used in the treatment of hypertension and as a sedative or tranquilising agent.
2. GLORIOSASUPERBALinn.
   (Liliaceae) A herbaceous climber.
   Tubers yield colchicine which is used for plant breeding work and in medicine.
3. COSTUS SPECIOSUS (Keen) Sm.
   (Zingiberaceae) A handsome plant 30-150 cm. high with long leafy stem. Tubers are used as anti-fertility drug.
4. MUCUNAPRURITAHOOK.
   (Leguminosae) A herbaceous twining annual. The seeds of this plant contain L-DOPA an anti-Parkinsonian drug.
5. PUERARIATUBEROSADC. (Leguminosae) A very large woody climber. Tubers are used as tonic, aphrodisiac, diuretic and galac-togogue. They have also shown anti-tubercular and anti fertility activity.
6. ADHATODAVASICA shrub.
   It is not wild in Bihar. It is grown in villages and sometimes as garden hedge. Leaves and roots are used in cough, chronic bronchitis, asthma and as expectorant.

7. CATHARANTHUS ROSEUS G. DON. yn. Vinca rosea. (Apocynaceae) A ḩerbaceous plant 30-60 cm. high with whi 3 or deep rose coloured flowers. Root and stem are medicinal and contain anti-cancerous properties. Occasionally grown in gardens or near temples for its flowers.

8. DATURA METAL Linn. (Solanaceae) A high shrub.
Occasionally grown near Lord shiva temples
Roots, leaves and seeds narcotic and poisonous in high doses and medicinal in small doses. Contains seopolamine, an alkaloid, used as pre-anassthetic in surgery and child-birth, in ophthalmology and in prevention of motion sickness.

9. DIOSCOREA FLORIBUNDA (Dioscoreaceae) A climber. It contains diosgenin used in preparation of birth-control pills.

10. SOLANUMKHASIANUMC.B.Clarke ; (Solanaceae)
A stout much branched under-shrub.
The berries contain solasodine, a commerci il source for steroids. The improved varieties of Iho planting material and the plantation technique may be obtained from Central Medinal Plant Organisation, Lucknow.

MEDICINAL PLANTS, THEIR OCCURRENCE IN INDIA & PARTS OF THE PLANT USED

The medicinal properties of a plant species varies from plant to plant. In some cases the entire plant is used whereas in other cases roots, bark, leaves, flowers, fruits etc. are used singly or in combination for treatment of various diseases. About 200 medicinal plant spec es are found in Bihar. A glance at the following table will give a gist of the medicinal plants, their occurrence and the various parts of the plant species used medicinally under different systems of medicines :-

I. Rhizome, tubers, bulbs, corms etc. used as medicine : -

1. Arundo donax Linn. (Arund) : A reed like grass. Occasional along ditches and streams in Bihar.

2. Asparagus racemosus Wilid. (Satawara) : A climber. F'equent in all districts of Jharkhand.

3. Costus speciosus (Koen.) Sm. (Keu) : A herbaceous plc.nt. common in moist localities in grassy and forest areas of Jharkhand.

4. Cyperu rotundus Linn. (Mutha) : A hecge plant. A common weed.

5. Curcuma amada Roxb. (Ama-haldi) : A herb. Common in the forests of Jharkhand & Uttaranchal.

6. Curcuma augustifolia Roxb. (Tikur) : A herb. Common in Jharkhand & Uttaranchal.
7. Dioscorea bulbifera Linn. (Pita-alu) : A climber. Frequent in Bihar & Jharkhand.
8. Dioscorea pentaphylla Linn. (Kanta-alu) : A climber. Freequent in Jharkhand.
9. Dioscorea hispida Dennst. Syn D. daenona Roxb. (Tain i) : A climber. Frequent in Uttaranchal and Santhal Parganas.
10. Gloriósa superba Linn. (Karihar) : A climber.

    Rare in low jungles of J & K.
11. Lasia aculeata Lour. Syn. L. spinosa Tnw. (Kanta-saru) : A herb.

    Common in Ligirda swamp (Saranda) and occasional alcng muddy streams in Nagaland.
12. Urginea indica Kunth. (Kandri, jangli-piaz) : A herb.

    Frequent on the fire lines in the forest and grassy areas in Assam & Jharkhand.
13. Zingiber cassumnar Roxb. : In damp shady forest of Jharkhand & South India.

II. Species whose roots are of medicinal use :-

1. Ardisia solenacea Roxb. (Gara-boi) : A large shrub along streams in Tamil Nadu.
2. Berberis asiatica Roxb. (Samlu) : A shrub. Rare, in Himalayan region.
3. Carissa spinarum Linn. (Karaunda) : A shrub. Frequent in scrub forest of Himalayan region.
4. Gardenia turgida Roxb. (Dharuk karhar) : A small tree. Gammon in dry forests of Meghalaya & Jharkhand.
5. Glycosmis pentaphylla Correa (Ban-Nimbu) : A shrub. Rare in Karnataka.
6. Hamiltonia suaveolens Roxb. (Selauli) : A shrub wit, foety leaves. Occasional in rocky places in J & K.
7. Hemidesmus indicus R. Br. (Ananta-mul) : A creeper. Co imon under shade and in moist localities
8. Inchnocarpus frutescens R. Br. (Kali-dudhi) : A creeper. Common specially in hedge in moist localities in all districts. Jharkhand & Bihar.

9. Indigofera pulchella Roxb. (Jirahul) : A large shrub. Common in Chatishgarh.
10. Leea robusta Roxb. (Galinil : A shrub. Frequent in Chatisgarh & Madhya Pradesh.
11. Morinda tincotoria Roxb. (Ach.) : A small tree. Valleys of Snghbhum, wild in Santhal Parganas and frequent in Chhoianagpur.
12. Premna herbacea Linn. (Bharangi) : A woody climber. On clay in open grounds, fire lines in Uttar Pradesh and Rajasthan.
13. Pueraria tuberosa DC. (Siali) : A woody climbe'. Frequent in Jharkhand along strems.
14. Rauvolfia secpentina Benth. (Sarpgandha) : A snrub. Rare in damp places in Gujrat & Rajasthan.
15. Smllax zeylanica Lam.
syn. S. macrophylla Roxb. (Atkir, rajdantani) : A climber. .Common in damper forests of Uttaranchal & J & K.
16. Smilax prolifera Roxb. (Atkir, rajdantani) : A clrnber. More frequent than S. zeylanica even in drier localities.
17. Vetiveria - zizaniodes (Lam.) Nash. (Khus) : A perennial grass. In damp and moist areas of Assam & Mizoram.

III. Species whose bark is used medicinally. :-

1. Acacia catechu Willd. (Khair) : A tree usually in dry forest. Common in Bihar & Jharkhand.
2. Adina cordifolia Linn. (Karam) : A tree. Frequent in Uttaranchal, Delhi & Uttar Pradesh.
3. Albizzia chinensis (Osbeek) Merr. Syn. A. stipulate Boiv. : Frequent in valleys of Jharkhand & Assam rare in frequently planted.
4. Anogiessus latifolia Wall. (Dhauwra, dheontha) : A tree. Common on the drier hills in Uttar Pradesh & Himachal Pradesh.
5. Ficus lacor Buch - Ham. Syn. F. infectoria Roxb. (Pakar) : A tree. Frequent in Jharkhand often planted.
6. Grewia tiliaefolia Vahl. (Dhamani) : A tree. Frequent in valleys in Jharkhand & Nagaland.
7. Holoptelea integrifolia Planch. (Chilbil) : A tree. Usually in valley, common in Uttar Pradesh, occasional in Mezoram rare in Megalaya.

8. Hymenodictyon excelsum Wall. (Bhurkund) : A tree. Especially near rocky nalas in drier hills in chotanagpur and Santhal Parganas.
9. Litsea polyantha Juss. (Meda Pojo) : A tree. F'equent in valleys, chiefly near streams and moist areas in Chhotanagpur.
10. Ougeinia oojeinensis (Roxb.) Hoch. : syn. O. chlbergioides Benih. (Sandan, Panjan) : A tree. Frequent in Chhotanagpur & Santhal Parganas.
11. Soymida febrifuga A Juss. (Rohan) - A tree. Common in Hazaribagh & Palamu, rare in Singhbhum.
12. Streblus asper Lour. (Seora) : A tree. Common in Purnea, rare in Chhotanagpur.
13. Saraca indica Linn. (Sita-ashok) : A tree. Valleys of Singhbhum, specially in the valley of Porahat. Often planted in gardens.
14. Terminalia a!ata. Syn. T.. tomentosa W. & A. (A?an) : A tree. Common as an associate of Sal in forests of Chhotanagpur and Santhal Parganas.
15. Mimosa pudica Linn. (Lajvanti) : A trailing shrub. Common in the damper districts in open places, way sides.
16. Oxa'is corniculata Linn. (Chalmori, amboti) : A herb. A common weed in gardens and lawns.
17. Phyllanthus niruri Linn. (Jar-amla) : An erect annual. Common as weed throughout the area.
18. Solanum xanthocarpum Sch. & Wend. (Kaateli, rangui kanta) : A trailing herb. Common along roadsides, wet places.
19. Vernonia cinerea Less. (Sahadevi) : A herb. A common weed of waste ground.
20. Adiantum lunulatum Linn. (A fern.) : Common in shady forests.
21. Salix tetrasperma Roxb. (Jatvetas, vait) : A sma I tree. Occasional along rivers and streams in all districts.

IV. Root and Bark : -

1. Bridelia retusa Spreng. (Kaj) : A tree. Frequent in the drier areas of Himalaya, usually near streams.
2. Mitragyna parvlfolia Korth. (Guri-karam) : A tree Frequent in the plains and valleys of Himalayan Region.

V. Root and Stem :-

1. Rubia cordifolia Linn. (Mangit) : A creeper. Neta hat (Palamu) and Parasnath (Giridih) only of Jharkhand State.

2. Saccharum munja Roxb. (Munj) : A perennial grtss. In open and along river beds in the forest. Common in Sone valley of Bihar.
3. Saccharum spontaneum Linn. (Kaans) : A perenrial grass. Usually along river-banks and swampy localities of Bihar
4. Tinospora cordifolia (Wild.) Miers. (Gurach) : A c reeper. Rare in rocky valley of Jharkhand.

VI. Root, bark and fruit :-

1. Hellcteres Isora Linn. (Aintha) : A shrub. Commor in tho valleys of Bihar & Jharkhand.
2. Oroxylum Indlcum Vent. (Sona) : A tree. Along streams and rivers, and the cool sides in Jharkhand & Assam.

VII. Roots and seed :-

1. Argcmone mexlcana Linn. (Spial kanta) : A prickly herb. Common on waste ground of Bihar, Uttar Pradesh
2. Mucuna prurita Hook. Syn. M. pruriens. (Kavanch) : A creeper. Frequent in dry jungles of Jharkhand.
3. Sida cordifolia Linn. (Kungyi) : A small shrub. In moist peaces, specially in scrub jungles of Jharkhand, Bihar & Uttar Pradesh.

VIII. Hoot, leaves, fruits & seeds :-

1. Annona squamosa Linn. (Saripha) : A small tree. Common in western Himalaya & North-East States.
2. Cassia fistula Linn. (Amaltas) : A tree. In all districts of Chhotanagpur Jharkhand & Bihar.

IX. Root, leaves and seeds : -

1. Abrus precatorius Linn. (Rati) : A creeper. Frequent .in Ch lotanagpur & Santhal Parganas, sometimes seen on brushes also.
2. Ageratum conyzoldes Linn. (Vchuinti) : A herb. A very common weed but chiefly in damp places.
3. Aıangium salvifolium syn. A. lamarkii Thw. (Dhela; Akolo) : A tree. Common in Jharkhand.
4. Capparis zeyianica Linn. Syn. C. horrida. (Ardanda) : A climbing thorny shrub. Common throughout the Jharkhand in hedges, thickets and along rocky nalas.

5. Cassia tora Linn. (Chakora) : An annual herb. In all districls; a common weed during the rains.
6. Cassia occidentalis Linn. (Kasandi) : A common weed. on waste ground during rains.
7. Vitex negundo Linn. (Nirgundi, sinduar) : A high shrub. Frequent in waste ground and hedges in all Jharkhand.

X. Species whose leaves are of medicinal value :-

1. Albizzia procera (Roxb.) Birth. (Safed siris) : A tree. In camp places in all districts of Bihar, West Bengal.
2. Cassia sophera Linn. (Kasaunda) : A shrub. In all districts of Jharkhand & West Bengal.
3. Gmclina arborea Linn, (gamhar) : A tree. Frequent on cool side of hills throughout Jharkhand.
4. Kydia calycina Roxb. (Pula) : A tree. On valleys and hill sleepes in Uttar Pradesh Jharkhand.
5. Nyctanthcs arbortristis Linn. (Har-singhar): A small tree. Frequent on the hills of Jharkhand & Uttaranchal. Also raised in gardens.

XI. Flowers used medicinally :-

1. Woodfordia fruiticosa Kuntz. (Dhawai) : A shrub. Common ir Jharkhand gregarious on clay soils.

XII. Fruits : -

1. Tamarindus indica Linn. (Imli) : A tree. Rare in forests but commonly grown in vilages. Naturalise near South India.
2. Terminalia belerica Roxb. (Bachera) : A tree. Occasionally found in valleys in Sal forests of Jharkhand & Uttaranchal.

2a. Terminalia chebula Retx. (Harra) : A tree. As above.

3. Tribulus terrestris Linn. (Chota gokhru) : A herb. Common along roadsides and pastures, specially in sandy soil throughout the area.

XIII. Seeds : -

1. Bucnnania lanzan Spreng. (oil and gum also) Piar : A tree. Common in Jharkhand & West Bengal.
2. Dolichos biflorus Linn, (decoction also) (Jungli kulthi) : A twiner. Rare in forest. Common pulse crop in Jharkhand.

3. Psoralia corylifolia Linn, (seed oil also) Babchi : A herb. Occasional in the sone catchment of Uttaranchal.
4. Centratherum anthelminticum (Willd) Kuntz. Syn. Vernonia anthelmintica Willd. (Sowraj) : A shrub. Frequent in Jharkhand, usually in shade near villages.

XIV. All Parts : -

1. Achyranthus aspera Linn, (chirchiri) : A shrub. Common in vaste ground near villages.
2. Andrographis paniculata Nees. (Kalmegh) : A herb. Com non specially on cleared stony places in the forests of Uttar Pradesh, West Bengal.
3. Bacopa monnieri (L.) Pennel syn. Herpestis monieri (L.) H.E & K.(Aindri, Barui) : Trailing herb.
4. Barleria cristata Linn. (Jhinti) : A shrub. Common in Jharkhand in rocky hills.
5. Barleria prionitis Linn. (Sairayak, Katsari) , A shrub. Frequent in Jharkhand, West Bengal, Andhra Pradesh.
6. Cassytha filiformis Linn. (Amar-beli) : A gr 3enish twining parasite.
7. Centella asiatica (Linn.) Urban. Syn. Hydrocotyle asiatica Linn. (Beng-seg, Brahni) : Trailing herb. Very common in wet places in Jharkhand.
8. Cuscuta reflexa Roxb. (Amar-bel) : A twining parasite. Frequent in Forest of Uttaranchal, West Bengal & Jharkhand.
9. Cynodon dactylon (Linn.) Pers, (Doob) : A common grass.
10. Eclipta alba Hassk. (Bhangra) : A herbaceous plant. Common in pasture lands and roadsides in Jharkhand.
11. Euphorbia hirta Linn. (Dudhi) : A herb. A common weed everywhere roadsides etc. in Andhra Pradesh Gujrat.
12. Leucas aspera Spreng. (Chota-halkusa) : A herb Common in waste ground and cultivation fields.
13. Marsilia minuta Linn. (Sunsuniya) : A herb. In wet places throughout the country.
14. Withania somnifera Dunal. (Ashgand) : An under shrub. Occasional in waste ground.

XV. Root, leaves and flowers :-

1. Ficus retusa Linn. (Kamrup, Buti-hesa) : A tree. Occasional in Jharkhand, Gujrat, Maharasthra.
2. Jasminum multiflorum And. Syn. J. pubescens Willd. (Chameli, Mogra): A shrub. Common in Gujrat, Maharastra, Andhra Pradesh, Assam.

XVI. Root, bark and flowers :-

1. Bauhinia purpurea Linn. (Koinar) : A tree. Frequent in all districts of Jharkhand often grown in villages.
2. Bauhinia Variegata Linn. (Kachnar) : A tre ; of dry rocky hills common in all areas of Uttar Pradesh, Maharastra, Jharkhand.

XVII. Root and leaves :-

1. Calotropis gigantea (Linn.) R. Br. (Akawan) : A shrub. Frequent in open waste ground throughout the country.
2. Calotropis procera (Alt.) R. Br. (Akawan) : A shrub. As above.
3. Clerodendron indicum (Linn.) Syn. C. Siphonanthus (Bharanghih) : A shrub. In moist places in Jharkhand.
4. Clerodendron serratum (Linn.) Moon (Baranci) : A shrub. Common in Jharkhand, West Bengal.
5. Dalbergia sissoo Roxb. (Shishum) : A tree indigenous in champaran. An important plantation species in Jharkhand.
6. Dalbergia volubillis Roxb. (Nari-Siris) : A shrub. Along ravines and nalas and eroded areas.
7. Elephantopus scaber Linn, (samdulan) : A nerb common in open ground.
8. Clssampelos pereira Linn. (Akanadi, ranu. red): A creeper. Frequent in open and rocky valley in Jharkhand.
9. Paedaria foetida Linn, (gandh-bail) : A climber. Occasionally along Koil river in Jharkhand.
10. Tephrosia purpurea (Linn.) (peas.) (Sarphonka) : A Shrub. Common in Jharkhand, Uttar Pradesh & West Bengal.

XVIII. Root, bark, leaves, flowers & fruit : -

1. Careya arborea Roxb. (Kumbhi) : A tree occasional in valley & moist places in Uttaranchal, Uttar Pradesh & North East States.

XIX. Root, bftrk. flowers and fruits :-

1. Aegle marmelos Corr. (Bel) : A tree. Frequent in Jharkhand Bihar, Uttar Pradesh, Maharastra and Gujrat
2. Bombax Ceiba syn. Salmalia malabaricum (Semul) : A tree. Frequent throughout the area especially in cool valleys of Dehredun.
3. Cleistanthus collinus (Roxb.) B & H.F. (Kangali) : A tree. Common on the hills of Himalayan region & Jharkhand.

XX. Root, bark, leaves, fruits and seeds :-

1. Azadirachta indica A. Juss. (Neem) : A large, tree. Frequent in Hazaribagh district but usually planted along roadsides.
2. Embilica officinalis Gaertn. (Aonla) : A tree Common throughout Jharkhand.
3. Ficus benghalensls Linn. (Bargad) : A troe. In all districts, wild and cultivated.
4. F. religiosa Linn. (Peepal) : A tree. In all districts, wild and cultivated.
5. F. racemosa Linn. Syn. F. glomerata Roxb. (Gular) : A tree. Frequent in all districts of Jharkhand.
6. Melia azedarach Linn. (Bakain) : A tree. Often plantec along roadsides and in plantations.
7. Michelia champaca Linn. (Champa) : A tree. ThalkobaJ and karampoda in Jharkhand. Often planted.
8. Pongamia plnnata (Linn.) Morr. (Kanni) : A tree. Largely planted along roadsides, villages etc. rare in the forest.
9. Zizyphus mauritiana Lamk. Syn. Z. Jujuba Lamk. Nor Mill (ber) : A tree. Frequent in Jharkhand. Largely cultivated for fruits and as lac-host in Jharkhand.

XXI. Bark, leaves and seeds :-

1. Albizzia lebbeck Benth. (Sin's) : A tree. Common in Jharkhand & West Bengal. Everywhere planted specially along roadsides.

XXII. Bark, leaves and fruits :-

1. Annona reticulata Linn. (Aanta) : A tree. Common in Pirnea. Often planted.
2. Cordia dichotoma Forst. F. Syn. C. myxa. Roxb. (Lasora) : A tree. Frequent along streams and moist localities.

3. Croton oblongifolius Roxb. (Putri) : A high shrub or sn all. Common in Jharkhand and Bihar & Uttar Pradesh.

4. Syzygium cumini (Linn.) Skeels. (Jamun) : A big tree. Common in Uttar Pradesh & Jharkhand Wild or cultivated.

5. Terminalia arjuna W. & A . (Arjun) : A big tree. Along nalas in Jharkhand & Uttar Pradesh. Also planted along roadsides.

XXIII. Bark & Flowers :-

1. Desmodium gangeticum DC. (Salpani) : A shrub. Common throughout the country.

2. Demodium pulchellum Benth. ex Baker (Bir kapi) : A shr ib. Common in Jharkhand, West Bengal.

3. Madhuca indica Gmel. (Mahua) : A tree. Common in West Bengal & Jharkhand but usually confined in the I'lls. Common in villages of tribal area.

XXIV. Bark and fruit : -

1. Acacia arabica Willd. (gum also) : A tree. Common along roadsides, railway lines and spoil banks.

2. Diospyros peregrina Gurke. Syn D. embryopteris Pers. (f.ale) : A tree. Seed oil also rare along streams in Jharkhand, frequent in Saranda Forest.

3. Randia dumetorum Lam. (Mainphal) : A large shrub. Freequent in Jharkhand & Uttar Pradesh.

4. Zizyphus oenoplia Mill. (Makai) : A thorny straggling shrub. Common in Jharkhand, West Bengal and Uttar Pradesh.

XXVI. Bark and leaves :-

1. Acacia pennata (Linn.) Wilid. (Arar): A thorny climber. Common in Jharkhand along moist places.

2. Anthocephalus indicus A. Rich syn. A cadamba Miq. (Ka-iam) : A large tree. Common in moist and damp places, specially in Jharkhand (Singhbhum). Usually planted.

3. Chloroxylon swietenia DC. (Bharhul) : A tree Frequent in Jharkhand & West Bengal. Common in Latehar (Palamu).

4. Erythrina varuegata Linn. Var. oriental's (Farhad) : A tree. Rare in forest commonly planted.

5. Lannea coromandeica (Houtt.) Merr. (Jhingan, dokeo) : A tree. Syn. L. grandis Engl. Common throughout the country.
6. Litsea chinensis Lam. Syn. L. sebifera Pers. (Garbijaur. Mcdha) : A tree. In valleys and cool aspects of hiils in Jharkhand.
7. Pterocarpus marsupium Roxb. (gum, wood also) (Paisar) : A tree. Frequent in Jharkhand & West Bengal specially on slopes near the nalas.
8. Vitcx pcduncularis Wall. (Nag bael) : A shrub. Frequent in moistor places in Jharkhand, West Bengal and Uttar Pradesh.

XXVII.Leaves & flowers :-

1. Clerodendron infortunatum Linn. (Bhant) : A shrub. Common in Jharkhand, West Bengal & Bihar.
2. Cochlospermum religiosum (Linn.) Alston. Syn. C. gossyp urn DC. (Gaigal) :

   A tree. Rare in Jharkhand but frequent on dry hills of West Bengal & Bihar.

XXVIII. Leaves & fruits :-

1. Garuga pinnata Roxb. (Jiga, kekar) : A tree. Frequent in valleys in districts of Jharkhand.
2. Putranjiva roxburghii Wall. (Jiyaputra) : Atree grown as garden Plant.
3. Solatium nigrum Linn. (Makoi) : A herb. A free Jent weed in Singhbhum in Jharkhand.

XXVIII. Bark & Seeds :-

1. Artocarpus lakoocha Roxb. (Barhar) : A tree. Occasional in Jharkhand.
2. Calastrus paniculatus Willd. (Malkagni) : A c! mbing shrub (oil also). Common in West Bengal, Jharkhand.
3. Holarrhena antidysenterica Wall. (Dudh-kortiya) : A large shrub. Common in Jharkhand, West Bengal and Bihar.
4. Schleicnera oleosa (Lour.) Oken. (kusum) : A karge tree (oil also). Frequent in Jharkhand. Planted as roadside avenues.
5. Somecarpus anacardium Linn. F. (Bhelwa) : A tree, (oil) Frequent in Jharkhand, West Bengal & Bihar.

6. Shorea robusta Gaertn. F. (resin also) (Saal) : A tree throughout Bihar except the Gangetic plain.

XXIX. Leaves & seeds :-

1. Bauhinia vahlii W. & A. (Mahulan) : A large cliriber. Common in Jharkhand, We3st Bengal and Bihar.

XXX. Gum only:-

1. Boswellia serrata Roxb. (Salai) : A tree. Comnon on dry hills of Jharkhand, Uttar Pradesh & West Bengal.
2. Bauhinia retusa Roxb. (Kanla, tewa) : A tree. Frequent on the northern slopes of Jharkhand.
3. Gardenia gummifera Linn. F. (Dekamali) : A tree. Rare in Jharkhandon dry soil with a clay soil.
4. Sterculia urens Roxb. (Keonjhi) : A tree. Frequ jnt in the dry rocky hills of Jharkhand, Gujrat & Maharastra.

XXXI. Bark, leaves, flowers, fruits :-

1. Butea monosperm (Lam.) kunhtze (gum also) P ilas : A tree. Common in Jharkhand and very common in Palamu & Gaya in Bihar.
2. Tectona grandis Linn. F. (Sagwaan) : A tree. In lantations only.
3. Stereospermum suaveolens DC. (Padhal) Adhkapan : A tree. Frequent in all districts Jharkhand.

XXXII. Glands, hairs of fruit : -

1. Mallotus phillipensis Muell (Kamla, rori) : A small tree. Throughout Jharkhand & Bihar & Gaya, specially in moist places.

XXXIII. Leaves oil:-

1. Cymbopogon martini (rusa-ghas) : A tall sweet scented grass. Chiefly on rocky hiils with sparse forests in Jharkhand & West Bengal.

XXXIV. Latex :-

Vallaris solanacea O. Mze. syn. v. heynei sprent (Kokur-botur) : A large creeper. In valleys of Jharkhand and West Bengal.

There are a number of medicinal species which either donot occur or are rare in our forest but they have been planted along roadside as avenues or as g irden/hedge plants in small or large compounds. They are mentioned below :-

a) Planted along road sides :-

1. Adansonia digitals Linn, (leaves, fruit) (Baobad, khurseni im i) : A tree. Planted in along road side in Jharkhand.
2. Adenanthera pavonina Linn. (Leaves, seed) : Pakta - chandan. A tree. Planted along road side in Jharkhand, West Bengal and Gujrat.
3. Ailanthus excelsa Roxb. (Bark, leaves) Ghorkaranj : A tree. Frequent in Jharkhand, West Bengal.
4. Alstonia scholaris R. Br. (Bark, latex) Chatwan : A tree. Rare in moist places of Jharkhand. Mostly cultivated.
5. Artocarpus heterophyllus Lam. Syn. A. integrifolia Linn. F. Kathal : A tree. Common in villagaes, planted along roadsides also.
6. Cedrela toona Roxb. (Tun) : A tree. Grown along roadsides in Jharkhand & West Bengal.
7. Lagerstroemia speciosa (Linn.) : Pers. Syn. L. flosreginae (Jam!) A tree. Rare in Jharkhand along the stream. Usually planted along the roadside or as a garden plant.
8. Pterospcrmum acerlfollum Mochkunel : A big tree. Along roadsides in Jharkhand, West Bengal and Maharastra.
9. Thesposia populnea Sol.ex. Corr. (Pan-paras) : A small tree. Planted in along road side Jharkhand.
10. Trewla nudlflora Linn. (R.) (Pani-garrhar) : A tree. As roadsides tree in Jharkhand, West Bengal and Maharastra.

b) Medicinal species as garden/hedge plants :-

1. Adhatoda vasica NCOS, (vasaka) : A shrub. Common , in villages as hedge.
2. Argyrcla speciosa swoot. : A handsome creeper. Cultivated in gardens.
3. Anacardium occidentale Linn. (Bark, root, fruit) Kaju : A tree. Large scale plantation in Jharkhand & Uttar Pradesh.
4. Datura metel Linn. (Dhatura) : A shrub.
5. Elaeorcarpus ganitrus (rudraksh) : A tree. Often planted in Purnea, Jharkhand, West Bengal and Orissa.
6. Euphorbia antiquorum Linn, (root, stem, latox) : A shrub. Grown as hedge in villages.
7. Euphorbia nerrifolia Linn, (root, latex) : A shrub. Common as village hedge.

8. Jatropha curcas Linn, (root, loaves, seed) Baghrandi : A shrub. Common in village hedges in Jharkhand, U.P. & West Bengal.
9. Justlcia gendarussa Burm. (leaves, stem) Nili-nargar.di : A shrub. Grown as hedge in Jharkhand State.
10. Lawsonia enermis Linn. (Mehndl) : A shrub. Frequent as hedge or garden plant.
11. Melia composite Willd. (Arangaka) : A tree. Planted in residential compounds in Jharkhand.
12. Moringa oleifera Lam. (Bark, seeds) Sohjan, munga : ,\ tree. Often planted for its flowers and fruits.
13. Mimusops eiengi Linn. (Maulsari) : A tree. Very freque illy planted in temples, gardens and roadside.
14. Plumbago zeylanica Linn, (chita) : A shrub. Usually a nong rocks Jharkhand cultivated for flowers.
15. Piurneria rubra var. acutifolia Bailey (Gulainch) : A sm. I! tree. In garden and temples.
16. Secbania grandiflora (Linn.) Pers. (Agastha) : A small :ree. Often planted for its flowers.
17. Sesbania sesban (Linn.) Merr. Syn. S. aegyphtica poii (Jaiyanti, jaint) : A small tree. Planted in gardens, sometimes on waste ground also.
18. Spondias pinnata Kurtz. (Amra) : A tree. Rare in the forests commonly planted in villages.
19. Santalum album Linn. (Chandan) : A tree. A few trees in Parash Nath in Jharkhand.
20. Thevetia peruviana (Pers.) Schum. Syn. T. neriifolia luss. (Kanair) : A small tree. Often planted for its flowers.
21. Vinea rosea. Linn. Syn. Lochnera rosea (Linn.) Reichb. ,'Sadabahar). : As under shrub grown for its flowers.

DEALERS, PRODUCERS AND EXPORTERS OF CRUDE DRUGS AND AROMATIC PLANTS IN INDIA

Name of the party	Items being dealt with
1	2
	NORTHERN ZONE
	1. PUNJAB
Ashok Kirana Bhandar, Chowk Passion, Amritsar.	Herbal crude drugs.
Bharat Agencies, 64, Mewa Mandi, Amritsar	- do -
Deepak Bawa & Company, Majith Mandi, Amritsar-143001	Aconite (Aconitum spp.); Chiretta (Swertia chirata) buch. Han Gentians roots (Gentians kurroo Royle); Picrorhiza kurroa Royle ex Benth; Rauvolfia serpentina (L.) Benth. ex kurz.; Saffron (Crocus sativus L.); Taraxacum roots (Taraxacum officinale webber); etc.
Dinesh Gurban Bawa Sons, Majith Mandi, Amritsar.	Herbal crude drugs.
Gyan Chand Ram Gopal, Majith Mandi, Amritsar.	- do -
Himalayan Traders, Katara Dulo, Amritsar-143001.	Angelica roots (Angelica spp.); Bach (Acorus calamus L.); Colchicum lutcum Baker; Heracleum c.andicans Wall. ex. DC; Iris root (Iris spp.).
Kashmir Ayurvedic Works, Azad Nagar, Putalighar, Amritsar-143001	Attis (Aconitum heterophyllum Wall.) Dalchini (CInnamomum tamaia Nees & Eberm.); Nagkeshar (Mesua ferrea L.); etc.
Krishna Chandra Brij Kumar, 539, Majith Mandi, Amritsar-143001	Celery seeds (Apium graveolens Linn.)
Krishna Kapoor & Co. Woolands, The Mall, Amritsar.	-do-
Krishna Pharmacy Pvt. Ltd., Hall Bazar, Amritsar-143001	Bawchi (Psoralea corylifolia L.); Croton (Croton tiglium L.); Datura (Datura spp.); Kalihari (Gloriosa superba L.); Kunth mithi (Saussurea lappa Clarke); etc.
Mehta Pharmaceuticals Pvt.. Ltd., Chheharata, Amritsar.	Herbal crude drugs.
Orient Traders, 615/VI, Bagh Jhanda Singh, First Floor, Amritsar-143001	Aconite (Aconitum spp.); Angelica (Angelica archaengellca L. and A. glauca Edgew.); Bawchi(Psoralea corylifolia L.); Berberis aristata DC; Colchicum seeds (Colchicum luteum Baker); Costus (Saussurea lappa Clarke); Peganum harmala L.; Valerian roots (Valeriana walllchil DC); etc.
Roshan Lal Sham Sunder, 502-51, Akali Market, Amritsar.	Aconitum heterophyllum wall; Calamus (Acorus calamus L.); Costus (Saussurea lappa Clarke); Cydonia (Cydonla vulgarls Pors.); Gentiana (Gentiana kurroo Royle); Harmal seeds (Peganum harmala L.); Liquorice (Glycyrrhiza globra L.), Soapnut (Sapindus trifoliatus L.); Valerian (Valerian wallichii DC); etc.
S.D. Mehta & Company, 51, Rani Ka bagh, Amritsar-143001.	Brahmi (Hydrocotýie asiatica L.); Costus (Saussurea Inppa Clarko); Glycyrrhira

	(Glycyrrhiza glabra L.); Kutki (Picrorhiza kurroa Roylo ox Benth.); Rasout (Berberis aristata DC.); Talispatri (Taxus baccata L./Abies webbiana Lindl.); etc.
Surun Singh Lachhman Singh, Katra Hari Sing, Amritsar.	Celery seed (Apium graveolens L.); Kamila powder :Mai!otus philipinensis Lam.) (Muoll); Valorian roots (Valcrinna wallichli DC.); etc.

2. HIMACHAL PRADESH AND JAMMU AND KASHMIR

Lahul Kuth Growers Co., (Opp. Marketing Society Ltd.), Himachal Pardosh, Lahul.	Hebal crude drugs.
P.S. Jamwal & Sons, Kachi Chowni, Jammu-180001	-do-
Shiekh Nazir Ahmed, Forest Lesse & Contractors, 15, New Iddgha Building, Park Road, Baramulla, Kashmir.	Ephedra (Ephedra spp.); Ferula roots (Ferula jaeschkeana Vatke); Horse chestnut (aesculus hippocastanum L.); Iris spp; Viburnum Seeds (Viburnum sp.); etc.

3. DELHI

Amar Kirana Co., 330 Khari Bawli, Delhi-6	Commission agent for herbal crude drug.
Anatram Lachmandas Agarwal, 6692/1, Khari Bawali, Delhi-110006.	Herbal crude drug.
Aruna Brothers, Post Box 352, New Delhi.	- do -
Asian Drug Co., 1244, Chah Rahat, Delhi -110006.	Aconite (Aconitumk spp.); Atropa acuminata Royle ex Land. Belladonna (A. belladona L./) Berberis (Berberis spp.); Costus (Isaussurea lappa Clarke); Digitalis lanata Ehrh., Parseley (Petroselinum crispum) (Mill.); Valerian (Valeriana walichii DC.) etc.
Assam Drug Co.. 1244, Chahraht, Delhi-110006	Herbal crude drugs.
Dewan Chand Dholari Das & Co., H-5/4, Model Town, Delhi-110009.	Belladonna (Atropa acuminata Royle ex. Lindl./ A. belladonna L.); Hubulus dana (Myrtus communis L. etc.
Dekkon International, Flat No. 12, Khan Market, New Delhi-110003.	Herbal crude drugs.
Drug & Alkaloid Co., P.B. No. 1297, Delhi-110006.	Gloriosa superba L., Iphigenia stellata Blatter; and other herbal crude drugs.
Himalayan Drugs, Herbs & Alkaloids Syndicate, C4/33A, Lawrence Road, New Delhi - 110035	Herbal crude drugs.
K-7, Green park extension, New Delhi-110016.	Acacia (Acacia spp.); Ambrette (Hibiscus abelmoschus L.) Ammi maus L.; Berberis (Berberis spp.); Coptis teeta Wall; Dioscorea (Dioscorea deltoidea W. ex Kunth.); Hydrocotyle asiatica L.; Valerian (Valeriana wallichii DC.) Vinca rosea L. (Catharanthus rose L.) G. Don; etc.
Mohandas Lalchand, 2445-46. Tiiak Bazar, Delhi-110006.	Various crude drugs, medicinal raw herbs and pharmaceutical raw materials.
Ravashi Industrial Corporation, 176/4, Baba Udham Singh Marg, Khurji, Delhi-110051.	Ayurvedic and Unani medicinal crude drugs; Henna powder (Lawsonia inermis L.); etc.

Rosma Overseas Pvt. Ltd. 4-6, Vandhana, 11-Tolstoy Marg, New Delhi-110001.	Herbal crude drugs.
Shri Rajvaidya Shital Prasad & Sons, 1331. Chandni Chowk, Delhi-110006.	Herbal crude drugs.
S. Chandra Enterprises, 34, Ferozeshah Road, New Delhi-110001.	- do -
Timex Agencies, 4 E/9, Jhandewalan Extension, Link Road. New Delhi - 1100035.	- do -
Zubex India, 8, Poorvi Marg, Vasant Vihar, New Dehli-110057.	- do -

4. UTTAR PRADESH

Balgovind Bhagwati Prasad, Nayaganj, Kanpur - 208001.	Herbal crude drugs.
Baijnath Santosh Kumar, R/o & P.O. Manekpur, Dist. Banda.	Aawanla (Phyllanthus emblica L.). Himalayan herbal crude drugs.
Bhesaja Sahkari Vikas Sangh, P.O. Chamoli, Dist. - Chamoli (U.P.).	Himalayan herbal crude drugs.
Brij Lal Bal Krishna, Nayaganj, Kanpur - 208001	Dioscorea (Dioscorea deltoidea Wall) ex. Kunth., Dioscorea Spp.); Hyoscyamus (Hyoscyamus nlger L.); Jatamansi (Nardostachys jatamansi D. Don. (DC.); Kasturi (Hibiscus abelmoschus L.); Valerian (Valeriana wallichii DC.); etc.
Desh Rakshak Avshadhalaya, Kankhal-249408 Hardwar.	Herbal crude drugs.
Doon trading company, Panetitwari, P.O. Premnagar, Dehradun-248007	Herbal crude drugs.
Ganesh Singh Dariyal, Jari Booti Vyapari, Munsiyari (Pithoragarh).	- do -
Garg Banaushadhi Bhandar, Vijay Garh, Aligarh.	Apamarga (Achyranthes aspera L.); Arni (Clerodendrum phlomidis L.F.); Kharetti (Sida cordifolia L.); etc.
Gulab Chandra Phool Chandra, Sahadat Ganj, Lucknow.	Herbal crude drugs.
Himalaya Herb Stores, Madho Nagar, P.Box No. 130, Saharanpur - 247001	Aconite (Aconitum spp.); Aristolochia bracteata Retz.: Belladonna (Atropa acuminata Royle ex. Lindl./A. Belladonna L.); Hyoscyamus niger Linn.
Himalayan Herbalist, P.O. Ranibagh, Nainital.	Charila (Parmelia spp.); Chaupattia (Marsilea spp.); Mallotus phillipensis (Lamks) Muell, Pippali (Piper spp.); Tager (Valeriana wallichii DC.); etc.
The Indian Herbs, Post Box No.5, Sharda Nagar, Saharanpur - 247001	Herbal crude drugs.
Inder Singh Rawat, P.O. Raturisera, Patte Barsari, Nakuri (Uttarkàshi).	Atees (Aconitum heterophyllum Wall) Kutki (Picrorhiza kurroa Royle ex, Benth.); Masi (Nardostachys Jatamansi (D. Don. DC.); etc.

Jagdeo Ram Madan Gopal, Kirana Merchant, Ahraura, Mirzapur, U.P.	Marorphali (Helicteres isora L.); Nagarmotha (Cyperus rotundus L.); Tejabal (Zoathoxylum alatum Roxb.; Z. rhetsa DC.); etc.
Indo HJerbolex & Chemolex Traders, House No. 124, Rajsadan, Dwarkapuri, Muzaffar Nagar-251001 U.P.	Ambrette seeds (Colchicum sp.); Gloriosa superba L. seeds; etc.
Lucknow Kirana Co., Subhash Marg, Lucknow.	Khutmi, Khubbaji (Malva spp.) : Makoi (Solanum nigrum.); Satawar (Asparagus spp.): Sugandhabala (Valeriana wallichii DC)
Mahesh Trading Co., 360/127, Mata Din Road, Sahadat Ganj, Lucknow.	Amma Haldi (Curcuma amada Roxb.) : Babunah (matricaria chamomila L.). Kapur kachri (Hedychium spicatam Ham. Ex. Smith) : Rauvolfia serpentina (L.) Benth ex kurz.: Satawar (Asparagus racemosus will etc.)
Mohan lal laxmikant, Safadarganj, Northern Railway, Barabanki.	Datura stramonium (L.): Gloriosa superba L. seeds/roots; kahu (Lactuca sativa L.) kusum (Carthamus tinctorius L.); Malkagni (Celastrus paniculatus willd.); Poppy (Papaver somniferum L.) etc.
Motilal Jain & Co. 16, Lakhi Bagh, Dehra Dun.	Hebral crude drugs.
Northern India pharmaceutical Lab., 12, Govind Market, Saharanpur-247001	Aconite roots (Aconitum spp.); Ammi majus L. Berberis roots (Berberis spp.); Gloriosa superba L. seeds: Rauvolfia roots (Rauvolfia serpentina L.) Benth. ex Kurz; etc.
Parvatiya Sahkari Bhesaja, Vikas evum Kraya Vikraya, Sangh, Pithoragarh, U.P.	Akhrothal (Dandasa) Juglans regia L.: Jatmansi (Nardostachys Jatamansi) (D. Don) (DC.); Kapur kachri (Hedychium spicatum Ham. Ex Smith); Crude drugs and medicinal herbs of alpine and temparate region such as; Kutki (Picrorhixa kurroa Royle ex benth.); etc.
Pattu Lal Bala Prasad, Sahadatganj, Lucknow.	Herbal crude drugs.
Ramji sahu, P.O. Ramnagar (Varanasi).	Amla (Embilica officinalis Gaertn.) Bahera (T., bellerica (Gaertn.) Roxb.); Hara (Tarminalia chebula Retz.); etc.
Pharmaceutical Crude Drugs Enterprises, Kosi Road, Ramnagar - 244715 (Naini Tal).	Herbal crude drugs of Himalayan region.
Sharda Brothers, Herb Supplies & Commission Agent, Kathgodam, Naini Tal, U.P.	Pakhan bhed (Bergenia ciliata) (Haw.) sternb.; Ritha (Sapindus trifoliatus L.); Sugandhabala (Valeriana wallichii DC.); etc.
Sahkari Vikas Sangh Ltd., Chamoli U.P.	Dagger phool (Parmelia spp.) Dioscorea Discorea deltoidea wild kapur kachari (Hedychium spicatum Ham. Ex. Smith); Kutki (Picrorhiza kurroa Royle & Benth.); Pati (Artemisia vulgaris Linn.); Vacha (Acorus calamus L.) etc.
Sobhan Singh, contractor, P.O. Harsil (Uttarkashi).	Dhoop (Jurinia macrocephala HK & Non DC.); Kutki (Picrorhiza kurroa Royle ex Benth.); Rudrawanti (Astragalus spp.); etc.
The Himalaya drug Co., 22-24, Mahatma Gandhi Road, Dehradun.	Herbal crude drug.
Trishul Aromatics & Alkoloids, Srinagar-246174	- do -

Shekharco Import Export, Post Box No. 256, Muzaffar Nagar - 251001, U.P.	- do -
Vishal Traders, Sahadat Ganj, Lucknow - 226003	- do -

5. BIHAR AND ORISA

Babulal Sarabhai & Co., Post Box No. 16. Khariar Road-766104, Orissa.	Bhilawan (Semecarpus anacardium Linn. F.) Dhaya (Woodfordia fruticosa kur) Marirphali (Helicteres isora), Nirmali (Stychnos potatorum) Vidayn (Embelia ribes Burm. F.)
Bharat Pharma, 3/9, Saicchi, P.B.No.21.	Herbal crude drugs.
Jamshedpur-834001, Bihar.	Acorus calamus L.;
Coalite Chemicals Pvt.Ltd. 2, Garstin Place, Calcutta-700001.	Berberis roots (Berberis spp.); Coptis teeta L.; Gloriosa superba L., etc.
Indo-Overseas Trading Co., Post Box No. 16, Khariar road, 766104, Orissa.	Herbal crude drugs.

B. EASTERN Z DNE

6. WEST BENGAL

ADCO Ltd., ADCO Nagar-712121, Hoogly, W.Bengal	Herbal crude drugs.
A.Damino & Co., 133. Canning Street, Calcutta-700072.	- do -
Ahmadalli Jafferjee Gandhi, 24, Pollowck Street, Calcutta-1	- do -
Arun & Co., Arun Chamber, P-38 Indian Exchange Place, Calcutta-1.	- do -
Ajit Kumar Kunda Bros, Laxman Chandra Rao, P-48/DC IT. Road, SCM 52, Calcutta-14.	- do -
Auro Impex Pvt. Ltd, 24-Ganesh Chand Avenue, Calcutta-700013.	- do -
Aurora Export Trading & Agency, 6-A, Saklat Place (3rd Floor), Calcutta-700013.	- do -
Bansidhar Dutt & Co., 126, Khengrapatty Street, Calcutta-1	- do -
Bumnipex Pvt. Ltd., 6, Church Lane, Calcutta-1.	- do -
Bansidhar Dutt & Co., 126, Khengrapatty Street, Calcutta-1.	- do -
Bengal Herb Stores, 2, Mullick Street, Calcutta-7	Herbal Crude Drugs.
Bhartmata Drugs & Chemicals Works, 49 & 49/1, Beniatolla Street, Calcutta-5	- do -
Bharat Pharma Chem., Block 'D', Chowringhee Mansion, 30, J.L. Nehru Road, Calcutta-700071.	Henna Powder (Lawsonia inermis) Crude drugs and medicinal herb.
Bheshaja Bhandar, 26 A, Karbala Tank Road, Calcutta-6.	Herbal crude drugs.

Bhogilal Mehta & Co., 52-A, Beni mandan Street P.B. No. 10031, Calcutta-700025.	- do -
Bihari Lal Hemraj, Cross Street, Bajaji, Calcutta.	- do -
Biyanis, 25, Ganesh Chandra Avenue, Calcutta-700013.	Ambrette (Hibiscus abelmosehus) Ammi majus L; Berberis roots (Berberis spp.) Copti teeta Wall.; Dioscorea deltoidea Wall. ex Ergot (Claviceps purpurea Fr. Tulasne); Heracleum candicans Wall. ex. Hydrocotyle asiatica L; Jalap root (Exogonium purga) Ñux vomica L.; Solanum khasianum Clarke; Valerian (Valeriana wallichi) etc.
Dabur (Dr. S.K.Burman) Pvt. Ltd., 142, Rash Behari Avenue,Calcutta-29.	Herbal crude drugs.
Doraco, 219/C.Old Chinabazar Street, Calcutta-1.	Herbal crude drugs.
Eastman & Co., 3, Southern Avenue, Calcutta-700026	Chiretta (Swertia chirata Buch Ham.); Colchicum Corm and seeds (Colchicum luteum Baker); Jatamansi (Nardostachys jatam (D.Don) DC; Podophyllum emod Wall. ex Royle); Rhubarb (Rheum spp.); Valerian (Valeriana wallichii) etc.
Excell Drug House, 18 B, Sukees Lane, Calcutta-700001.	Ayurvedic and unani crude drugs.
Ghose Bros. (Perfumers), 59, Ezra Street, Calcutta-700001	- do -
Global Enterprise, 34/1, Puddapukur Road, Calcutta-700026.	- do -
Giga Exports, 60, Sir Harirm Goenka Street Calcutta-700007.	Chiretta (Swetia chirata Buch Ham.); Galangal (Alpinia galanga (L) Kutki (Picrorhiza Kurroa Royle Benth); Rose petals (Rosa Sp.); Safflower (Carthamus tinctored Linn.); Tukmaria (Occimum basilicum Linn. etc.)
Hill Green Co. of India, 12-B, Netaji Subhash Road, Calcutta-700001.	Acenite (Aconitum ferox Wall) Belladonna (Atropa belladonna acuminata Royle ex Lind) Berberis (Berberis aristata) Colchim luteum baker.; Captis teeta Wall.; Ephedra (Ephedra Spp.); etc.
Harris Chunder Bose & Co. Pvt Ltd., 33, Canning Street, Calcutta-1	Herbal crude drugs.
Indian Drugs (Crude) Distributors 12-B Clive Row, Post Box No. 2386, Calcutta - 700 001	Atees (Aconitum heterophyllum Wall.) Belladonna (Atropa belladonna) L./A. acuminata Royle ex Lind.); Berberis (Berberis aristata DC / B. Spp.) Chiretta (Swertia chirata Buch-Ham. Ors spp.); Indian squill (Urginea Indica (Roxb.) Kunth.; Kutaja (Hoiarrhena antidysenterica L.); Nagkesher (Mesua ferrea L.); Psyllium (Plantago ovata L.); Solanum xanthocarpoum Schard. & Wdndi; Vasaka (Adhatoda Vasica Nees); etc.
Indo world Trading Company Ltd., 10, Armanian Street, Calcutta-1.	Herbal crude drugs.

Jai Trading corporation, 11, Brabourne Road, Calcutta-700001.	- do -
Jerambhai Export Pvt. Ltd., 25, Princess Street, Calcutta-700013.	- do -
Kanial Ramkumar, 173, Harrison Road, Calcutta-7.	- do -
Kaviraj N.N.Son & Co. Pvt. Ltd., 18/1, Lower Chitpur Road, Calcutta.	- do -
Krishna Traders, 20, Harichandra Mullick Street, Calcutta-700005.	- do -
Mahesh Laboratories Pvt. Ltd 30/4, Canal East Road, Calcutta-1	Herbal crude drugs.
Merchants & Traders Pvt. Ltd., 32, Armencences Street, Calcutta-700001	- do -
Miller & Co. (Calcutta) Pvt. Ltd., 24, Netaji Subhash Road, G.P.O. Box No. 2567, Calcutta-700001.	- do -
Minex Agencies, 71, Ganesh Chandra Avenue, Calcutta - 700013	- do -
Mira Traders, Amber Tala Street, Caloutta-1	- do -
M.S.Vanda & Co., 67, B.R.B.B. Road, Canning Street, Calcuita-700001.	Crude drugs and medicinal herbs; Amla (Phyllanthus emblica) Gaern.) Chiretta (Swertia chirata Buch-Hum.); Myrobalans (Terminalia chebula Retz. T.bellirica Gaertn. Roxb.); etc.
Pharmachem International, 8, Camac Street, Calcutta,	Herbal crude drugs.
Pharma Impex, 10, Middleton Road, Calcutta-700081.	- do -
Ram Sohanlal, 3, Mallick Street, Calcutta-700007. Wall, ex Kunth.); Eswarmul (Aristolochia indica L	Chiretta (Swert a chirata Buch-Ham.); Dioscorea deltoidea Jatamansi (Nardostachys Jatamansi (D.Don) DC.; etc.
Robinsomar & Co., 2, Misson Row Extension, Calcutta-1	Herbal crude drugs.
S. Bhattacharjee & Co., Circular Road, Serampur, W. Bengal.	- do -
S.Lai, 10/1, Rowland Road, Calcutta-700020.	Herbal crude drugs.
S. Sharnsudding & Sons, 119-B, Circular leach Road, Calcutta-700020.	- do -
United Chemicals & Allied Product 10, Clive Road, Calcutta-700001.	Belladonna (Atropa belladonna L.) Datura (Datura metel L.): Hyoscyamus niger L.; Ulat-Kambal (Abroma augusta L.) etc.
Universai Carbon Co., 46, Ezra Street, Calcutta-700001.	Herbal crude drugs.

7. ASSAM / MEGHALAYA

Essential Oil Industries, P.O. Sepon, Dist. Sibsagar, Assam - 785673.	Herbal crude drugs.
Gogoi Citronella & Eucalyptus Estate, cordoloi nagar, Tinsukia - 786125.	Annato seeds (Bixa orellana L.) Rouvolfia conescens Lino.; Solamum berris.
Chranji Lal Raut, 12th, Milo, P.O. Kukramara, Sadiya, Distt. Dibrugarh, Assam.	Herbal crude drugs.

Gobind Prasad Raut, 8th Mile, P.O. Mandil Sadiya, Distt. Dibrugarh, Assam.	- do -
Sarju Singh, P.O. Gauripur, Distt. Golapara, Assam.	- DO -
Shillong Co-Operative Marketing Society Ltd., Bara Bazar, Shillong-2, Meghalaya.	- DO -
Shiv Punda Kundu & sons, 168-B, Cotton Street, Calcutta-7	- do -

C. CENTRAL AND SUTHERN ZONES
8. ANDHRA PRADESH / MADHYA PRADESH / KARNATAKA

Andhra Trading Co., 34, Kadalaikkar Street, Virudhunagar-626001.	Dagad phool, stone flowers (Parmelia spp.) Elva (Aloe spp.); Senna (Cassia augustifolia Vahl) Elava
A.S.R. Company, Punalur-626001. Kerala.	- do -
Babulal Sarabhai Co., C/o Kathiawar Stores, Abid Road, Hyderabad-500001.	Ayurvedic and Unani crude drugs
Bharat Drugs & Extracts, 13, Jail Road, Dewas City-450001 (M.P.)	Herbal crude drugs.
Bhogilal C.Shah & Co., 8/1, South Tukogang, Nabhc iip, Indore-452001.	Indigenous herbal curde drugs.
Chittory Venkatraz, Near Park, Pithapuram-533450 (A P.)	Rauvolfia canescens L.Roots.
Dholokia Crude Drugs, 5-2-96 (137), R.P. Road, Secunderabad-500003.	Herbal curde drugs.
Herbisindia, 5-1-624, Troop Bazar, Hyderabad-5000C1.	- do -
Jain Brothers, Nemuch- 458441 (M.P.) indica (L.) ; Dill seeds (Anethum graveolens);	Asgandh (Withania Somnifera); Ashok Chal (Saraca Sandal (Santalum album L.) etc.
Kothari Brothers, Neemuch -458441.	Herbal crude drugs.
Madhukar Ganpat Karche, Agricultural Entrepreneur, Post Box No.63, Rathi Bhawan, Buldana Road, Malkapur, C.Rly.,443101.	Ambrette (Hibiscus abelmoschus L.)O Asgandh (Withania Somnifera Dunel) Datura seeds and leaves (Datura Spp. Gloriosa seeds (Gloriosa superba L.) Rauvolfia (Rauvolfia
Serpentina (L.) Benth. ex kurz); Solanum khasiamum G.Don	Clarke berries; Vinca (Catharanthus rosesz (L.)
Mahavir Jari Buti, Ayurved Bhawan, Rajeshvvari Road, Shivapuri (M.P.), Via Jhansi.	Herbal crude drugs.
Shroff Channabasappa & Sons, 174-175, Avenue Road, Bangalore-2	- do -

9. TAMIL NADU/KERALA

General Merchantile Overseas Corporation, 314, Mint Street, Madras-600001.	Ambrette seed (Hibiscus albelmoschus L.); Cinchona bark (Cinchona spp.); Colchicum seed (colchicum luteum Baker); etc.
International Trade Links, 11/12, Anna Pillai Street, Madras-600001.	Herbal curde drugs.

Intra Catholic Centre, Armenian Street, Madras-600001 serpentina Benth. ex Kurz); Vinca rosea (Catharantnus	Gloriosa superba L.Seeds; Sarpagandha (Rauvolfia roseus (L.) G. Don.
J.K. Drug Stores, 163, Nainiappa Naick Street, Madras-3	Herbal crude drugs.
K.D. Shah Enterprises. 49/15, Sir C.V. Raman Road, Madras -600013.	- do -
Kothari Phytochemicals (International), Thanich-Choyam Nagar Madurai-624221.	- do -
M.K. Industries, No. 1, First Cross St., United India Colony, Madras-600024.	Herbal crude drugs.
M.K. Natraja Pillai, 49, Emperor Street, Tuticorin-628001.	- do -
Mutual Traders. Badrian Street. Madras-600001.	- do -
M.M. Abdul Hamed & Sons, Post Box No. 12, Alwaye -683101	Vinca rosea Leaves and roots (Catharanthus roseus (L.) G. Don
P.S. Sankarling Nadar, 50, P.S.S. Nadar Street,	Galangal (Alpinia galanga (L.) wild.); Senna leaves (Cassis angustifol Vahl);
Roddy Venkata Roa, Kakinada (South India.)	Herbal crude drugs.
Vin Rose, Blue Hill Farmers, Coimbatore-641040.	- do -
Cochin Trading Corporation, 17/220 A, Chullikkal, Cochin-5 (Kerala).	- do -
Makkar Pillai, Alwaye, Travancore (Kerala.)	- do -
Money Corporation, T.C. 27/979, Sreekandeswaram East Road, Trivandrum - 605023.	Bala (Sida cordifolia L.); Brahmi (Hydrocotyle asiatica L.) Indian squll (Urginea indica (Roxb.) Kunth,); etc.
P.G. Monoy, Pijoy School Building, Near Bank of Cochin, Munnar - 658612 (Kerala).	Ganda Puro (Gaultheria Fragrantessima Wall.) Lycoodium (Lycopodium clavatum wild cherry (Prunus spp.) etc.
Shell India, Thayakal P.O., P.S. No. 1, om -688530,	Herbal crude drugs.

D. WESTERN ZONE

10. GUJARAT AND RAJASTHAN

Balisana Isabgol, G.I.D.C. Industrial Estate, Balisana, 384110, (C.N. (Gujarat)	Psyllium busk (Plantago ovata Fors) and other curde drugs.
Deepan Trading Corporation, 1st Floor, A-47, Maskati Market, Ahmedabad - 380002	Herbal crude drugs.
Gulab Chand Laduram, Attar, Naya ZBazar, Ajmer (Rajasthan)	- do -

11. MAHARASHTRA

A. Amrit Lal & Co., 406, E.R. Road, Pydhonie, Bombay-3.	Herbal crude drugs.
A. Kader Aii & Co., 189, Samuel Street, Khoja Galli, Bombay-3	- do -
Adeo Corporation, Dada Manzil, 2nd Floor, 67/69, Mohammaadali Road, Bombay- 400003.	- do -

Agarwal 7 Sons, 54, Mahavir Darshan, Bhandari Street, Bombay-400003.	- do -
Alim Pharmacy, E.R. Road, Pydhonie, Bombay-3.	- do -
All India Drug Supply Co., 11, Dariyasthan Street, Vadagadi (Masjid Bunder Road), Bombay -400003.	Aconite (Aconitum spp.); Centella asiatica (L.) Urban; Colchicum luteum Baker seeds; Ephedra (Ephedra spp.); Gloriosa superba L.; Parseley (Petroselinum crispum Nym. ex Hill; Safflower (Carthamus Tinctorius Valerian roots (Valeriane walli DC.); etc.
All India Medical Corporation, 185-Princess Street, Bombay-2	Herbal crude drugs.
Amsar Pvt. Ltd., 2, Hormuz Mansion, 72V. Deshai Road, Bombay-400026.	Aloes (Aloe barbadensis Mill.); hyscyamus niger L.; Ipecac Cephaelis ipecacuanha (Stokes) Ball; Estracts of Aconite (Aconitum spp;) etc.
Amul Trading Co., 12-14, Kazi Sayed Street, (Shak Gully). Bombay - 9.	Herbal crude drugs.
Ashwini Kumar Niranjan & Bros., 194, Samuel Street, Bombay-3.	- do -
Asian Trading Corporation, 38/40, Veer Vithaldas, Chandan Street, Bombay-400003.	- do -
D.Jamnadas & Co., 207, Samuel Street, Vadgadi, Bombay-3	- do -
Diwan Enterprised, Navapur -425418, Maharashtra. Eastern Drug House, 9, Premji Ladha Building, Bhandarkar Road, Bombay-19. Shertaliay, Ernakulam (Kerala).	- do -
Fortrad, The Foreign Trade Co, of India, 12 Walkashwar Road, Bombay-400006.	Herbal crude drugs.
Franco-Indian Pharmaceutical (P.) Ltd., 20, Dr. E. Moses Road, Bombay - 400001.	- do -
Ganesh Ashadhi Bhandar, 229, Kalbadevi Road, Bombay-2 Gopal Brothers, 180/81, Samuel Street, Bombay-400009	- do -
Hargovind Das Ramji & Sons, 194, Samuel Street, Bombay-400008.	- do -
Harbal Stores, P.O. Box No. 5047, 23 Khadak Street, Bombay-400008.	- do -
India Kirana Stores, 392, Ebrahim Rahematulla Road, Pydhoni, Bombay-400003.	- do -
Jadavaji Lallu Bhai &.Co., 247, Kalba Devi Road, Bombay -400009. chirata Buch. Ham.); Gloriosa superba L,; Henna	Aconite (Aconitum Spp.); Ambrette seed (Hibiscus abelmoscus Calamus roots (Acorus calamus L.); Chiretta (Swertia (Lawsonia inermis L.); Valeriana wallichii DC.; etc.
J.Nagindas & Co., 36, Kazi Sayed Street, Mandvi, Bombay-400003.	Herbal crude drugs.
Kika Bhai Gulam Hasan, 210, Samuel Street, Vadgadi, Bombay-400009.	Herbal crude drugs.

K.Uttam Lal (Exports) Ltd. Bhagwan Bhavan, 1st Floor 196/198, Samuel Street, Bombay-400009.	- do -
K.K. Traders, Bajaj Ward, Gondia (M.S.)	- do -
Manish Manufacturers & Suppliers, 503 Budhwar Peth, Room No. 34, Poona-2.	- do -
Silviculturist, Maharashtra State, Poona	- do -
Mahesh Kumar & Co., 215, Samuel Street, Bombay-3.	- do -
Moh. Ibrahim Pharmacy, E.R. Pharmacy Pydhonie, Bombay - 3	- do -
Manilal & Lallu Bhai & Co., P.B. No. 2008. Kalbadei Road, Near Narain Mandir, Bombay-400002.	Ayurvedic and Unani herbal crude drugs.
Mohan Kumar & Co., 185/187 Samuel Street Bombay-9.	- do -
Nathubhai Cooverji & Co., Arna Bhawan, 87/C, Broach Street, Bombay-400009.	- do -
N.Nimatlal & Co., 332/334, Narsinath Street, Bombay-400009. Naroji Anandaji, 15/17, Raghunath Road, Bombay-3	- do -
Navneer Lai Savailal, 64, Mudi Bazar, Mandavi, Bombay-3.	- do -
Niyogi & Co., 352/354, Samuel Street, Vadagadi, Bombay-400003.	Ayurvedic and Unani herbal crude drugs.
P.Khadilal Khandesia & Co., 167, Old China Bazar, Calcutta-700001.	- do -
Pharmachem International, 361, Maulana Azad Road, Bombay-400004.	- do -
Premji Haridar & Co., Bhanushali Chambers, 166-170, Sant Tukaram Road, Bombay-400003.	- do -
Rama Agency, 200 Kazi Sayed Street, Bombay-400003.	Ambreted seeds (Hibiscus abelmoschus L.); Babchi (soralia corylifolia L.); Calamus roots (Acorus calamus L.); Mesua ferrea L.; Solanum dulcamara L; etc.
Santosh Ayurvedic Drug Supply Co., 33 Daryansthan Street, Majit Bhandar, Bombay-400003.	Herbal crude and ayurvedic drugs.
Shamji Amreshji & Sons, 392, E, Rahimmtala Road, Pydhonie, Bombay-3.	- do -
Swami Corporation, 4, Hari Niwas, C. Road, Church Gate, Bombay-400020.	Abroma augusta (L.) L.f.;' Berberis bark (Berberis Spp.); Chiretta (Swertia chirata Buch. Ham.); Dill seed (Anethum graveolens L.); Ephedra vulgaris Hook. f.; Poppy shells and seeds (Papaver Somniferum.) etc.
Swastik Traders, 394, Kulh Bazar, Mamaya Chambers, Bombay-400009.	Herbal crude drugs.

E. NORTHERN ZONE

12. DELHI

Durg & Alkaloid Co., 4127, Nayz Bazar (2nd Floor), Delhi-110006.	Navin Bharat manufacturers, D-373, Defence Colony, New Delhi - 110024

Himalayan Drutgs, Herbs & Alkaloids Syndicate, C4/33-A, Lawrance Road, New Delhi-110035.

Indian Drugs & Alkaloid Co., 282/2A, Marg No.2, Than Singh Nagar, Anand Parbat, New Delhi-110005.

Jai Hind Trading Corpn., 28, New Rohtak Road, New Delhi-110005.

Kadmi International, Kashi House, 7-A, Connaught Place, New Delhi-110001.

Timex Agencies, 4E/9, Jhandewalan Extension, Line Road, New Delhi - 110055

13.PUNJAB

Dinesh Gurbase Bawa Sons, Majith Mandi, Amritsar.

Himalayan Traders, katra Dulo, Amritsar.

Krishna Kapoor & Co., Woodlands, The mall, Amritsar.,

Orient Traders, 615/VI, Bagh Jhanda Singh, First Floor, Amritsar.

14. UTTAR PRADESH

Shekherco, Post Box No. 256, Muzafarnagar-251001.

V.V. General Traders, I, Sutharashahi Kundanpura, Muzafarnagar - 251 002.

F. EASTERN ZONE

15. WEST BENGAL

Arun & Co.,
'Arun Chamber'
P-38, Indian Exchange Place, Calcutta - 700001

Expo. International,
20, Old Court House Street,
Calcutta - 700007

Giga Exports,
60, Sir Hari Ram Goenka Street,
Calcuta - 70007

Global Export Pvt. Ltd.,
55, Stephen House,
4 B, B-D. Bagh East,
Calcutta - 700001

Hill Green Co.,
12, Netaji Subhash Road, Calcutta - 700001

Hiralal Gajdharlal,
24, Maktaram Babu Street,
Calcutta - 700007

Indian Marine Service Pvt. Ltd.,
6/1, Lurdsay Street,
Calcutta - 700016

Jai Trading Corporation,
11, Brabourne Road,
Calcutta - 700001

Jiwanram Sheoduttraj,
Block 'D' Chowringhee Mansion,
30 Jawaharlal Nehru Road, Calcutta - 700016

Minex Agencies
71, Middleton Road,
Calcutta - 700016

Organon (P) Limited
Himalaya House,
38, Chowringhee Road,
Calcutta - 700016.

Pharmachem International,
7, Camac Street, Calcutta - 17

Radharam Sohan Lal,
3, Malik Street,
Calcutta - 700007

Tolaram (India) Limited,
68, Nalini Sett Road,
Calcutta - 700070

16. MADHYA PRADESH / ANDHRA PRADESH / KARNATAKA

Bharat Drugs & Entracts,
13, Jail Road, Dewas City - 455001 (M.P.)

Shroff Channabasappa & Sons,
Avenue Road, Bangalore - 560002

Shree Swastik Industries,
Manasnund, Raipur - 492001

Sri Srinivasa,
Fish Meal Products & Whole Dry Fish Suppliers.
Vunduro, Via (Samalkot),
E.G. Distt., Andhra Pradesh.

17. TAMIL NADU AND KERALA

Asian Exporting Corpn.,
Tuticorin - 1,
Tamil Nadu.

Cochin Trading Corpn.,
17/220 A., Chullikkal,
Cochin - 5, Kerala.

Canes Corporation,
New Colony, Tuticorin

General Merchantile Overseas
Corporation,
140, Mint Street,
Madras - 600001.

Kothari Phytochemicals
(International)
Nagari, Thanichchiyam - 624221. Madurai (T.I.)

M.K.Nataraja Pillai,
49, Emperor Street,
Tuticorin - 628001

Methuswamy, S.P.,
Great Cotton Road, Tuticorin -1.
Ponnu Saw Mills,
316, South Cotton Road,
Tuticorin -01.

Ponselvan Traders,
213, South Cotton Road, Tuticorin -1.

R.M.K. Industries,
No. 1, First Cross Society,
United India Company,
Madras - 600004.

Murugan & Bros.
78, South Raja Street,
Tuticorin - 1.

Murugasan & Bros.
972/2, North Third St.,
Padukottai - 628001

Nadar, P.P.M. Thangayaiah,
78, South Raja Street, Tuticorin - 1.

Nadar, P.S. Sonkaralinga
50, P.S.S. Nadar Street,
Tuticorin - 1.

Palanichamy, V.M.K.P.,
T.R. Naidu Street,

Pandian, J.R.,
North Cotton Road,
Tuticorin -1.

Pillay, S.P.M. Pirama Nayagam,
South Raja Street, Tuticorin
Thenammal & Co.,
54, P.S. Nadar Street,
Tuticorin -1.

V.C.H. Venkataswamy,
Perira Street, Tuticorin.

V.S. Arulanadam & Sons,
35, Gin Factory,
Tuticorin -2.

G. WESTERN ZONE
18. MAHARASHTRA

Ashish Overseas Corpn.,
34, Sarvodaya Industrial Estate,
Mahakali Road, Andheri (East), Bombay - 400 093

Biddle Swayer & Co. Pvt. Ltd.
25, Dalaal Street
Bombay - 400 009

Bisco Co.
Rawji Manasion,
Shak Galli,
12/14, Kazi Sayed Street, Bombay - 400 009

Export Enterprise,
34, Bhagwan Bhavan,
196/98, Samuel St.,
Bombay - 400 009

John Trading Corporation,
Post Box No. 5503
Bombay - 400 014

Gautam Export Corpn.,
506, Surat Sadan,
5th Floor, Surat Street, Bombay - 400 009

Hembot & Ocer,
509, G.B. No. 7
100/102, Dakastreeyed St., Bombay - 9.

Herbal Stores,
P.O. Box No. 5047,
23, Khadak Street,
Bombay - 400 009

International Traders,
Ramesh Chambers,
2nd Floor, 14, Garidbat Street,
Vadgadi, Bombay - 400 003

P.L. Associates.
Botwala Building (Top Floor),
8, Horniman Circle, Fort Bombay - 400 001.

Kamalkant Chotalal,
106, Bhandari Street,
Bombay - 400 003

Laxmidas Haridas Tanna & Co.,
Gulabi House,
3rd Floor,
111/115, Kazi Sayed St., Bombay - 400 009.

L.Nanalal Bros.,
P.O. Box 5022,
257, Narshi Natha St.,
New Anant Bhavan,
3rd Floor, Room No. 303, Bombay - 400 009.

Monji Vishram,
306, Shaikh Menon Street,
Bombay - 400 002.

Premji Haridas & Co.,
Bhanushali Chambers,
166-170, Sant Tukaram Road, Bombay-400009

Sierra Enterprises,
163, Atlanta,
16th floor, Nariman Point,
Bombay - 400 021.

Sigma Trading Co.,
Nafees Chambers,
3rd Floor,
Lokmanya Tilak Marg,
Bombay - 400 001.

The Fairdeal Corporation Pvt. Ltd.
66, Lakshmi Building,
Sir, P.M. Road, P.B. No. 1925,
Fort, Bombay - 400 001.

22. EXPORTERS OF PSYLLIUM HUSK AND SEED

Abhyudaya Industries
Highway, Khali, Sidhpur.
Ashok Industries,
Highway, Bhandu, Distt. Mehsana.

Balisana Isabgol,
C.I.D.C. Estate, Balisana, Dist. - Patan.

B.K. Industries,
C.I.D.C. Estate,
Palanpur.

Indo Exports,
Bindu Sarower Road, Sidhpur.

Jai Industries,
Highway, Khali, Sidhpur.

J.P. Mills, Sidhpur.
Kantilal Joitaram Patel,
Kokosi Road, Sidhpur.

Keshavlal Vithaldas patel,
S.T. Road, Sidhpur.

K.V. Patel & Co.,
S.T. Road, Sidhpur.

Modi Industries,
Sidhpur.

North Gujarat State,
Isabgol Industries,
Highway, Unjha.

Patel Corporation,
Highway, Khali, Sidhpur.

Dashrathlal Ramjibhai Patel,
Hari Cotton Mills Compound, Sidhpur.
Giriraj Enterprise,
Highway, Unjha.

Girdharilal Vithaldas Patel,
S.T. Road, Sidhpur.

Hindustan Trading,
Corporation,
Sidhpur.

Rajendra Brothers,
Highway, Khali, Sidhpur.

Sanjay Traders,
Highway, Khalli, Sidhpur.

Satpal Kamal & Sons,
Bindu Sarowar Road,
Sidhpur

Sidhpur Isabgol Processing Co.,
Sidhpur.

Swastik Industries,
Deesa Highway, Palanpur.

Sweti Export Corporation,
Deesa Highway, Palanpur.

The Golden Sat Isamgol Factory,
Highway, Fakosi Char Rasta,
Sidhpur

Vadilal Vithaldas Patel,
S.T. Road, Sidhpur.

23. EXPORTERS OF SENNA

A. Meeran,
M/s Tropical Traidng Co.
Tuticorin - 1, Tamil Nadu.

J.R. Pandian,
M/s PSPK Jayarajapandiyan,
Tuticorin - 1, Tamil Nadu.

H.L. Damani,
M/s G. Dass & Company (Pvt.) Ltd.,
27, W. Great Cotton Road,
Tuticorin -2. Tamil Nadu.

K. Muthuswamy,
M/s S.P.M. Sons,
Tuticorin -1.

M.L. Khetewar,
M/s Kothari Plantations,
29, New Colony, Tuticorin - 1, Tamil Nadu.

P.A.V. Sundaram,
M/s P.V.A. Sundaram,
166, New Colony,
Tuticorin - 1, Tamil Nadu.

K.M. Abdul Kadher,
M/s K.M. Abdul Kadher,
Tuticorin -2,
Tamil Nadu.

P.S.S. Ganesan,
M/s P.S. Sankaralinga Nadar,
North Cotton Road, Tuticorin -1, Tamil Nadu.

S.V. Ponnaiah,
M/s Jawahar Industries,
Tuticorin - 1, Tamil Nadu.

T.P. Srinivasagam,
M/s P.P.M. Thangaiah Nadar,
South Cotton Road,
Tutirocin - 1. Tamil Nadu.

H. EASTERN ZONE

Bunimpex Private Ltd., 6, Church Lane, Calcutta - 700 001.	Essential Oils.
Calcutta Chemicals Co. Ltd., 35, Panditia Road, Calcutta - 700 029.	- do -
Doraco, Perfumers, 219, C. Old China Bazar St., 1st Floor, Calcutta - 700 001.	- do -
Ghose Bros., (Perfumers) 50, Ezra Street, P.O. Box No. 627, Calcutta - 700001	- do -
Howrah Chemicals Works, 148, Mahatma Gandhi Road, Calcutta - 700 007	- do -
Jay Bee Tex, 20, Waterioo Street, 1st Floor, Calcutta - 400 069.	- do -
Paul & Sons, P.O. Box No. 902, Calcutta - 700 001.	- do -
Shankar Chandra Lodh, 60, Sir Hariram Gounka Street, Calcutta - 700 007.	- do -
Zenith Enterprises 19, R.N. Mukherjee Road, Calcutta - 700 001.	- do -

I. NORTHERN ZONE

Amar Bharat Industries Ltd., 1152, Main Street, Paharganj, New Delhi - 55.	Essential Oils.
Aroma Chemicals, Moradabad Road, Chandausi - 202412.	- do -

Asgar Ali Mohd. Ali, Hina Building, Phool Wali Gali Chowk, Lucknow (U.P.)	- do -
Beni Ram Mool Chand, Post Box No. 32, Kannauj - 209 725 (U.P.)	Calamus Oil, Celery Oil, Cinnamon Oil, Citronelia Oil, Dill Seed Oil, Sandal Wood Oil, Valerian Oil etc.
Budh Sen Sindhi Nath, Kannauj - 209725 (U.P.)	Essential Oils.
Devi Prasad Sunder Lal Khattri, Shyam Sundar Bhavan, Kannauj - 209725 (U.P.)	Palmarosa Oil, Sandalwood Oil and vetiver oil.
Deepak Menthol Corporation, A 5/3, Mayapuri Industrial, Area, Phase - II, New Delhi - 110064	Cardamom Oil, Mentha Oil, Peppermint Oil.
E.S. Chandra Enterprises, 34, Ferozeshah Road, New Delhi.	Essential Oils.
Essential & Sandalwood Oil India, 41, Industrial Estate, Kalpi Road, P.B. No. 358, Kanpur - 12 (U.P.)	- do -
Gupta Brothers, Sadar Bazar, Delhi - 1100 06	- do -
Herbs & Chemicals, Rly. Station Sarkara, V. & P.O. Sultanpur, District Naini Tal (U.P.)	- do -
Hindustan Monarck Pvt. Ltd., 20/1, Asaf Ali Road, New Delhi - 110 002	Essential Oils.
Indian Fragrance & Chemicals Works, Kannauj - 20975	Calamus Oil, Costus Oil, Peapermint Oil, Tagates Oil etc.
Jagat Aroma Oils Distillery, Kannauj - 209725 (U.P.)	- do -
J.K. Sugandhi Works, Tikait Raj Talab, Lucknow - 22604 (U.P.)	- do -
Jagat Narain Kapoor, C/o M/s Manualal Raj Narain, P.O. Saraimiran, Kannauj - 209725 (U.P.)	-do -
Kulu Aromatics, P.O. Shamshi - 175126, Kulu (H.P.)	Essential Oils.
Lala Banarsidar Khattri, Sabzi Mandi, Kannauj - 209725 (U.P.)	Nagarmotha Oil, Palmarosa Oil, Vetiver Oil; etc.
Lal Ji Kedar Nath Khatri, Nadan Mahal Road, Lucknow - 4 (U.P.)	Mentha Oil, Nagarmotha Oil, Peppermint Oil, Spice Oil, Vetiver Oil

Lucknow Stores, Chaura Bazar, Ludhiana (Punjab)	Essential Oils
Makhulal Ayodhya Prasad, Kannauj - 209725 (U.P.)	Essential Oils.
Malayagiri Sandalwood Oil Distillery, P.B. No. 354, Kanpur (U.P.)	- do -
Manohar Dass Sewak Ram, Kannauj.	- do -
Mathura Prasad Shri Prakash, Kannauj (U.P.)	- do -
M.L. Ram Narain, Perfumers, Kannauj - 209725 (U.P.)	- do -
Mohd. Ayub - Mohd. Yaqub, Kannauj - 209725 (U.P.)	- do -
Mohd. Idrish Nazrul Rehman, Perfumers, Kannauj - 209725 (U.P.)	- do -
Mannalal Sons & Co., Hina Building, Kannauj - 209725 (U.P.)	- do -
Navin Bharat Manufactuers, D-373, Defence Colony, New Delhi - 110024	Lemongrass Oil, Mentha Oil, Palmarosa Oil, Sandalwood Oil, vetiver
Neeru Enterprises, Rajdwara Road, Rampur - 244901 (U.P.)	Essential Oils.
Perfumers (India) Pvt. Ltd., Kaiserbagh, P.B. No. 87, Lucknow (U.P.)	Citronella Oil, Mentha Oil, Palmarosa Oil, Perfumed water, Sandalwood Oil, vetiver oil,
Perfumers (India) Pvt. Ltd., Kaiserbah, P.B. No. 87, Lucknow	Citronella Oil, Mentha Oil, Palmarosa Oil, Perfumed water, Sandalwood Oil, Vetiver Oil.
Perfumers and Essential Oils Co., Rambagh, Chandganj Gardens, Lucknow - 226006 (U.P.)	Citronella Oil, Mentha Oil, Palmarosa Oil, Perfumed Water; Sandalwood Oil, Vetiver Oil.
Pragdutt Dindayal, Perfumers, Kannauj - 209725 (U.P.)	Essential Oils
Radha Sales Corporation, 548, Fasil Road, Lahori Gate, Delhi - 6.	- do -
Ram Krishna Gupta & Bros., 33/107, Gaya Prasad Lane, Kanpur - 208001 (U.P.)	Sandalwood oil and other essential oils and perfumed water.
Richardson Hindustan Ltd., P.O. Bilaspur, Distt. Rampur (U.P.)	Essential Oils
S. Chandra Enterprises, 34, Ferozshah Road, New Delhi - 110001	- do -
Saran Brothers, Kannauj - 209725 (U.P.)	Nagarmotha Oil, Palmarosa Oil, Perfumed water, Sandalwood Oil, Vetiver Oil.

Shaivi Industries, Chowk, Lucknow - 226004 (U.P.)	Citronella Oil, Cyprus Oil, Jatamanshi Oil, etc.
Standard Essential Oils Distellery, 84/113, Affim Kothi Crossing, Near Juhi Rly, Under Bridge, Kanpur - 208003 (U.P.)	Calamus Oil, Cardamom Oil, Celery Oil, Costus root Oil, Dill seed Oil, Fennel Oil, Ginger Oil; etc.
United Perfumes Pvt. Ltd., 66/188, Bhustaoli, Kanpur - 208001 (U.P.)	Essential Oils.

J. WESTERN ZONE

B.C. Patel & Co., No.1, Devkaran Mansion, 2nd Floor, 79, Princess Street, Bombay - 400 002 Br.	Essential Oils
B.D. Shah & Co., Damodar Buildings, 105, Princess Street, Bombay - 400 002.	- do -
Bhagat Fine Chemicals, 164, Fanaswadi (Sitaram Poddar Marg), Harlalka Bhavan, Bombay - 400 002.	Bargamot Oil, Mentha Oil, Peppermint Oil,
Bonanza Export Corporation, 18 BM Jolly Maker Chambers, No.1, 1st Floor, Nariman Point, Bombay - 400 021	Essential Oil
Calcutta essence Supply Co., Mulchand Mansion, Shamaldas Gandhi Marg, Bombay - 400 002.	Essential Oils (Aromatic Chemicals & Resinoids).
Ethical Products (India) Pvt. Ltd., Advani Chambers, Sir Phirozshsh Mehta Road, P.O. Box No. 1244, Bombay - 400 001	Cannaanga Oil, Garanium Oil, etc.
Eastern Aromatics, Dutt Mandir Road, Bhandup, Bombay - 78 NB.	Essential Oils.
Goldensun Manufacturing Co., Renvir Building, 70, Princess Street, Bombay - 2	Bergamot Oil, Clove Oil, Eucalyptus Oil, Lime Oil, Mentha Oil, Peppermint Oil etc.
Gordhandas Tokersey, 106, Kazi Sayed Street, Bombay - 3.	Essential Oil,
Hindustan Lever Ltd., Hindustan Lever House, Backbay Reclamation, Bombnay - 400 020.	- do -
Impex India International D-52/499, Gandhi Nagar, Bandra (East), Bombay - 400 051.	- do -
Industrial Perfumes Limited, Hay Bunder Road, Tank Road P.O., Bombay - 400 033.	- do -
Industrial Export Import Co., 65, Modi Street Fort, Bombay - 400 001.	Essential Oils (aroma Chemicals, oleoresins and resinoids, etc.).
Mazda Exports, Appeoy House, 4th Floor, 130, Bombay Samachar Marg, Bombay - 400 023.	Essential Oils

Mehta Corporation Pvt. Ltd., Rajesh Mills Premises, Fergusson Road, Bombay - 400 013.	- do -
M.C. Devar & Co., Sai Bhavan, 143, Princes Street, Bombay - 400 002.	- do -
Naarden Chemicals Works (India) Ltd., Plot No.6, Saki Vihar Road, Bombay - 400 072.	- do -
New India Aromatic Co., Devkaran Mansions No. 3, 3rd Floor, 45/47, Princes Street, Bombay - 400 002.	- do -
N.R. Shah & Company, 613, Parekh Market, Paper Mill Compound, Girguan Off., Kannedy Bridge, Bombay - 400 004.	Essential Oils
Prabhat General Agencies, 195, Kalvadevi Road, Bombay - 400 002.	- do -
Premium Exports, 84, Bombay Samachar Marg, Bombay - 400 002.	- do -
R.K. Shapurji & Co., P.O. Box No. 1065, Prospect Chambers, 319, Dr. Naoroji Road, Bombay - 400 009	- do -
Sanjaya Kumar Shankar Lal & Co. 88/89, Surat Street, 208, Surat Sadan, Bombay - 9.	- do -
Scientific Products, 54, Poddar Chamber, Parsi Bazar Street, Bombay - 9.	- do -
S.H. Kelkar & Co. Ltd., Lal Bahadur Shastri Marg, Mulund, Bombay - 88.	- do -

K. CENTRAL & SOUTH ZONES

A. Maschmeiger Jr. (India) Pvt. Ltd., G.S.T. Road, Madras - 44 (Tamil Nadu)	Essential Oils.
Abdul Hameed & Sons, Post Box No. 12, Alwaye - 683101 (Kerala)	Ohaulmoagra Oil, Oinnamon Oil, Oitronella Oil, Eucalyptus citridora oil, Lemongrass oil, Palmarosa Oil, vetiver Oil.
Bharat Sandal Oil Distrillers, P.B. No. 106, Salem - 636001 (South India).	Sandalwood Oil (Agmark).
Biochemical & Synthetic Products Ltd. Sanat Nagar, Hyderabad - 18 (Andhra Pradesh).	Essential Oils
Devi Industries 19, Azad Street, Gandhi Nagar, Madurai - 625020 (Tamil Nadu)	Ajawain Oil, Cardamom Oil, Calery Oil, Coriander seed oil, Cumin oil, Dill seed oil, Galangal root oil, Ginger oil, Pepper oil, Turmertic oil.
D.V. Deo, P.B. No. 135, Pandit Road, Corner Cochin - 682002 (Kerala)	Black pepper oil, Cinnamon leaf oil, Citronella Oil, Davana Oil, Eucalyptus Citriodora Oil, Ginger Oil, Gingergrass Oil, Lemongrass Oil, Palmarosa Oil, Sandalwood Oil, Vetiver Oil.

East Indian Sandal Oil Distrillers (P) Ltd. Krishnagiri Road, Kuppam (Andhra Pradesh)	Essential Oils.
Kangundi Industrial Works Pvt. Ltd. Krishnagiri Road, Kuppam (Andhra Pradesh).	- do -
Mackar Pillay & Sons Pvt. Ltd. Post Box No. 18, Alwaye (Kerala).	- do -
M.L. Krishnan, No. 12, 'Sreegayam', Ranga Rao Road, Shankara Puram, Bangalore - 4 (Karnataka)	- do -
Mysore Essential Oil Industries, Kuppam (Andhra Pradesh)	- do -
N. Krishnaswamy, Sakthi Extracts Co., 13/23, Main Road, Mettupalayam - 641301, Coimbatore (T.Nadu)	- do -
New Pratap Chemicals India, Vellody Buildings, Kallai Road Calicut-673002 (Kerala).	Eucalyptus citriodora oil, Lemongrass oil, Sandalwood oil, Vetiver oil.
Paul Chaly, P.O. Pangarapilly, Ernakulam District, (Kerala)	Essential Oils
Puchiri Exports, Punchiri Buildings, Jetty Road Alleppey (Kerala)	- do -
Seth Sons, Post Box No. 195, Jaipur-302001 (Rajasthan)	- do -
Salem Essential Oil Co., P.B. No. 106, Salem-636001 (Tamil Nadu).	Sandalwood Oil (Agmark).

NORTHERN ZONE

DELHI

Drug & Alkaloid Co., Post Box No. 1297, 4127 Naya Bazar, (2nd Floor), Delhi - 110006.	Catharanthus roseus (L.) G. Don; Hyoscyamus nioger L.

PUNJAB

Bharat Automobiles & Fertilizers/Traders, Pharwahi Bazar, Barnalal.	Japanese mint (Month arvensis L. Subsp. haplocalyx Briquvar. piperasceus Holmes).

JAMMU AND KASHMIR

Drug Farms & Factories, DRL (CIMP), Canal Road Jammu-Tawi.	Belladonna (Atropa belladonna Bergamot mint (M.citrata Ehrh.) Ergot (Claviceps purpurea (Fr.) Tulsane); Hops (Humulus lupulus L.); Japanese mint (Mentha arvensis L.) Peppermint (Mentha piperita L.);

UTTAR PRADESH

Aroma Chemicals, Maradabad Road, Chandausi.	Japanese mint (M. arvensis L.).
Bhavana Chemicals 8, Civil Lines, Rampur.	- do -
Central Institute of Medicinal & Aromatic Plants, P.O. Faridinagar, Lucknow - 226010.	Bergamot mint (M. citrata Enrh.); Citronella Java (C. Winterianus Jowitt); Japanese mint (Mentha arvensis L.) Balmarosa (Cymbopogon martini Stapf variety motia); Feppermint (Mentha piperita L.).

CIMAP Regional Centre, P.O. Dairy Farm, Nagia-263149. Rly, Stn. Pantnagar, Dist. Naini Tal (U.P.)	Ambrette (Hibiscus abelmoschus L.) Bergamot mint (Mentha citrata Ehrl Citronella Java (Cymbopogon Winterianus Jawitt). Ergot (Claviceps purpures (Fr.); Tulsane); Japanese mint (Mentha arvensis L.) Stapf var. motia); Peppermint (Mentha piperita L.)
Herba & Chemicals (Mentha Distillation Unit), Rly. Stn. Sarkara, V. & P.O. Sultanpur, Distt. Naini Tal.	Japanese mint oil (Mentha arvensis L.)
Himalayan Herb Stores, Madho Nagar, Post Box No. 130, Saharnpur - 247001	Hyolscyamis niger Linn.
Joshi, M.D., Mentha Factory, Chapparrah, Distt. Rampur.	- do -
Kanpur, P.L. Kundeora, Distt. Nani Tal	- do -
Mengal Singh & Sons, Amroha, Distt. Moradabad	Cinnamomum tamala Near & Eberm Oil.
Meghalaya Essential Oils & Chemicals Ltd., P.O. Clutterbukganj-243502	Japanese mint oil (Mentha arvensis)
Mentha & Allied Produceres, Sambhal, Distt. Moradabad.	- do -
Menthol & Allied Chemicals (India), Opp. Central Bank of India, Rampur-244901.	- do -
Miracle Menthol, Sambhal, Distt. Moradabad.	- do -
Mukhi Farms & Orchards, P.O. Behat, Saharanpur.	- do -
Orgamatic Chemicals, Sambhal, Distt. Moradabad.	- do -
Puranlal Essential Oil Plantation (Industries), P.O. Doiwala, Distt. Dehra Dun.	Palmarosa (Cymbopogon martinil Stapf var. Motia).
Puri I.M. Peru, Madras, Distt. Naini Tal	Japanese mint oil (Mentha arvensis L.)
Qasim, S.A. 179, Lukerganj, Allahabad - 211001	Ambrette seed (Hibicus abel-moschatus L.); Catharanthus roseus (L.) G. Don Citronella java (Cymbopogon) Winterianus Jowitt); japanese mint (Mentha arvensi etc.
R.A. Chemicals, Sayadwara, Badaun.	Palmarosa (Cymbopogon martinii Staff var. motia); Peppermint oil (Mentha piperita)
Richardson Hindustan Ltd., mentha Research & Development Centre, P.O. Bilaspur, Distt. Rampur (U.P.)	Japanese mint oil (Mentha arvensis L.).
Sambal Chemicals, Sambhal, Distt. Moradabad.	- do -
Shanti Industries, mentha Distiller, Rajdwara Road, Rampur	- do -

EASTERN ZONE
ASSAM

Assam Aromatic Enterprises, R.K. Bose Road, Jowitt) P.O. Dhubri - 783301.	Citronella Oil (Cymbopogon Winterianus
Bhuyankhat Udyaog, P.O. Borhat, Assam. Jowitt)	Citronella oil (Cymbopogon winterianus
Bosabari Tea Estate, Post Box No. 27, Jorhat - 785001.	- do -

Dilli Tea Estate, P.O. Doom Dooma, Assam.	- do -
Essential Oil Industries, P.O. Sepon - 785673 Distt. Sibsagar (Assam).	- do -
Harishpur Tea Estate, H.O. Chowdidinghee, P.O. Dibrugarh (Assam).	- do -
Harlalka Manufacturing Enterprises, Dhubri (Assam).	- do -
Itakhooli Tea Estate, P.O. Tinshukia, dibrugarh (Assam).	Citronella Oil (Cymbopogon winterianus Jowitt)
Jallan Enterprises and J.P. Agro Plantations, Jallan House, G.F. Road, Golaghat (Sibsagar) Assam - 785621.	Citronella Oil (Cymbopogon winterianus Jowitt) Dioscorea composita hemsl.; D. floribunda mart. & Gal.; Eucalyptus citriodora Hook.
Joonktolle Tea Estate, P.O. Barbam (Assam).	Citronella Oil (Cymbopogon winterianus Jowitt)
Kamrup Essential Oils, P.O. Palasbari, Kamrup (Assam)	- do -
Koranganj & Azizbagh Tea Estate, Sibsagar (Assam)	- do -
Moran Tea Co., Sapon Tea Estate, P.O. Moran (Assam)	- do -
Nait Co. Ltd., Post Box No. 11, G.S. Road, Ulubari, Gauhati - 781007 (Assam)	- do -
P.S.I. Co-operative Farm, Indian Air Force, Jorhat (Assam)	- do -
Silviculturist, Assam P.O. Gauhati (Assam).	Rauvolfia canascens L.
Sonajuli Tea Estate, P.O. Borjuli (Assam)	- do -
Warren Tea Group, P.O. Hoolgrijan, Distt. Dibrugarh (Assam).	- do -

CENTRAL ZONE
ANDHRA PRADESH & MADHYA PRADESH

Aromatic Chemicals & Oil Co., Senthramapuram P.O. Vishkapatanam, Distt. Andhra Pradesh.	Citronella Oil (Cymbopogon Winterianus Jowitt) Japanesse Mint (Mentha Arvensis)
Buche, Vinay Kumar, H.No.2-2-2003, Ballakunta, Hyderabad - 500044	- do -
Chief Conservator of Forests, Maharashtra, Poona.	Linaloe (Bursera delpechiana Poiss)
Chief Conservator of Forests, Andhra Pradesh, Hyderabad.	- do -
General Manager, The M.T. State Agro-Industries Development Corporation Ltd., Babai, Hoshangabad, (M.P.)	Citronella Oil (Cymbopogon winterianus Jowitt) Lemongrass Oil (C. Flexuosus Nees ex Steud); Palmarosa Oil (cmartinil Stapf var. motia).
Herbsindia, 4-1-624, Troop Bazar, Hyderabad	Catharanthus roseus (L.) G.Don.
India Seeds corporation, P.B. No.62, Aurobindonagar Vellore - 524003 (A.P.)	- do -
Rao, G. Ramachandra, Principall, Samatha Type Institute, Main Road, Chirala-523155.	- do -
Rao, Srinivasa, Eish Meal Products & Whole Dry Fish Suppliers, Vunduro, Via Saranlkot, E.G. District (A.P.)	- do -

Srikrishna Agencies, Chilakaluriopet, Distt. Guntur (A.P.)	- do -
Veeraghavalu, N. Bosenagar, Jandrapeta, P.O., 523165 (A.P.)	- do -
Vepeon Oil & Chemicals Pvt. Ltd., Vijay Villa, Lallguda, Secunderabad (A.P.)	- do -

SOUTHERN ZONE
KARNATAKA

Chief Conservator of Forests, Karnataka, Bangalore.	Linaloe (Bursera delpechiana Poiss);
CIMP Regional Centre, NAL (W.T.C.), Belur, Bangalore - 560037.	Cathararthus roscus (L.) G.Don; Citronella Oil Java (Cymbopogon) winterianus Jowitt); Davana Oil (Artemisia pallens wall Eucalyptus citriodora Oil; (Eucalyptus citriodora Hook); Geranioura Oil (Pelargonium Graveolens (L.) her. ex Air. Palmarosa Oil (Cymbopogon martinii stapf var. motia)
Kadamane Estate Co., Kadamane D.O. 573127, Hassan District, Karnataka.	Citronella Oil (Cymbopogon winterianus Jowitt).

TAMILNADU / KERALA

Abdulrasheed, Sheejamanzil, Anchal P.O., Quilon District, Kerala.	lemongras oil (Cymbopogon flexuosus (Nees ex steud.) W. Wats.
K.V. Joseph, Karimathotil, neyyasseri, Thodupuzha, idukki Distt. Kerala.	palmarosa oil (Cymbopogan martinii Stapf var. motia.).
Mathew mathai, Kuruvithadam, Moovattupuzha, Ernakulam, Distt. Kerala.	- do -
Mathai Varkey kadilikandam Valara P.O., Iduki Distt. Kerala.	- do -
M.J. Sebastin, St. Philominas Estate, Kothamangalam Ernakulam Distt. Kerala.	- do -
Michael Francis, Poovathinkal., Thidangannad P.O. Thodupuzha, Idukki Distt. Kerala	Palmarosa Oil (Cymbopogan Martinii Stapf Var. Motia.)
N.K. Kuriakose, Narinmattamthil House, Mukkandy, Mukut, P.O. Tellicherry, Cannanore, Distt. Kerala	- do -
O.C. Abraham, Ottakkadasseril, Oonnukal P.O., Ernakulam, distt. Kerala	- do -
P.M. Ninan, Palamattathil House, Kannathadi P.O., Vellathooval, Idukki Distt. Kerala	- do -
Parameswaran, Thazhathukudy, Nellimattom, Ernakulam, Distt. Kerala.	- do -
P.K. Sasieharan, Padnalayam Neriyamangalam Ernakulam, Distt. Kerala.	- do -
P.K. Padmanabhan, Puthenpurkal, Neriyamangalam Ernakuilam, Distt. Kerala.	- do -
P.M. Kurmaran, Palakkulathii Pindimana, Ernakulam Distt. Kerala.	- do -

P.M. George Joseph, Kothamangalam Ernakulam Distt. Kerala.	- do -
P.S. Luckose, Poondikulam Bharanganam, Kottayam, Distt. Kerala.	Apanese Mint (Mentha Arvensis)
Sasikumar, T.G., Tokekallel House, Konnathadi Central P.O., Vellathooval, Idukki District, Kerala	Japanese Mint (Mentha Arvensis L.)
Shahudden, M., Palavila Veedu, Agasthancode, Anchal P.O. Quilon District. Kerala.	- do -
V.K. Narayanan Namboodiripad, Varikkasserimana, P.O. Kizhur, Palghat, Distt. Kerala.	- do -

WESTERN ZONE
MAHARASHTRA

Silviculturist, Maharashtra State, Pune - 411001.	Dioscorea Floribunda Mart. & Gal.
Bharat Vinca Producers, 9/E Dhavalganga C.H.S. Ltd., Bandra (W.), Bombay - 400 050. Badaun Distt.	Vinca (Catharanthus roseus (L.) G. Don).

Roophel Khand Mentha Works, Badaun.
Barcilly Distt.

Essential Oils & Chemicals, Kuhadapoer.
Parsu Pura Mentha Factory, Clutterb Akganj
(Near Soyabeed Facality Bareilly. Dehra Dun Distt.

Puran Lal Essential Oil Plantation (Industry),
P.O. Virbhadra, Lucknow Distt.

Central. Industitute of Medicinal & Artomatic Plants.
P.O. Faridinagar, Lucknow - 226010. Moradabad Distt.

Aroma Chemicals, Chandausi.
Agarwal Mentha Factory, Ferojpur.

Bharat Chemicals, Sambhal.
Dalpat Pur Mentha Factory, Dalpat Pur,

Dharma Enterprises, Sambhal.

Delite Chemicals, Ferojepur.

Friends Chemicals, Ferojepur.

Gupta Chemicals, Sambhal.

Haji Gaus Mohd, Asafpur.

Haji Kidaytula, Sambhal, Joya Road, Moradabad.

Haji Matloob Husain, Amroha.

Jagadish Moha (Bajarwala), Ferozepur.

Jain Enterprizes, Sambhal, Joya Road, Moradabad,

Janta Chemicals, Ferojepur.

Jain Aromatics, Delpat Pur

Kapoor Chemicals, Dalpat Pur.

Kissan Mentha Factory, Sambhal

Kissan Mentha Udyaog, Amroha.

Kissan Industries, Sambhal.

Khan Mentha Factory, Sambhal.

Kumar Chemicals, Sambhal, Kussam Industriesd, Sirswan Doraha.

Managal Singh Sons, Amooha.

Mentha & Allied Products (Nanda), Sambhal.

Mentha Rasyan Shals, Sambhal, Joya Road, Moradabad.

Miracles Mentha Oil Distillary, Sambhal.

Orgamatic Chemicals, Sambhal.

Pakwara Mentha Factory, Delhi Road, Moradabad.

Sambhal Chemicals, Sambhal.

Shiromanni Chemicals, Sambhal.

Super Mentha Factory, Ferojepur.

S.S. Chemicals, Dalpat Pur.

United Chemicals, Ferojepur.

Z.I. Khan, Sambhal., Distt. Nanital.

Aromatic Herb & Chemicals, Sultanpur-224713

CIMAP Regional Centre, P.o. Dairy Farm, Nagla-263149

Govt. Fruit Utilization & mentha Factory, Kunda Chauraha.

I.M.Puri, Peru Madara.

Kapson Enterprises, Ramnagar Road, Kashipur.

Mehrotra, Kundeora.

Narula Mentha Factory, Kunda.
Pilibhit Distt.

Gita Aromatics, P.O. Bisalpur (NER) - 262201.

Rampur Distt.

Abdul Hasan, Kun Khera.

Allied Products, Lalpur.

Agrawal Mentha Factory, Menhdipur.

Ajwani & Chemicals, Singan Khera.

Ashu Drugs, Rampur.

Bhawana Chemicals, Lalpur.

Bhawana Chemicals, Benazir.

Bhawana Chemicals, Kdandia.

Bhawana Chemicals, Nerpat Nagar.

Bhawana Chemicals, Karimpur.

Cefearoma, Rampur.

Chaudhri Mentha Factory, Sohria.

Chandausi Mentha Factory, Tanda-badli.

Eskay Chemicals, Rampur.

Gupta Mentha Udayog, Koili, Bilaspur.

Guramail Mentha Works, Rampur

Gupta Mentha Factory, Singan Khera.

Gupta Mentha Factory, Hakimgaj.

Gupta & Co. Suar. Harishanker
Arya Mentha Works, Lalawalabagi Bilaspur.

Herbs & Chemicals, Patti Sultanpur.

India Chemicals, Bilaspur.

Ishrat Chemicals, Rampur.

J.S. Bolaria Mentha Works, Smodia.

Kanhayia Lal Mentha Factory, Singan Khera.

Kasturba Mentha Factory.

Kamal Mentha Works, Kosi.

Khan, Mentha Factory, Shahzad Nagar.

Kier Highness, Khas-Bagh (N.Party), Rampur.

Kisan Mentha Factory, Suar.

Khod Tel Udyog (Ras Rasayan), Khempur.

Lal Ji Mentha Factory, Azim Nagar.

Lal Ji Mentha Factory, Khempur.

Mangal Sen & Sons (India Chemicals) Bhot, Bilaspur.

Mint India Co., Parchai.

Mentha & Allied Products (Nanda), Chamraua.

Moradabad Mentha Factory, Munda ZPandey.

Mini Enterprizes

Nadim Mentha Factory, Thuna Pur, Bilaspur.

Neem Enterprises, Rajwadra Road, Rampur.

Progressive Enterprises, Bazarkhata, Rampur.

Rampur Industries.

Raj Mentha Industries, Near Chamraua.

S.S. Chemicals, Shadikemaraioah.

Sanbhal Chemicals, Khod.

Sonkara Mentha Factory, Sankara, Bilaspur

Shanti Industries, Rajdwara Road.

Singh Months Factory, Charattadnagar.

Singh Works, Nasingaj (Near Saidnagar).

Vivek Mentha Udyog, Khaprailwala.

Y.D. Joshi Mentha Factory, Chapparrah.

NORTHERN ZONE

Arya Vastu Bhandar, 46, Dispensary Road,
Dehradun, U.P.

Govind Singh S/o Pahelwan Singh,
Village Laron, P.O. Kachnev, Dist. Jhansi, U.P.

Ashok Kumar Gupta S/o Sri Ram Kishore, 38-8 Dube Chowk, Manranipur, Jhansi-284004 U.P.

Bhadur Singh Villayal, Saurah Siroli, P.O. Thano, Dhra Dum, U.P.

Bechan Prasad, Contractor, Forest Division V.P.O. Nau Garh, Distt. Varanasi, U.P.

Bhagwan Das S/o Sri Gore, Lal Jain Gram Gona, P.O. Narhat Distt. Lalitpur, U.P.

Bhaiya Lal S/o Shri Fadali Villaye Lakhjar, Post Office Bhadawara, Distt. Lalitpur, U.P.

Brijpal Singh S/o Sri Chaul Singh, Village Aurangabad P.O. Khas, Distt. Saharanpur, U.P.

Daya Ram 901/9, Lakhi Bagh, Dehra Dun, U.P.

Giris Chandra Agrawal, Mohalla Kalluganj, Najibad, Distt. Bijnor, U.P.

Mani Ram Mishra, Railway Road Rishikesh, U.P.

Mahesh Chand S/o Amar Chand Jain V.P.O. Suraj, Distt. Lalitpur, U.P.

Madanpur Sahakari Shjarma, Sanvida, Samiti Ltd. Madanpur, P.O. Didoria, Distt. Lalitpur, U.P.

P.S. Jamwal Kachii, Chowi, Jammu-Tawi J & K.

Parshu Ram S/o Gopal Das, V.P.O. Narhatr, Distt. Lalitpur, U.P.

Parol Gram Udyaog Sewa Mandal, V.P.O. Parol, Distt. Pillibhit, U.P.

Gupta Trading Co., Matadin Road Sahadat Ganj, Lucknow U.P.

Harvans Singh S/o Sri Kahan Singh, Nadipura, Distt. Kalitpur, U.P.

Hari Prasad Agrawal, Forest Contractor Minor Produce, Ramnagar, Nani Tal. U.P.

Ikhadar Alam Khan S/o M.O. Asaf Husain Imali, Jue Wali Rampur, U.P.

Harnam Das & Sons, Majit Mandi, Amritsar Punjab.

Jhanshed Ali S/o Sri Babbi Bix, Village Aneki, P.O. Khas, Distt. Nainital, U.P.

Kishori Lal S/o Sri Bhola, Bannakhera, Bazpur Distt. Nainital, U.P.

Kishori Saran S/o Munna Lal, P.O. Katra, Distt. Jhansi, U.P.

Kunji Lal S/o Sri Ram Sahu, V.P.O. Walawehat Distt. Lalitpur.

Sri Sahid C/o Najaf, Vill. Shat, Manshr, P.O. Buggawala, Distt. Saharanpur, U.P.

Srimati Lareti Bai, W/o Rajju, Magar Para Magarwara, Distt. Jhansi, U.P.

Shri Shiv Bahadur Singh C/o Chandrika Singh, Contractor, V.P.O. Ramgarh, Distt. Mirzapur, U.P.

Sheikh Nazir Ahmed, Forest Lesse, Baramula, J & K.

Shri Shiv Ram Contractor S/o Shri Alam Singh, R/o Raipur, P.O. Raipur, Distt. Dehra Dun, U.P.

SOUTHERN ZONE

Ram Kumar S/o Shri Chhedilal, M.O. Dube Bishalpur, Distt. Pilibhit, U.P.

Ramesh Chandra S/o Shri Umrao Singh, Ward No.2, Tanakpur, Distt. Nainital, U.P.

Raja Ram S/o Sri Laxman Singh, Ward No.2 Tanakpru, Distt. Naini Tal, U.P.

Ram Murti Lal S/o Shri Kedar Nath, Civil Line, South Pilibhit, Distt. Pilibhit, U.P.

Ram Prasad S/o Sri Laxman, P.O. Bhadawara, Distt. Lalitpur. U.P.

Sahakari Bhesaja Vikasevum Karya Vikraya Sangh, Uttarkashi, U.P.

C.M. Poulose, P.O. Chedleth, South Wynad Taluk, Kerala.

C.A. Chennappan S/o Chennamoorthy, Chettiyar, Harpur, Tamil Nadu.

A.N. Arthanari Mudaliar, Forest contractor, Adhiyur, Kerala.

A.S. Akbar Forest Contractor, P.O. Denkanikotta, Distt. Dharmpuri, Kerala.

A. Venkatachalam S/o Appu Chinne, Marandahally, A.P.

Bashu Dev Mulchand, At Shirpur, Distt. Dhulia, Maharashtra.

C.P. Mathew, Chirkkaran House, Mission Quarters, Trichur, Trivandrum Kerala.

K. Siddhan, Kottapatty, Agraharam, Tamil Nadu

K.M. Kuppuswamy, Keerapatty, Harur, T.K., Tamil Nadu

K.P. Rangaswamy Gounder, Forest Contractor, P.O. Kothpur, Via Natrampalli, N.A., Tamilnadu

E.Rangaswamy S/o Elayaperumal Gounder, Kerapatty Harur, T.K., Tamil Nadu

J.K. Md. Sahib, forest Contractor, Veppanappali, Distt. Krishnagiri, Andhra Pradesh.

Kurumba Girijan Service, Co-operative Society, P.O. Andvai, Mukkali (Palghat), Kerala.

K.P. Moideenkutty, Mill House, Mannarghat (Paighat Distt.), Kerala.

K. kunhabdulla, P.O. Adukkath, Kuttiady, Distt. Calicut, Kerala.

Kathirval & Co., Forest Contractors, Satyamangalam, Kerala.

Kuppuswamy S/o Matha Naidu, Bilianore Agraharam (Via) Indore, M.P.

K.M. Subramanium S/o Muthaia Mudaliar

Kottapatty, Harpur, T.K., Tamil Nadu.

Malampuzha Giri Jan Service Co-operative Society Ltd., N.P. No. P. 669, Malampuzha, Palghat, Kerala.

Murya Lal S/o Sri Puran Chand, V.P.O. Barayatha, Distt. Sagar, Madhya Pradesh.

M.P. Shah of Silvasa, Dadar and Nagar Haveli.

Mahendra Kumar R. Popat of Gondia, Dadar and Nagar Haveli.

M.Nallagounder S/o Muthugounder, P.O. Vannipathy (Via) Mathrandahally, Tamil Nadu.

Murugayyan Chettiar, Vedakkepurathu House, P.O. Pottassery, Palghat Distt., Kerala.

M. Narayanan, Malayil House, P.O. Koddodi, Calicut, Kerala.

M.P. Abdurahiman, Stall No. 11 N.V. Market, Calicut-2, Kerala.

M.Kandaswamy S/o Mariappan Modiyoor, P.O. Noolahalli, B.Agraharam (Via), Distt. Dahrmapur, Tamil Nadu.

M.S.A. Yasin Sahib, Forest Contractor, Ambur (North Arcot Distt.) Tamil Nadu.

Sholayar Girijan Service, Co-operative Society Ltd. No. P. 496, P.O. Varagampadi, Via Thadakam Palghat, Kerala.

S.Sankarapandi S/o Ehocklingam Pillai, Vettaikerampudur, Pollachi, Tamil Nadu.

K.V. Mathew, Lilavihar, Mannarghat, Kerala.

M. Kader Moideen C/o T.S.A. Rahmattullah, No. 138, Madurai Road, Pegambur Dindigul, Distt. Madurai, Tamil Nadu.

Kurein Poulose S/o Kurien, Kaungampillai House, Kothemangalam, P.O. Chelad, Alway R.S., Kerala.

Lalit Kumar B. Patel of Silvasa Dadar & Nagar Haveli.

Manu Bhai Daya Bhai of Silvasa, Dadar & Nagar Haveli.

M.P. Paul, Manjali House, P.O. Varandarapilly, Distt. Trichur, Kerala.

M.Hamza Melathil House, Nallipuzha, Mannarghat, Kerala.

M. Abdul Wahab, Forest Contractor, Dennikotta,

P.O. Dharmapuri Distt., Kerala.

N.A. Singravelu S/o Arunachalam Pillai, Palviduthi P.O., Vaiyampatty (Via), Trichy Distt., Tamil Nadu.

Natar Lal P.Patel of Silvasa Dadar and Nagar Haveli.

The President, Pachamalai Co-operative Marketing Society, To-Sengattupatty, Puthur & P.O. Thuraiyur Talauk, Distt. Tiruchi Tamil Nadu.

P.K. Venkatarama Gounder S/o Kaliappa Gounder, Mulluvadi, Pennagaram, Tamil Nadu.

P.S. Gulab Johan S/o P.K. Silar Sahib, Pennagaram, Tamil Nadu.

P.B. Pasupathy S/o Obasuvaraj Chetty, Pennagaram, Tamil Nadu.

P.Kadiri Nadiri Naidu S/o Narasimha Naidu, P.O. Achulvadi., Harur, T.K., Tamil Nadu.

P. Kesavankutty S/o K. Ramakrishnan, Mukrangad, Kottekkad, Palghat, Kerala.

Mr. V.Kanthappan Chettiar, Cane Merchant, Pottassery, Marghat, Kerala.

V.M. Moidu Hajee, Valiaparambil House, Mukkarikunnu, Distt. Trichur, Kerala.

Mr. V.P. Subramanium, Merchant, Big Bazar, Palghat, Kerala.

M. Murughyan Chettiar, Cane Merchan t, Pottassery, Mannarghat, Kerala.

Satyamangalam, Range Hill, Thekkady Girijan Service Tribes Co-operative Marketing Society Ltd. Satyamangalam, Kerala.

S.A. Saleem, Forest Contractor, Satyamangalam Kerala.

S.A. Ibrahim S/o Akbar Sahib, Bairanaickanpatty, Harur Taluk, Tamil Nadu.

S.Eranavelu, Merchant, V.R.T. Office, Vettaikaranpudur, Pollaichi, Tamil Nadu.

Tiru Podur Maniappa, Forest Contractor, Maragatta Village, Noganur & P.O. Denkanikotta, T.K., Distt. Dharampuri, Kerala.

Thekkady Girijan Service Co-operative Society Ltd., No. P-651, P.O. Sethmadadi, Chittur Taluk, Kerala. T.O. Lona, Thekkath House, Varandarappilly, Distt. Irichur, Kerala.

V.A. Venkitachala Iyer, 25/26, Vadakkanthara Village, Palghat, Kerala

V. Govindan, Forest Contractor, Kothapadi Village, Pennagaram Sub-taluk, Distt. Dharampuri, Kerala.

V.P. Mohammed, Puthenpurayil P.O. Chedleth, Via Sultan Battery, South Wynad, Kerala.

V.P.K. Kulanthaivel Gounder S/o Perumal Gounder, P.O. Kalapathy, Manapari Taluk, Tiruchirappali, Distt. Tamil Nadu.

V.Venkatarama Gounder Forest Contractor, Maharajagadi P.O., Krishanagiri Taluk, Distt. Dharmpuri, Kerala.

V. Govindon S/o Venkatarama Chettiar, Bocharam Patty, P.O. Koothadipatti, Dist. Dharamapuri, Kerala.

Description of Medicinal Plants

Abies (Tourn) Linn.
Pinaceae

Abies spectabilis (D.Don.) Spach (Syn. /I. webbiana L.) "Fir" :

A hardy evergreen coniferous tree, large, tall and pyramidal in shape, sometimes attain height of 54m. Cones, erect and oblong. A native to Himalaya, commonly occurring in chakrata region, at a height of 2500 m to 4500 m above the sea level.

It requires Cool humid climate, good depth of soil essential for the optimum growth of the plant. This species of Plants are propagated by seeds.

Uses : The leaves are used as carminative, expectorant and stomachic.

Abrus Linn.
Leguminasae (Fabaceae)

Abrus pecatorious Linn : "Rosary Pea" or "Weather Plant" :

A perennial plant with multiple branches, deciduous climber with a maximum height of 6 m. Flowers are pale purple in colour. Seeds are scarlet and black. The plant species ia a native to Tropics, commonly occurring in Chakrata region up to a height of 1500 m above the sea level.

A Compost mixture consisting of two parts of loamy soil, one part each of leafmould, cowdung and sand is ideal. Plants are to be watered freely during dry season. This plant species is propagated by seeds or cuttings of firm shoots.

A plant contain glycyrrhizin and the seeds taxalbumin "abrin".

Uses : Leaves, seeds and roots are useful. Leaf juice is sweet in taste and used to treat hoar senses, aw leaf juice mixed with oil is applied on painful swellings of the body. Seeds are used as beads for rosaries and also is used as a substitute of liquorice.

Abutilon Mill.
Malvaceae

About one hundred and fifty species of this genus are present, mostly shrubs, undershrubs or herbs. Angled or palmately lobedleans. Axilary or terminal inflorescence.

Abutilon hirtum G. Don. :

An erect herb up to 100 cm in height. Flowers are yellowish orange in colour. This is distributed in many parts of country.

A compost mixture consisting of two parts loamy soil, one part well rotten cow manure and sand is good for this species. Sunny position and moderate watering year round are preferable. Plants are propagated by seeds or by shoot cutting.

User : A plantpossesses diuretic, demulcent and laxative properties and is also used to treat fever..

Abutilon indicum (L) SW. :

A shrub with golden yellow flowers; plant attain an ultimate height of 95 cm. and occurring in various parts of the country.

Uses : Most of the plants are useful, and used for the same purpose as that of A. hirtum. A stems of this plants on retting, yield a fibre used for making ropes.

Acacia (Tourn) Linn
Leguminasae (= Mimosaceae)

It has about nearly eight hundred tropical and subtropical species of this genus. About twenty five species are native to India, mostly trees or shrubs, erect or climbing in habit of growth and are usually armed. Most of the spec ies yield tannin, gum and timber.

Acacia catechu Wiled. "Catechu Tree" (V. Khair) :

An evergreen weedy perennial native to India, commonly found in Chhindwara District of M.P. at Chimtipur in Sandy loam soil and sunny situation. It is Propagated from seeds or cuttings of half ripened shoots.

Uses : The bark decoction of this plant about 50 ml is taken in the morning and evening for three days in stomachache. It is administered in case of gum trouble and chronic diarrhoea.

A. sinuata (Lour) Merr. :

This plant is available in the tropical jungles throughout India especially in the Deccan Peninsula. The pods of this plants are extensively used as a detergent for washing hair, silk and woolen fabrics. The powdered pods are used in place of soap while taking an oil-bath, since it does not leave the skin dry.

Uses : Seeds of A. cibaria F.V. Meull, A dictyophleba F.V. Meull, A. greggi A. Gray, A. longifolia Willd., A. nilotica Delile, A. oswaldi F. Muell are eaten.

Gums of A. abyssinica, A. dealbata Link., A. decora Rchb., A pycnandra Benth, A. senegal Willd, A. macrotachya, A. pycantha Benth and A. rivalis J.M. Black are useful. Leaves of A. concinna DC and A. suaveolens are also eaten. A campylacantha species yield good adhesive mucilage and suitable for mucilage. Bark of A. caesia is used as a substitute for soap for washing hairs.

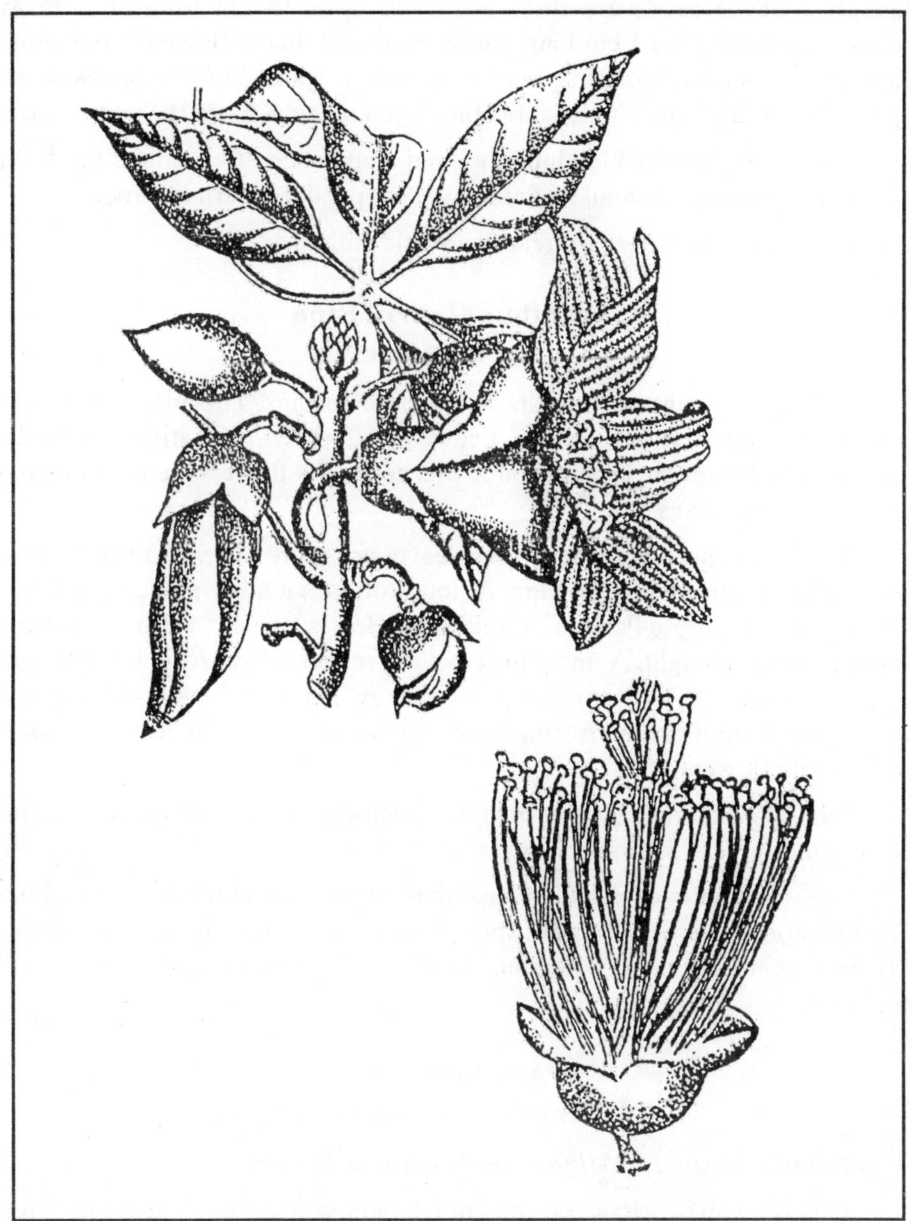

Fig. 1. Semul, Rakat-senbal (Bombax ceiba Linn. Bombacaceae)

AcampeLinn
Orchidaceae

Acampe papillosa L. :

This is an epiphytic orchid with erect stem which is up to 45 cm tall, densely leafy. Leaves are thick recurved upto 15 cm long and fleshy. Inflorescence are upto 5 cm long, rarely branched, many flowered and dense. Flowers are 15 cm across, fragment, dark yellow spotted brown in colour and lips are white. It is native to Eastern Himalayan range, Khasia Hills and Assam.

This plants require high humidity and abundance of water and light. It is grown in a mixture of chunks of charcoal, chopped tree fern and moss.

Uses : Roots are used to treat rheumatic patients.

Acanthus (Tourn) Linn
Acanthaceae

The genus Acanthus consists of Fiftyfive species. This plant species are erect or twining herbs or shrubs. Leaves are toothed, pinnatified and often spinous. The leaves are ornamental in this genus. Its flowers are in terminal or pseudoaxillary spikes.

A can thus caroli-alexandri, Linn, leaves are spiny which grow upto 50 cm height and produce white flowers. A. longifolius is a taller plant, upto l.30 ni height and bear purple flowers. A. millis L. "Bear's Breech" is a lilac flowered herb of medium height. A. montanus leaves are spiny, bear rosy white flowers up to 95 cm tall. A. perringi also possess spiny leaves, which is a dwarf herb with rose coloured bloom. A. spinosus L. grows to 1.25 m tall, leaves are spiny with purple flowers.

Sandy loam soil and warm sheltered position are prefered by these plants; propagated by root cuttings or by division of roots.

Uses : All parts of plants of Acanthus contain mineral salts and its juice. The infusion of the flowers and leaves, have aperient and digestive properties. Crushed Acanthus has soothing and healing effects when applied to wounds and sores.

Achillea Linn
Compositae (= Asteraceae)

Achilla millefolium L. "Milfoil" or "Common Yarrow" :

This is a hardy herbaceous perennial plant with aromatic leaves, attain a height of 15 m. Leaves are alternate, lanceolate-oblong, more or less hairy, deeply divided, usually simple, a very narrow band running down either side of the mid rib; graceful and lacy. Stems are simple and without branches. Flowering stems are erect, stiff, upto 50 cm tall, covered with hairs. Flowers are white, appear in dense cluster. Flower heads are small, stalked, numerous, forming flat topped corymbose inflorescence at the end of the stem and branches. Fruits are compressed, shining, smooth, with no crown or pappus.

This species is widely distributed in Europe, Western Asia and North America. The plant is common in temperate and alpine areas in Chamba, Kangra, Mandi, Kullu, Shimla, Lahaul-Spiti and Kinnaur between 2200 to 4800 m elevation in Himachal Pradesh.

Its require ordinary rich soil and open sunny situation. Plants are propagated by division of tubers.

The leaves, flowering tops and herbs when fresh possess an agreeable and aromatic odour. It has bitterish, aromatic and slightly astringent taste.

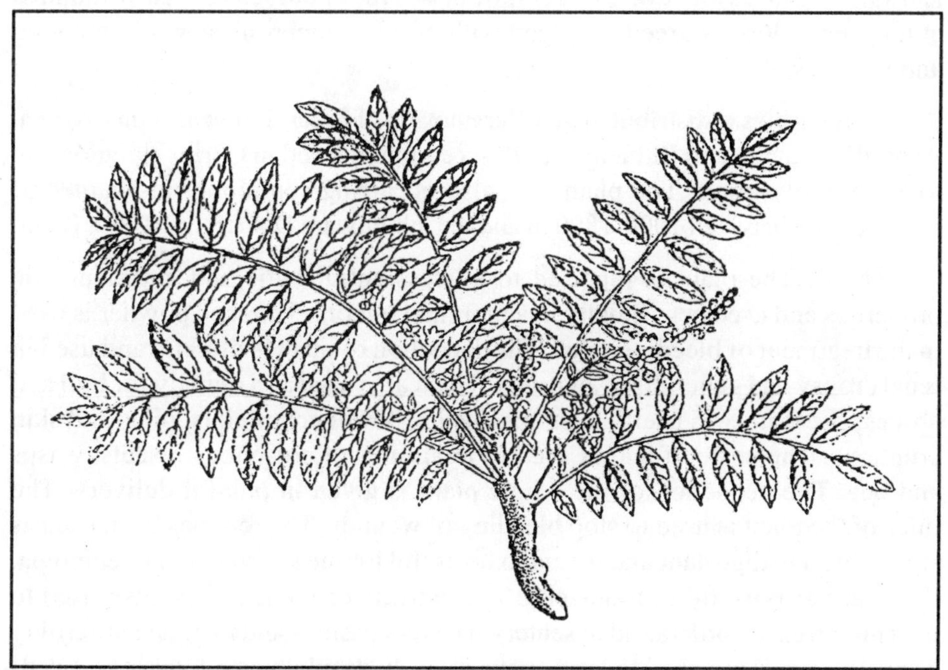

Fig. 2. Salai, Indian olibanum tree (Boswellia serrata Roxb. Burseraceae)

The bitter principle present in this plant, achilleine; an acid known achileic acid and tannin. For its volalite oil, achilleine and tannic acid are useful.

Uses : Entire plant in full bloom is harvested for medicinal purpose. It is a mild aromatic tonic and astringent. Poultice made from plant is applied to cure skin rash. The leaves are chewed in acute toothache. The tea made from the plant is used to treat tuberculosis, disordered stomach, fever and headache. The plant is used as tonic and stimulant and also as diuretic.

Achyranthes Linn
Amaranthaceae

This genus have about fifty six species under this genus and four species are native to India.

Achyranthes aspera Linn :

This is an erect herb or undershrub which attains a maximum height upto 1.4 m. Stem is stiff and not much branched. Branches are pubsescent, terete and striate. Leaves are few, thick, elliptic or obovate, pubescent, usually rounded at the apex. Flowers green or tinged with purple, numerous, appear on a long and woody spike.

This species is distributed in different parts of India, Tropical Asia> Africa, Australia, Ceylon and Baluchistan. It is a common weed in Garhwal regions up to 1000 m elevation. The plants are also occurring commonly in Damtee of Utter Khashi between 1000-1800 m and in Chhindwara District of M.P. at Rater.

Uses : The plant is reported to possess antidiabetic and antirheumatic properties and used beneficially in abdominal tumors. The seed powder is used in the treatment of bleeding piles. The decoction of plant is diuretic and used in renal dropsy and generalised anasarca. The leaf is used as a remedy for boil and abscess. Leaf juice is useful in stomachache, bowel complaints, piles and skin eruptions. The paste of leaf is used to treat bites of poisonous insects, wasps and bee. The decoction of the whole plant is given in painful delivery. The juice of the plant is used to stop bleeding of wounds. The root has been used as stomachic and digestant and is said to be useful for the treatment of pneumonia. Roots are also useful in toothache. The extracts of the roots are also used to treat menstrual disorders and dysentery. The root paste is said to be an antifertility drug. The seeds are used to treat snake bites, hydrophobia and itchings. Seeds are emetic and also used as a brain tonic.

Achranthes bi dent at a Bl. :

This species is distributed in temperate zones of Garhwal Himalaya, mostly herb.

Uses : Plant is diuretic and astringent.

Aconitum Linn
Ranunculaceae

This is a large genus consisting of about four hundred species. Thirty five species are native to India in Himayalayan region. The tuberous roots of several Aconitum species are used in medicine and number of them are known to contain highly toxic alkaloids.

Aconitum atrox (Brul) Mukherjee (V. Meetha) :

This is found in nature from Garhwal to Nepal between 3500 to 4800 m elevation, Chyangseel of Utterkashi, Rishi Ganga Valley and Dudhaloi.

Uses : A tuberous root has medicinal properties. The root paste is applied on snake bite. For relieving cough and cold a little amount of root powder mixed with tobacco is taken. Roots are diuretic, diaphoratic and antiperiodic.

Aconitum balfouri Stapf. (V. Mitha Bish, Teelia) :

This is a perennial herb with tuberous roots. The flowers are hooded and foliage delphinium-like .

The plants occur in Kumaon and Garhwal in the shady slopes and forest edges, between 300-4200 m altitude.

Rich soil and partly shaded situation are ideal for the growth of these plants. Propagated by seeds and division of roots.

Uses : Root is used against neuralgia and rheumatism.

Aconitum chasmanthus Stapf. ex Holmes (V. Mohri) :

This plant is distributed in North-Western Himalaya between 2200 to 4500 m multitude.

The species tuberous roots are the source of the drug aconite used externally for rheumatism, and internally to relieve pain and fever.

Aconitum falconeri Stapf.:

This is an erect biennial herb upto Hm tall. The flowers are hooded and foliage delphinium-like.

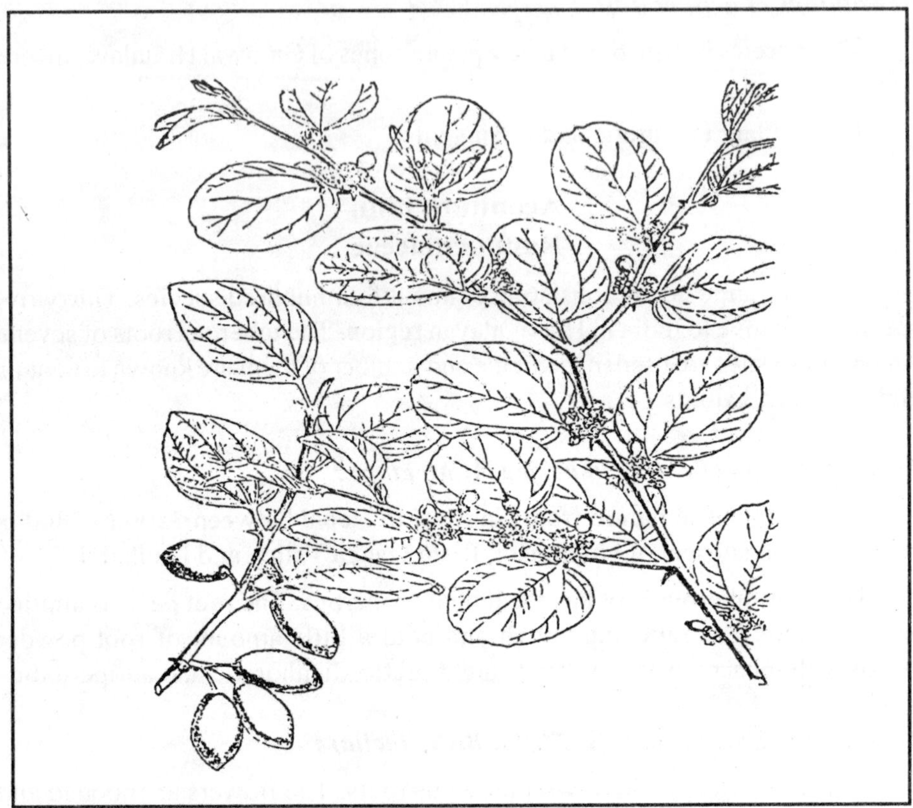

Fig. 3. Baer, Indian Jujube (Ziziphus mauritiana Lam. Rhamnaceae)

This species is native to sub-alpine and alpine zones of Himalayas, found in Garhwal Hills between 3000 to 4000 m. altitude.

These plants can be grown in ordinary soil in partly shaded location and are propagated by division.

Uses : Leaves, flowers and tuberous roots are used in medicinal preparations. This plant is known as antidiabetic, antipyretic and diuretic. The roots of these plants are poisonous. Various preparations of flowers and roots are used for the treatment of skin diseases, rheumatism, neuralgia, fever and diarrhoea.

Aconitum heterophyllum Wall. ex. Hoyle :

This species is a perennial herb with tuberous roots.

It is found commonly in Donpur of Utterkashi between 3000-3800 m, and in alpine Mountains and grassy slopes.

Uses : Roots are useful for medicinal preparations. Roots are non-poisonous and are used as appetizer, astringent, carminative and antihelminthic. It is used in diarrhoea, dysentery and gastric pain. For stomach ache a small dose of root powder is given. This is also used to cure cold, for which a small quantity of root powder is mixed in tobacco.

Aconitum violaceum Jacq. ex. Stapf. :

This is a small biperennial herb with tuberous roots.

It is seen in alpine sunny slopes and also in Changseel of Utter Kashi Between 3200-3800 m; Bhagirathi valley to Kumaon between 3000 to 4500 m altitude.

Uses : This is given to infants in the cold, cough and stomach ache by mixing root powder with mother's milk. The roots are also useful in piles.

Aconitum vulparia Rchb. :

A juice seueezed from the fresh leaves and roots of the plants have medicinal properties. The active ingredient is an organic alkali called aconitine.

These are used in natural form, alcoholic tincture, alcoholatine, tablets and liniment. It is administered for headaches, sciatica, toothache, acute and chronic articular rheumatism, gout and also for dysentery.

Acorus Linn
Araceae

Acorus calamus Linn :

This plant is an aromatic marshy herb with creeping root stocks. Leaves are distichous and ensiform. The flower spathe is continuing the ensiform peduncle.

The species is found throughout India, wild or cultivated. In nature, it is seen in moist places between 1700-2500 m elevation, at Arakot of Utter Kashi.

Uses : In Kwashiorkor disease of children, small pieces of roots are tied around the neck. The rhizomes are used as a carminative, stimulant and tonic. An essential oil is extracted from the rhizomes which is utilised in perfumery.

Actaea (Tourn) Linn
Ranunculaceae

Actaea spicata Linn :

A hardy herbaceous perennial upto 60 cm tall with horizontal roots. Leaves are ovate, lanceolate, three lobed and entire. Flowers are white in colour. Berries are black. There is one variety in it, known as "arguta" which is taller.

The species is native to Britain and also occur in Garhwal Hills between 2500-3500 m. altitude and also other parts of temperate Himalaya.

Ordinary organic rich soil and sunny situation are ideal. Plants are propagated from seeds or by division of roots.

Uses : Leaves, berries and roots are useful. Dried powdered plants are used as insecticide. The berry contain essential oil which is toxic. The roots possess violent emetic and purgative properties. This is also known as nerve sedative.

Actiniopteris Link
Actiniopteridaceae

Actiniopteris radiata (Swartz.) Link.:

It is a green house evergreen fern with a spreading segmented fronds.

It is native to India and found throughout the Aravalli hills between 400-1000 m.

The compost mixture consisting of equal parts of well decayed organic manure, loamy soil, charcoal, potsherds and silver sand is preferable. For its cultivation good drainage and clean pots are needed. Moist atmosphere and adequate watering are beneficial for these plants. These are propagated by spores.

Uses : The plant is used in bronchitis and gynaecological disorders. The plant ash is administered in burns, wounds and scalds. The dry leaves are used in tuberculosis.

Adhatoda (Tourn.) Mill.
Acanthaceae

Adhatoda zeylanica Medic. :

An evergreen shrub with large lanceolate leaves. Flowers are in dense, short spikes. Colour of the flower is white with few purplish markings. Fruits are capsular.

It is commonly found in Chhindwara district of M.P. at Chhindi and Chimtipur and also occurs throughout the plains and sub-mountaneous region of India.

Uses : The plant parts are boiled in water and used for bath in the treatment of body inflammation and bodyache. The leaf decoction is administered in cough and chronic bronchitis.

Adiantum Linn
Palipodiaceae (Adiantaceae)

Adiantum capillus-veneris :

It is a hardy temperate zone fern upto 15 cm tall. There are numerous variety in it.

These plants are also occurring at Mount Abu, Alwar, Gwaparnath of Rajasthan.

Uses : The compost mixture consisting of two parts of organic manure, one part each of loamy soil and silver sand, mixed with charcoal is used for this species. Propagated by spores sown in fine sandy soil with leaf mould, and kept in moist and shades under glass or polythene cover. These plants are grown under glasshouses in humid situation. The decoction of leaves is taken for acute bronchitis and fever. Tea made from the leaves is widely used in coughs resulting from colds, nasal congestion or catarrh. This is taken also along with tea in abnormal stoppage of the menses.

Fig. 4. Khair, Cutch tree (Acacia catechu Willd. Mimosaceae)

Adiantum incisum Forsk :

It is Commonly found in Aravalli Hills.

The leaf powder is mixed with butter and used for controlling the internal burning of the body. Also used in cough, diabetes and skin diseases.

Adiantum lunnulatum Brum :

It is a fern, frond attain a height of 40 cm with long rooting at apex.

Commonly found at Mt. Abu, Gwaparnath, Ajmer and Menal of Rajasthan. The plant is native to Tropics.

It is used in cough, asthma, fever, leprosy and hair falling.

Adonis Linn
Ranunculaceae

There are about twenty five species of this genus.

Adonis aestivalis Linn :

A herbaceous plant with much divided leaves. Flowers are scarlet with a dark-purple eye and appear at the ends of the branchlets.

It is distributed in nature in Western Himalaya.

The plant is known as cardiac stimulant.

Adonis capensis "Adonis Flower" :

This acts as cantharides.

Leaves a: Infusion of their seeds was good for colic.

Adonis vernalis Linn :

It is a herbaceous perennial, 50 cm tall called false hell above. Plants produce yellow flowers.

Native to Southern Europe.

It require fairly rich soil and open sunny situation. Propagated by seeds and division of roots.

It has abundant acrid principles and a dangerous plant; regarded as abortifacient.

Aegle Correa ex Koen.
Rutaceae

Aegle marmelos (L.) Corr. :

It is a tree with numerous thorns, possesses leaves with 3 or 5 leaflets. Flowers are 2-6 cm across, greenish white and sweet scented. Produce attractive edible fruits. Leaves and fruits are aromatic, widely distributed in tropical and subtropical area of India.

It is native to Indo-Malayan region.

Good sandy loam soil, sunny situation and warm humid climate are suitable for cultivation of these plants. Plants generally propagated by seed. Generalty called "Bael' (Ve.Bel).

Uses : The fruit is used as drink and medicine. The unripe fruit is powdered and taken with water in dysentery. The bark decoction of about 20 ml is taken thrice a day for one week in intermittent fever. The fruit is valuable for its mucilage and pectin. Leaves, fruits and the roots have antibiotic properties.

Aerva Forsk.
Amaranthaceae

Aerva lanata (L) Juss. :

A plant is native to tropical Africa and Asia. Distributed in Chhindwara district of M.P. at Rater and also throughout tropical India as weed. Locally name as chilarbhaji.

Uses : A root paste is rubbed on the forehead in headache.

Aesculus Linn.
Hippocastanaceae (Sapindaceae)

Aesculus rubicunda Linn :

It is a tree with large winter buds, covered with resinous scale leaves. The buds expand rapidly in spring. Flowers are large and irregular.

The plant is native to North America and Europe. It is locally called Horse drestrout.

Uses : The bark of this plant contains esculin, fraxin and tannin. Esculin is odourless which is bitter in taste and it is used as a substitute for quinine.

The bark is also prescribed in intermittent fever and periodic neuralgia. The oil extracted from seeds of these plants are externally applied for rheumatic and gouty problems.

Fig. 5. Maharuk (Ailanthus excelsa Roxb. Simaroubaceae)

Aethusa Linn.
Umbelliferae

Aethusa cynapium Linn :

This is a poisonous weed resembling parsley.

It is distributed in Europe, North Africa and West Asia; Locally called Fool's parsley or "Ass Parsley".

Uses : The leaves of these plants are externally used and are known as emollient and resolvent. The crushed plant is mixed with resolvent flours to make poultice which is used for gout and sciatica. These plants are also used for whooping cough and congestions of the abdominal viscera.

Agathosma Willd.
Rutaceae

Agathosma betulina, Pillans :

This plant is native to South Africa. It constitutes diosmin, calcium oxalate, diosphenol, d-limonene, dipentene, and 1-menthone. Leaves are medicinal.

Uses : It is used as diuretic and urinary antiseptic. It is called round Badu.

Agathosma crenulata Pillans
(Barosma crenulata Hook; = B. Serratifolia Willd) :

This is a native to South Africa. Medicinal properties are same as that of A. betulina. It is locally called oral Buchu.

Agave Linn
Agavaceae

Agave sisalana Perr. :

It is a perennial succulent plant, forming stem upto 1.5 m tall. Leaves are in dense rosette, stiff and spreading in all directions, flat or slightly concave, sword shaped, 1.5 m to 1.78 m long, 6.5 to 7.9 cm wide and 3.5 to 4.9 m thick, margins are smooth or bear a few small teeth; epidermis leathery, grey green with conical terminal short spine. Inflorescence emerges from the centre of the rosette, 7 to 8 m tall; flowers are 7.5 cm long, tulip-shaped, green, ill smelling. Flowers appear in clusters at the forked tips of the 35 to 45 horizontal branches. The plant bloom once in six to nine years, sometimes after twenty years and die thereafter. Fruits are few only if any, 6 to 7 cm long and seeds are12 mm long. Nearly 2200 to 3500 bulbils develop after the flowers are shed. Bulbils fall to the ground and take root which develop into new plants. This species is grown in plantations and yields the sisal hemp.

Distribution :

This plant is distributed from Yucatan. It is commercially cultivated in Mexico, Brazil, Haiti, East Africa, South Africa, Venezuela, Madagascar and India. In India, this species is cultivated in W. Bengal, Bihar, Assam, Maharashtra, Karnataka and Chennai.

This plant is best grown in humid tropical climate and well drained open textured soil. Sisal is propagated by bulbils and suckers. Nearly 3000 plants can be accommodated per hectare. Soil should be well fertilized and weeding is necessary only in initial years. After three years leaves are harvested and continued upto seven to eight years. Normally, three to four years old plant attain a height of 1.8 m and bear 118 to 125 leaves. All leaves are harvested annually retaining only 22 to 29 central leaves. With the appearance of flower stalk, they are quickly cut to 1.4 m height so that the leaves do not shrivel and developed to a usable size. The plantation is ploughed down when 72 to 85% of the plants develop flower stalks.

Uses : Soon after the harvest of leaves, these are tied in bundles and within twenty-four hours these are processed and fibre is extracted. The extracted fibre is washed, dried, brushed, graded and baled. From one hectare of land nearly 2200 kg of fibre can be obtained per year. From about 1000 Kg of leaves 85-95 kg of dry fibre can be extracted. From the pulp waste hecogenin, tigogenin and other sapogenins are extracted, which is a source of cortisone for pharmaceutical industries. Cortisone drugs are useful for the treatment of rheumatoid arthritis, rheumatic fever, sciatica, skin diseases and certain allergic diseases. Sisal fibre is largely used in cottage industry level and also for making ropes, twine, socks etc. Cellulose is manufactured from sisal. The fresh pulp waste is used as cattle feed. This pulp waste is also used as fertilizer in sisal plantation.

Ageratum
Compositae (Asteraceae)

Ageratum conyzoides Linn :

This is a half hardy annual up to 60 cm tall with pale blue fragrant flowers, it has scaly pappus.

This plant is native to tropics and found as weed in different parts of India. It is also a common weed in Ceylon. Commenly known as Floss of lower or Goat weed.

Organic rich soil and sunny situation are ideal for these plants. Propagated by seeds or by cutting of young shoots. Pinching is done periodically to prevent flowering.

Uses : Leaves of these plants are used to treat sores and cuts.

Ailanthus Desf.
Simaroubaceae

Ailanthus altissima (Mill) Swingle :

It is a tall tree of about 25 m height. Leaves are lam long with 16-20 small leaflets and these are teethed near the base. This species is native to China. Commenly name as Tree of Heaven.

Rich, light soil and moist situation are ideal for these plants. Generally propagated by root cuttings.

Uses : Bark, trunk and roots are medicinal. To treat intestinal tapeworms, stomach troubles and dysentery, powdered bark is administered. The root bark is used to cure epilepsy, heart trouble and asthma.

Ainsliaea DC.
Compositae

Ainsliaea aptera DC. :

This is an erect perennial herb with white flowers, cylindric and appear before the leaves during March-April. In autumn also flower appear but remain unopened. Commonly occurring in the finest shades in association with Valeriana jatamansi throughout Himachal Pradesh between 2000-3200 m altitude.

Uses : In acute stomach ache, the powder of the root is used as an emergency medicine. It is also used as wormicide, and against diarrhoea.

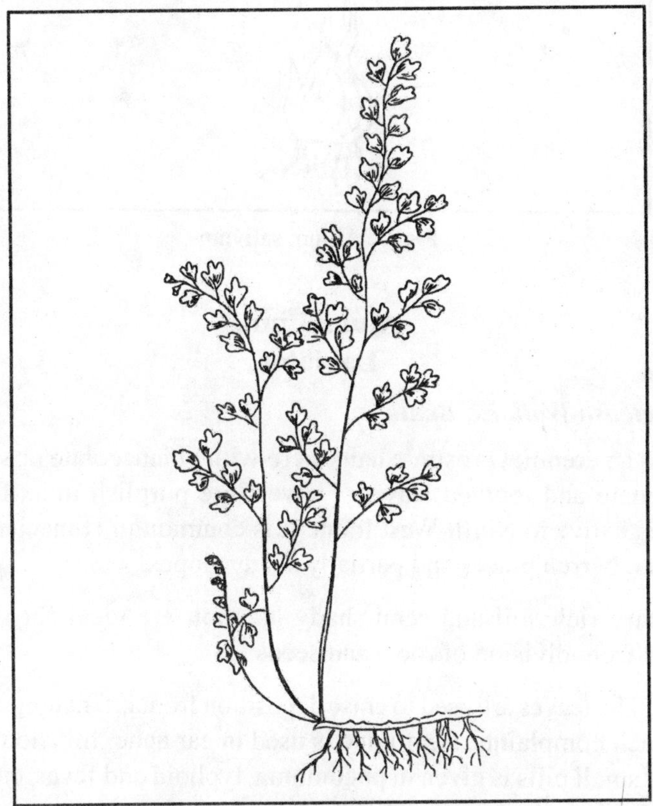

Fig. 6. Adiantum capillus-veneris

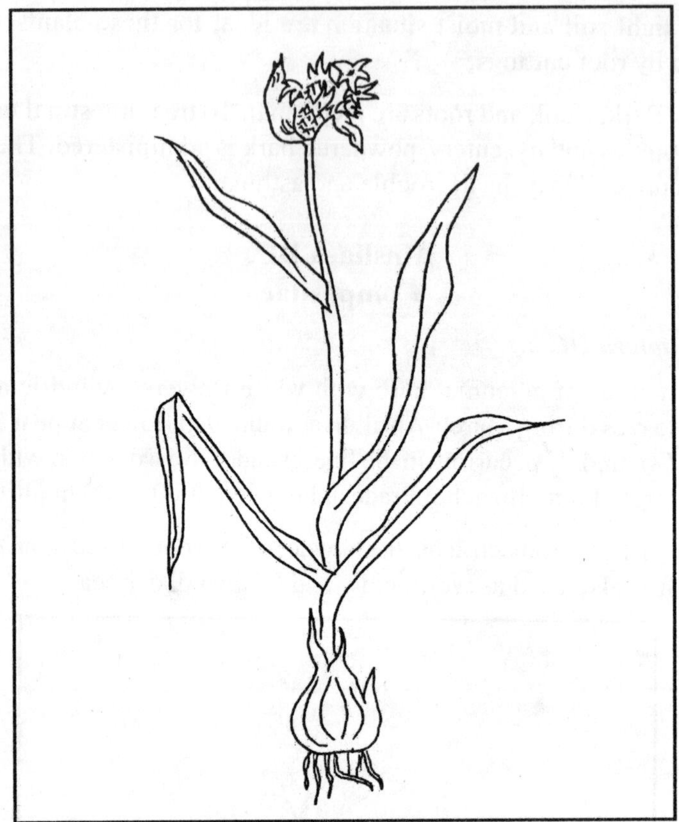

Fig. 7. Allium sativum

Ajuga Linn.
Labiatae

Ajuga bracteosa Wall. ex. Benth. :

This is a perennial prostrate hairy herb with oblanceolate or subspatulate, obtuse, sinuate and toothed leaves. Flowers are purplish in axillary whorls. The plant is native to North-West India. It is common in Himachal Pradesh in field bushes, barren places and partially shady slopes.

Ordinary rich soil and semi-shady location are ideal for these plants. Propagated from division of roots and seeds.

Uses : The leaves are used to erase deposition from infant tongues, suffering from stomach complaints. Leaf juice is used in ear ache. Infusion of leaves in the form of small pills is given in pneumonia, typhoid and fever, bronchitis and gastric pains. The plant is considered a bitter tonic.

Aletris Linn.
Liliaceae

Aletris farinosa Linn. :

This is a perennial plant. Leaves are flat, long and grass like. Flowers are tubular, yellowish white.

This species is native to East Asia and North America.

They prefer moist damp boggy location for proper growth. Commenly called white tube stargrass.

Uses : Leaves and roots are medicinal. Powdered leaves are used to treat sore breast and sore backs. Root preparation is sedative and toxic, and these are used to treat rheumatism, stomach pains. Such preparation is diuretic..

Allium Linn.
Family - Liliaceae

Allium sativum Linn. "Garlic" :

This is a spice crop, widely cultivated in India. It has a compound bulb, greyish white in colour and sub-globular in shape about 4-6 cm in diameter with 9-25 cloves. Entire bulb is surrounded by 3-5 whitish papery membranous scales. The cloves are attached to a flattened circular, wood axis with numerous roots.

It is native to South Europe and Central Asia and widely cultivated in India. Commenly called Garlic.

For cultivation of these plants, well drained loamy soil, rich in humus is good. It can be grown in monsoon and winter season. They are propagated from cloves. During soil preparation about 10 tonnes of FYM is incorporated with the soil. These are cultivated either by dibbling or furrow planting. Around 500 kgs of planting material is required per hectare. The bulbs are planted at 5 cm depth and 15 cm apart. A fertilizer mixture consisting of 250 kg urea, 400 kg superphosphate and 100 kg muriate of potash is applied per hectare. When the leaves turn yellow or brown and begin to dry, bulbs are harvested. Average yield of 8000 kg bulbs/hectare may be expected. Plants are affected by thrips, maggots, caterpillar, tobacco worm and mite. Diseases like smut, smudge, leaf blight and rust are also reported.

The active principle is allin. A number of antibiotic principles like allistatin I, allistatin II and garlicin are isolated from garlic. It is used in medicine for both human beings and animals. It is a flavouring agent and one cannot do without it in the kitchen. In medicine it is used on a decoction, in the form of tincture, juice, syrup or poultice. It is boiled in milk and administered as an anthelminthic and expectorant. The garlic pulp is regarded as rubefacient and even vesicant. Importance of garlic for the treatment of hypertension, diabetes, helminthiasis, chronic colitis and gastritis, angina pectoris, bacterial and fungal infections, amoebiosis, arterioselerosis, rheumatoid arthritis and cancer were reported. Garlic is a stimulant and cures digestive disorder, lungs and respiratory tract diseases. A paste made from it can relieve pain cause by the sting of scorpion.

Allium govanianum Wall. :

It is a perennial herb with aromatic leaves.

Frequently occur in eastern hills and is also found in Garhwal between 3200-3800 m elevation.

Uses : The leaves are used in wounds and cuts.

Alnus Mill.
Family - Betulaceae

Alnus serrulate (Ait) Willd.:

It is a small tree upto 9 m height. Bark of this plant is aromatic, smooth and greyish in colour. Leaves are oval, leathery, margins teethed upto 4.5 cm long. Flowers are catkins.

It is native to North Temperate region of Indo-China. Commenly Called Hazel Alder.

Swampy and coastal areas are suitable for growing of these plants.

Uses : Stem bark are medicinal. The tea made from bark is a blood purifier and used to treat diarrhoea. Decoction of bark is useful to treat syphilis and malaria, while a decoction of catkins is used in bloody urine. Dried bark steeped in water is used to cure eye infection. The bark is also used to treat wounds and skin ulcers.

Aloe Linn.
Family - Liliaceae

A large genus comprising of species native to South Africa, tropical Africa to North Africa, Abyssinia, Eritrea, Arabia, Mediterranean Region, Atlantic Islands and Madagascar. These are perennial succulent plants where leaves are arranged in rosettes. A few species of these are cultivated commercially in different countries of the world for latex production. Dried powdered latex and mucilaginous pulp of the leaves are used in the pharmaceutical and cosmetic industries.

Aloe barbadensis Mill. :

Its Stem is short and forming offsetts. Leaves are rosulate to 50 cm long and 8 cm wide, rounded on the reverse, grey green. Inflorescence are simple, tall and raceme is elongated. Flowers are cylindrical, yellow in colour. There are several forms in this species.

It is native to Cape Verde Islands, San Antonio, Canary Islands, Medeira and growing wild in Southern Mediterranean regions. It is grown in Egypt, West Indies, Tropical America and in 'die Netherlands. This species is naturalised in India. Locally known as curaced or Barbodos Aloe.

Uses : The leaves contain anthraquinone glycosides which are collectively termed as "aloin". The aloin content of A. barbadensis is 35%.

Aloe ferox Mill. :

It is a large growing perennial succulent plant with stems varying from 2 to 4 or more in height. Leaves are 55-65 numbers in a dense rosette, lanceolate, sword-shaped, upto 1 m. long and 20 cm wide at base. The leaves are broad, glaucous, and spiny in margins, lower and upper surfaces. The leaf epidermis is smooth, thick and rubbery, mucilaginous pulp within. The flowers are borne in cylindrical, terminal racemes on central flower stalks upto 1 m tall. Raceme erect, densely many flowered upto 90 cm long. Flowers are pale red in colour.

It is native to Cape Province and Natal. The aloin content is 15%.

The dried latex contains aloinosides A and B.

Aloe Sruticosa Lam. :
A. arborescence Mill :

This plant attains a height upto 5 m in the native habitat. Shoots developing from the lower part of the stem frequently become layered. Leaves are in dense rosettes upto 70 cm long, grey green with spiny, waved margins. Inflorescence is simple, raceme conical-cylindrical. There are number of varieties in it which differ in leaf colour. Flowers are tubular, red in colour.

It is distributed in Cape Province; Natal to Transvaal.

Aloe perryi Baker :

This is a small plant, leaves are in a dense rosette, lanceolate, 15-25 numbers per plant, 35 cm long and 7 cm wide, gradually tapering, blue green in colour.

Inflorescence is upto 70 cm tall, stout, simple, raceme oblong-cylindrical, many flowered. Flowers are red tipped with green.

It is native to Isle of Socotra and cultivated in Arabia and East Africa. Aloin content of this species is 30 to 35%. Commenly called socotrine or zanzibar Aloe.

The plants are propagated for commercial cultivation from offshoots or suckers. It requires sandy calcarious soil with adequate drainage. Plant grow luxuriantly with moderate irrigation during dry season and generous fertilisation. Leaves attain a full size within three years of planting. Mature leaves are harvested with care so that undue leakage of latex can be avoided.

The leaves are cut and the juice which exudes is collected and evaporated to a dark viscous material which is later solidified.

From the subepidermal longitudinal cells of the cut aloe leaves, very bitter yellow latex exudes which contain aloin. It has been reported that aloe contain resinotannols with cinnamic or p-hydroxy-cinnamic and other aromatic acids. The mucilaginous pulp is composed of glucose, mannose, traces of arabinose, galactose, xylose and 98% water.

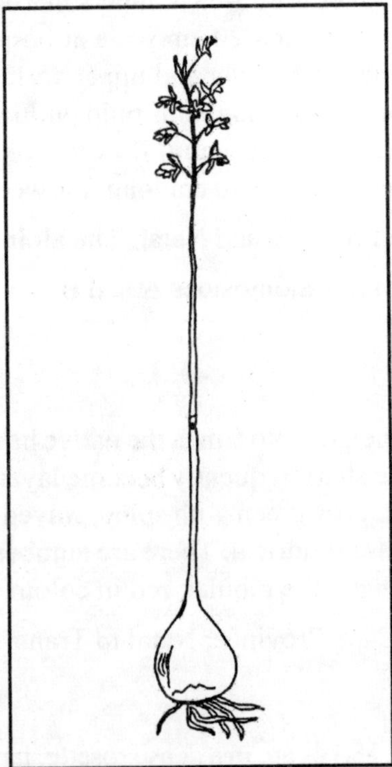

Fig. 8. Aplectrum hyemale

Fig. 9. Arctium lappa

Uses : The resinous juice from the leaves has been used medicinally since ancient times for a wide variety of ailments. It is used as stomachic tonic, purgative and emmenagouge. The fresh latex is taken in a very small dose as purgative. The mucilaginous pulp is said to possess biogenic stimulators and wound healing hormone activity. Hence, the peeled fresh gel is used to treat inflamed eyes, skin inflammations, sores and burns. This pulp is used into many cosmetic preparations like cream, lotions and ointments for softening, soothing or moisturizing the skin. The fresh juice is administered for minor burns, sunburn and other emollient uses. The pulp is taken internally for curing ulcers and intestinal ailments, sore throat etc. As an ingredient in compound tincture benzoin, aloe is used.

Alpinia Roxb.
Zineiberaceae

Alpinia galanga (Linn.) Willd. :

It is a herbaceous perennial plant which attains a height upto 2 m. Leaves are long narrow. Flowers arise in compound dense branches, greenish white in colour. Fruits are orange red, 13 mm long and contains 3-6 seeds. Rhizomes are slightly aromatic, deep orange-brown in colour, pungent and bitter.

It is native to Eastern Himalaya and the Western Ghats.

These plants require compost mixture of equal parts of leaf mould, well decayed cowdung manure and loamy soil. Propagated by division of roots. Commenly known as Greater Galangal.

Uses : The green rhizomes contains 0.04% essential oil. It consists of methyl-cinnamate, cineol, some camphor, and probably d-pinene. Leaves also yield volatile oil.

Dried rhizome of this plant constitute the drug. It is used as tonic, deodorant and disinfectant. It possesses a stimulant aroma like ginger. It is used to treat bronchial catarrh, respiratory complaints and rheumatism. The drug is a depressant of the cardio-vascular system. The rhizomes are also carminative and stomachic.

Alstonia R. Br.
Apocynaceae

Alstonia scholaris R. Br. :

A tall evergreen tree, may attain a height of 25 m with a tall trunk and spreading whorled branches, having bitter milky juice. Leaves are upto 20 cm long, are in whorles of 5 to 7, spreading, shortly stalked, lanceolate or obovate oblong, mid rib very prominent beneath, leathery. Bark grey colored, uneven and rough, with many lenticels on the younger branches. Flowers are small, greenish white in colour, shortly stalked, in small clusters, combined into compact long peduncled rounded paniculate cyms. Fruits are cylindrical, pendulous narrow, slender upto 60 cm long, hanging in pairs and forming dense clusters. Seeds are dorsally compressed, numerous, pale orange.

It is distributed in the humid region of India/ Ceylon, Burma, Java and Philippines, Western tropical Africa, New Guinea and Australia.Locally known as “Devel tree”.

Uses : The bark has bitter taste, not acrid and devoid of any odour. An alkaloid has been detected from the bark which is known as diamine.

The dried bark of the tree constitute the drug. The bark is regarded as a valuable antiperiodic and tonic. It is effective in chronic diarrhoea and dysentery. It is useful in malarial fever and brings down the temperature. It is also used to treat skin diseases; debility after fever and other exhausting diseases. The wood is used for making wooden slates, furniture, pencil etc.

Althaea Linn
Malvaceae

***Althaea officinalis* Linn. :**

An annual herb, sparsely branched, and the plant height varies from 40 to 160 cm. Leaves are ovate or heart shaped greyish-green velvety, alternate, covered with fine star shaped hairs. Flowers in very short, dense, axillary panicles, rarely solitary, of a delicate uniform bluish colour. The tap root is greyish-yellow, corky externally and white fibrous within.

It is native to Siberia and Western Asia, Southern and Eastern England and from Denmark through Central and Southern Europe to North Africa. The plant is commercially cultivated in Belgium, France, Germany and Yugoslavia. This species is naturalised in India and occurring from Punjab to Kashmir.

They grow in ordinary loamy soil and sunny situation. Plants grow better when the seeds are sown directly in situ and extra seedlings are removed to maintain a distance of 60 cm either way.

The tap root is slightly scented, sweetish and astringent. Dried tap root contains mucilage, starch, pectin, sugar, tannin and asparagine. For medicinal use, roots should be dried thoroughly otherwise the roots will quickly decompose and develop sour odour and taste.

The flowers and roots of these plants are emollient. From the dried root infusions, fluid extracts, syrups and tincture are prepared for medicinal use. The flowers and roots are used in a decoction for all inflammation. The decoction of root is prescribed for asthma, bronchitis, hoarseness, pleurisy, dysentery, muscular and nerve injury. The leaves are recommended to treat very acid urine, for chronic phlegmasia of the alimentary canal, constipation, irritation of biliary system, nephritis, cystitis and dry cough. Concentrated decoctions are made from minced leaves that can be applied in poultices. The decoction of leaves and roots is applied in the form of lotions, formulations, eye wash and rectal infections, whenever there is need to soothe irritated inflamed parts. The infusion of flowers is used in chest complaints. The roots are used in a decoction in cases of poisoning, gastric and intestinal inflammations and toothache. This is used as a gargle for gum troubles and sore throats. The powdered root is applied as a poultice on abrassions and eruptions externally. It is used as a demulcent for children as is used in poultice, lozenges etc.

Amaranthus Linn
Amaranthaceae

A genus of over fifty five species and about twenty one species are native to India.

Amaranthus hybridus Linn
A. paniculatus Linn "Smooth Pigwood"

An erect branched annual herb may attain a height of 3 m. Leaves are petioled, ovate with wavy margins, rough hairy, upto 16 cm long, dull green in colour. Flowers appear in terminal clusters, greenish or red. Tap root is long fleshy, red or pink in colour,

It is distributed in tropical and temperate regions.

About 30 kg of seeds are required for a hectare of land. Seeds are mixed with 8-10 parts of sand and broadcasted in well prepared beds. The seeds may be sown any time of the year. A fertilizer mixture of 85 Kg N, 30 kg P_2O_5 and 30 kg K_2O per hectare is recommended. The leaf yield is around 15 tonnes from one hectare. The crop is of 3 months duration.

Uses : Leaves have medicinal value. It is used to treat intestinal haemorrhage, ulcers, diarrhoea, dysentery, piles and excessive menstrual flow. It has a cleansing effect, and helps to reduce tissue swelling. The plant is a blood purifier. The leaves are also eaten as vegetable.

Amaranthus tricolor Linn. :

A half hardy annual plant. Leaves are carmine and yellow. There are two varieties in it e.g. splendens and salicifolius.

It is native to India, though it occurs wild in other countries. Found in Chhindwara district of M.P. at Rater and also distributed widely in the different parts of the country. Commenly known as Matabhaji.

Moderately fertile, well drained soil and sunny situation are ideal for these plants. Propagated by seeds. It is a warm season crop and grows best in the rainy season. The seeds are sown during March-July in the plains and May-July in the hills. The seed rate/hectare is 3 to 4 kg depending upon the varieties. The average yield of leaves/hectare is about 8600 kg.

Uses : The root decoction of about 70 ml is used twice a day for seven days in ulcers of the throat and mouth, and in pus formation. Its tender leaves and shoots are cooked like spinach. It is a good source of vitamins A, B, C and iron. The seeds are eaten by some tribes.

Amaranthus retroflexus Linn. :

A fuzzy weed 40 cm tall. Leaves are dull green with long stalks. Flowers appear in long spike; densely packed. Commenly known as Red rood Pigweed.

Uses : Leaves are rich in saponin. They are used to stop excessive menstrual flow, internal haemorrhaging and diarrhoea..

Amaranthus gracilis Desf. :

An erect herb upto 80 cm tall with green flowers.

The species is distributed in different parts of India and sporadically occurs on the Garhwal hills between 1000-1800 m altitude.

Uses : The plant has emollient action and tender shoots are eaten as vegetable.

Amaranthus spinosus Linn. :

This is a prickly annual herb, erect in nature with greenish flowers

It occurs in different parts of India in waste land and cultivated fields.

Uses : The root of this plant is diuretic and also used in colic pain and leucorrhoea. The leaves are eaten as vegetable.

Ammi Linn.
Umbelliferae

Ammi majus Linn. :

A biennial plant, constitutes principal commercial source of Xanthotoxin.

This is native to Europe and Africa and successfully tried its cultivation in Dehra Dun, Jammu and Punjab. This species is also cultivated as a winter ornamental and has run wild in North India. It is grown as annual crop and the seeds are sown in September. Commenly Known as Bishop's weed.

Well drained, loamy soil rich in organic matter is good for this crop. During vegetative phase they require cool climate, thereafter dry and warm weather is preferable. The seeds can be sown directly in the final site or seedlings are transplanted. Planting the transplanted seedlings in line makes interculture operations easier. Seeds are sown in a well-fertilized, porous and raised beds. About 1.5 kg of seeds are required per hectare for transplanted crop. After sowing, seeds are covered with fine layer of sieved FYM or leaf mould. Seeds germinate within 12.-14 days after sowing. Seedlings are transplanted in the final site during October-November when the seedlings attain an age of 60-70 days. The land is prepared by giving 5 to 6 cross ploughings and mixing 60 tonnes of FYM per hectare. The land is finally levelled and divided into small plots of convenient sizes. Around 80 Kg N, 50 kg P_2O_5 and 40 kg K,O per hectare is recommended. Nitrogen is applied in two split doses, 40 kg at transplanting and remaining 40 kg during branching of plants. The crop may require 3 weeding and first irrigation is given just after transplanting. Irrigation at regular interval is needed in dry and hot weather. Seeds are harvested during April-May when the seeds in most of the umbels turn light brown. After harvesting, the crop is threshed manually in the threshing floor. Around 12 quintals of seeds can be obtained from 1 hectare of land.

Uses : The seeds contain about 0.4% xanthotoxin which is commonly used in the treatment of leucoderma and also in formulating sun-tan lotion.

Ampelocissus Planch.
Vitidaceae

Ampelocissus rugosa (Wall.) Planch. :

A climbing shrub with tendrils on peduncles.

Commonly found in pine forest of Gamari, Utter Kashi between 1000-2000 m.

Uses : The sap of the hollow stem is sucked for oral apthae and rashes.

Anacardium Linn
Anacardiaceae

Anacardium occidentale Linn. :

An evergreen tree widely cultivated in the tropics for its edible nuts.

In India, cashew is commercially cultivated. Other major cashew producing countries are Madagascar, Kenya, Mozambique, Tanzania and Brazil. Commenly known aas Cashewnut.

The plants grow well under hot and humid climate of the West Coast. Plant grows in different soils, except heavy clay and alkaline soils. Deep loamy soil with good drainage is well suited. Plants are propagated by seeds and air layering. Seedlings are raised individually in polythene bags, and 2 to 3 months old seedlings are transplanted. Land is prepared by ploughing, cross ploughing and harrowing. Pits are dug of 60 cm depth and diameter at 8 m apart. Pits are filled with 15 kg of FYM. Seedlings or rooted air layers are planted in the centre of the pits during monsoon, one in each pit. During September-October, after heavy rains, NPK fertilizers are applied depending on the age of the plants. Plants are attacked by tea mosquito, stem borer, leaf miner, thrips, bark borer, wild silk caterpillar, slug caterpillar and aphids. Cashe.wnut is attacked by Dieback which is a serious one. A combination spray of Metacid and copper-oxy-chloride is suggested to control insects and diseases. The plants begin to bear from 4-5 years onwards. Main season of harvest is from February-May and around 1000 kg of cashew nuts can be expected from a hectare of land.

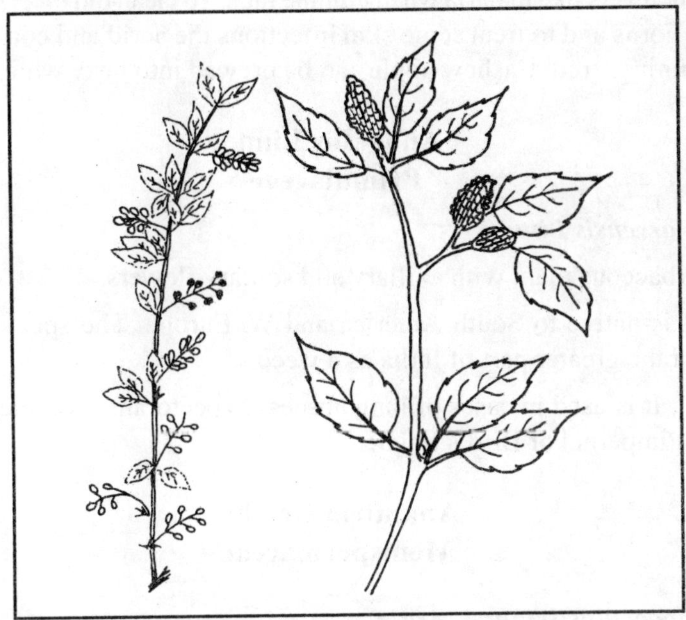

Fig. 10. Berberis vulgaris, Betula utilis

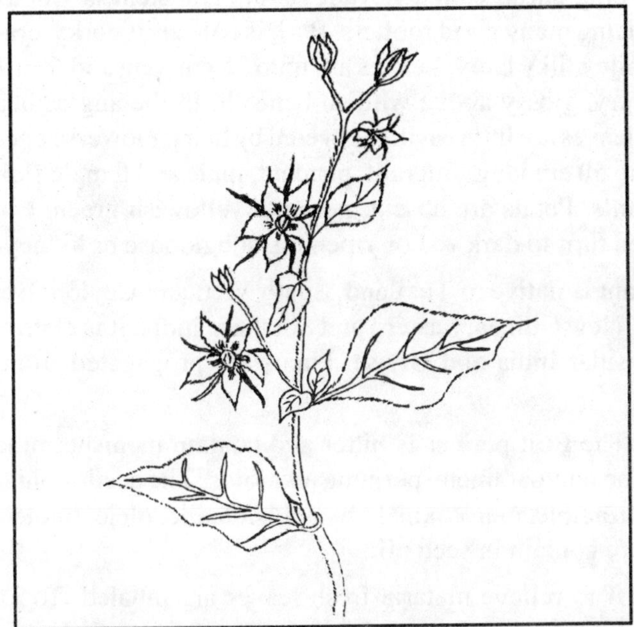

Fig. 11. Borago officinalis

Uses : The kernels of cashew are rich in unsaturated fats, carbohydrate, minerals and vitamins. The cashew apple juice contains vitamin C 260 mg/ 100 gm, sugars 10 to 12.5% and 0.35% malic acid. Tree bark are used for some intermittent fevers like malaria when quinine fails. To clean old ulcers, to remove warts and corns and to treat some skin infections the acrid and corrosive fresh juice is administered. Cashew apple can be brewed into tasty wine.

Anagallis Linn.
Primulaceae

Anagallis arvensis Linn. :

A herbaceous plant with axillary and solitary flowers.

Plant is native to South America and W. Europe. The species is found throughout the greater part of India as a weed.

Uses : It is used in cases of lung abscess expectorant. Commenly known as scarlet Pimpernel or Bird's tingue.

Anamirta Colebr.
Menispermaceae

Anamirta cocculus Wight & Arn. :

A climbing shrub called Levant Berry. The stem is woody upto 10 cm thick, producing many acrid rootlets. Bark is pale gray, corky, cracked with age and with white milky latex. Leaves are upto 25 cm long and 6 cm broad, broad-ovate, leathery, glossy above whitish beneath. In the angles between the big veins of the leaves are little cavities covered by hairs. Flowers appear in branched panicles upto 50 cm long, fragrant, pendent, male and female flowers borne on separate plants. Petals are absent, perianth yellowish green. Fruits are white initially, then turn to dark-red on ripening, subglobose or kidney shaped.

The plant is native to Thailand, South Vietnam, Ceylon, Burma, Malaya, Philippines, New Guinea, Eastern and Southern India. It is distributed in Khasi Hills, Peninsular India and Orissa. Plants are propagated from seed or root cuttings.

Uses : The fruit pericar is bitter and contain menispermine and contain menispermine and paramenispermme alkaloids. The seeds contain very bitter, crystalline principle, picrotoxin. Fatty acids, stearic, oleic, linoleic, trisaturated glycerides are contain in seed oil.

An snuff to relieve malaria fresh leaves are inhaled. To extract Guinea worms leaf juice is administered. Fresh fruit juice is applied on foul ulcers and scabies. The seeds which contain picrotoxin is useful in the treatment of epilepsy and to overcome night sweating in cases of phthisis, paralysis of pharnyx and lower limbs. Picrotoxin is said to counteract morphine and chloral poisoning; used as a fish-poison.

Anemone Linn.
Ranunculaceae

Anemone pulsatilla Linn. :

A herbaceous plant known as Wind flower & Pasque Flower with rhizomes. Leaves are radical, lobed or divided. Flowers appear on scapes. Honey secreted by the long tubed purple flowers which is mainly visited by bees.

It is native to Europe.

Uses : It has acrid burning taste. The roots are very bitter and it acts as an aperient and stimulant in minute doses. Strong dose is poisonous. The root and leaves are used in a powder and in an infusion. To prepare homoeopathic medicine and used for toothache, colic and measles, this plant is used.

Andira (Lam.) Juss.
Leguminasae Fabaceae

Andira araroba Aguiar :

It is native of Brazil. Goa powder or Araroba from trunk of tree is used in medicine.

Uses : It contains chrysophanol and anthranol. It is applied on skin diseases; an irritant on eyes and nose.

Andrographis Wall.
Family : Acanthaceae

Andrographis paniculata Wall. ex Ness. :

This is a much branched erect annual herb. Branches are four angled. Leaves are lance-shaped. Flowers appear in large spreading sparse branches, individual flowers are small, upto 1 cm long, rose colored. Fruit is capsular upto 2 cm in diameter.

It is native to Tropical Asia and occurring in the plains of India. Commenly known as Kalmegh or Indian Chirettar or koriyat plant.

Uses : Leaves and stem constitute the drug. It is a bitter tonic and possess antityphoid and antibiotic properties. It is used to treat liver and digestion complaints, general weakness, worms, fever, dysentery and excessive gas formation in stomach.

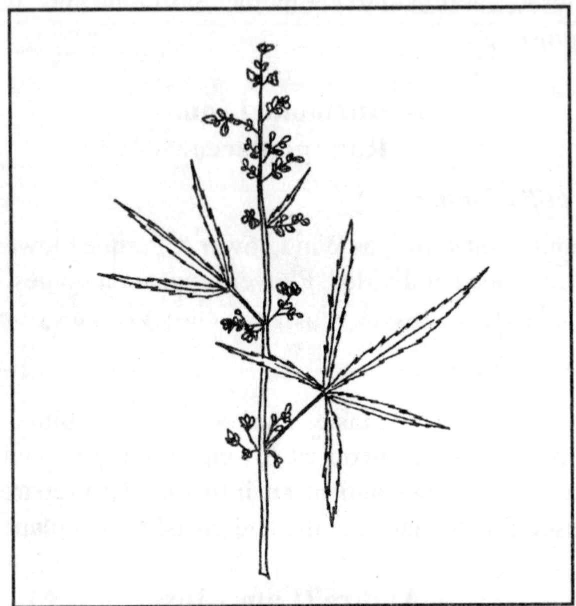

Fig. 12. Cannabis sativa

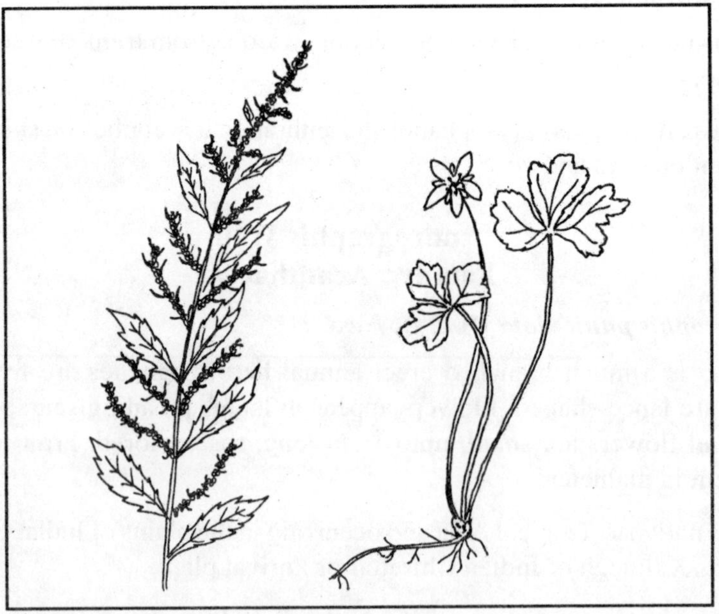

Fig. 13. Chenopodium ambrosioides, Coptis groenlandica

Androsace (Tourn) Linn.
Family : Primulaceae

Andrographis paniculata Wall. ex Ness. :

A small, hairy tufled herb with long stalked rounded leaves. Flowers pink and white appear in irregular umbel during April to September.

The species is occurring sporadically in all parts of H.P. between 1500-2800 m, elevation, in moist partially shady slopes. Commenly known as Rock Jasmine.

Compost mixture consisting of sandy loam soil, leaf mould and sharp grit is ideal. Propagation is by seeds and division of roots.

Uses : Leaves are eaten with gur and ghee for correcting menstrual flow for the age groups of 25-30 years of women.

Angelica Linn.
Family : Apiaceae (Umbelliferae)

A Angelica glauca Edgew.

A large aromatic herb with hollow stems. The tap roots are short and thick. They have aromatic odour and a warm, pungent, bitter-sweet taste.

It is distributed in W. Himalayas, occasionally found in alpine scrubs and forest edges in India between 3400-3800 m elevation.

Deep moist loamy soil and shady situation are ideal for these plants. It is propagated by seeds. Commenly known as Long Wort.

Uses : Roots are used as drug in wounds and gastric pains. This aromatic plant in known as a stomachic, carminative, stimulant, sudorific. It is used in dyspepsia anorexia, spasms, flatulent colics and bronchitis. It is also used for flavouring confectionary and liquors and the root candied for cake decoration.

Angelica officinalis Hoffmann.

A young stem, roots and fruits are used for flavouring foods. The fresh roots contain 1.2 to 1.3% essential oil and fruits 1.2 to 1.3%. The main constituent of essential oil is fi-phallandrene. Commenly known as American Angelica.

Uses : The root and fruits are stimulant, expectorant and diaphoretic. The stem, roots and fruits are used for flavouring wines, liquors and confectionery.

Anthemis Linn.
Family : Compositae (Asteraceae)

Anthemis nobilis Linn. :

A hardy perennial. Foliage is fully cut strongly scented. Flower heads are terminal, pedunculated, white in colour.

The species is native to Germany, England, Belgium and France.

Dried flowers are medicinal. It constitute azulene, dihydrocinnamic acid, apigenin and angelic acid. Commenly known as common charnomile.

Uses : An infusion of this plant is used in cases of fever, hysteria, nervousness and dyspepsia. Lotion is used to treat neuralgia, earache and toothache. It is used for making chamomile tea.

Anthyllis Linn.
Family : Leguminasae (Fabaceae)

Anthyllis vulneraria Linn. :

This is a hardy perennial plant, grows to a height of 30 cm. Flowers are yellow.

This species is native to Britain.

It grows in ordinary rich soil and partly shaded location. Propagated by seeds, cutting or division of roots. Commenly known as wound or cat's claws.

Uses : Flowers are used for making Swiss tea. Plants are used in infusions or poultices for bruises and wounds.

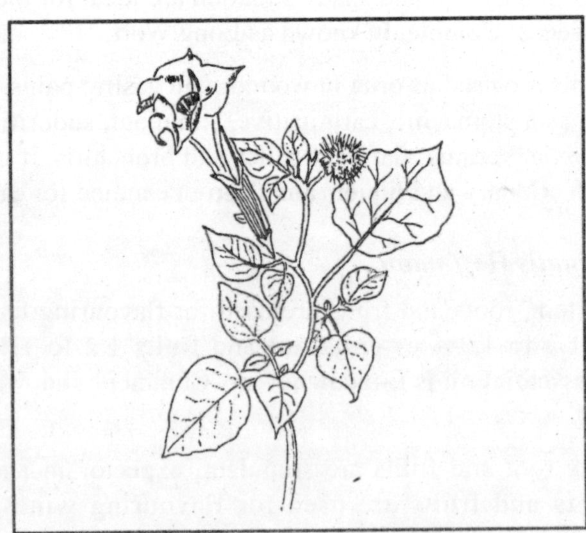

Fig. 14. Datura meteloides

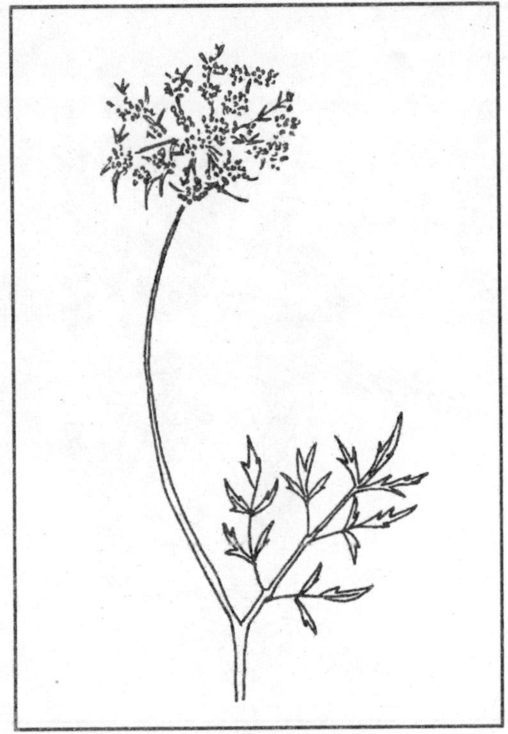

Fig. 15. Daucus carota

Apium Linn.
Family : Umbelliferae

Apium graveolens Linn. :

This is grown as annual in India which is a herbaceous plant with erett stem. Leaves are compound pinnate with long stalks. Flowers are greenish white in colour, appear in compound umbels. The fruits are formed from two compressed carpels, enclosing the seeds, very small dark brown cremocarp, with pungent taste and agreeable odour. Roots are adventitious.

It is native to Europe, India, North and South Africa.

Uses : The leaves, stalks, seeds and its oil is used. The leaves and stalk are used fresh or after blanching. The essential oil of seed is used for flavouring food and pharmaceutical industries. Stems of these plants are used to make a very good preserve for windy colics. The root is used as a aperitive and diuretic and administered for jaundice, nephritic colic and obstructions in the urinary passages. The leaves and stalk are cooked as vegetables. It is used in soups, sauces, purees. The seeds are also used as spice.

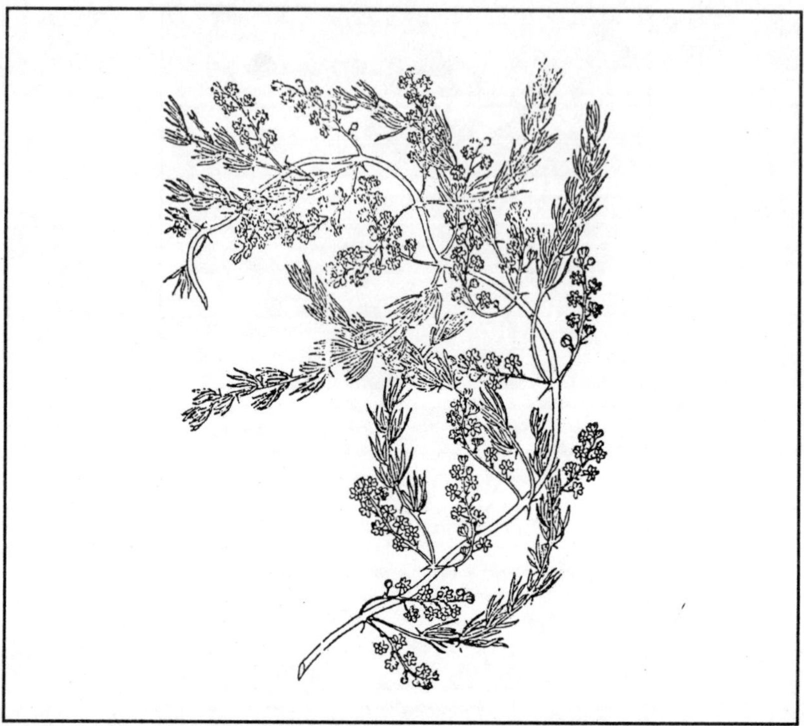

Fig. 16. Satawar (Asparagus recemosus Willd. Liliaceae.

Aplectrum (Nutt) Torr.
Family : Orchidaceae

Aplectrum hyemale (Muhl) Torr. :

It is an orchid species Commenly known as Puttyroot. Plant arises from a series of bulk-like roots with a single pointed leaf attached at the base. The flower appear in a spike, purplish at base, lip is white.

It is native to Temperate North America.

The plant is grown in a humus rich soil and in a shady area.

Uses : The bulb is medicinal and used for bronchial problems.

Apocynum Linn.
Family : Apocynaceae

Apocynum cannabinum Linn. :

A perennial plant 2.5 m tall, and produce milky juice. Leaves are oval with short stalks. Flowers are borne on terminal clusters, whitish green in colour. Pods are 15 cm long, slender and pointed.

It is native to North America and Mexico.

Uses: Whole plant, roots and rhizomes have medicinal properties. A decoction of roots is laxative, and powdered roots induce vomiting. Plant is considered a heart stimulant. The plant steeped in water is used to treat indigestion, inflammation of intestines, intestinal worms, fever, asthma, pneumonia and dysentery.

Arachis Linn.
Family : Leguminasae (Fabaceae)

Arachis hypogaea Linn. :

This is an annual herb. Flowers are dense and appear in axillary spikes. This species is native to Brazil and grown throughout tropics and subtropics. It is widely grown in India.

Seeds are used for fixed oil, which constitute glycerides of arachidic, stearic, lignoceric, oleic acid etc.

Uses : The oil serves as a vehicle for injections. It is an ingredient in ointments and liniments. Oil is used as a substitute for olive oil. The nuts are used for roasting or salting and for preparation'of peanut-butter. Filtered and refined oil used for cooking.

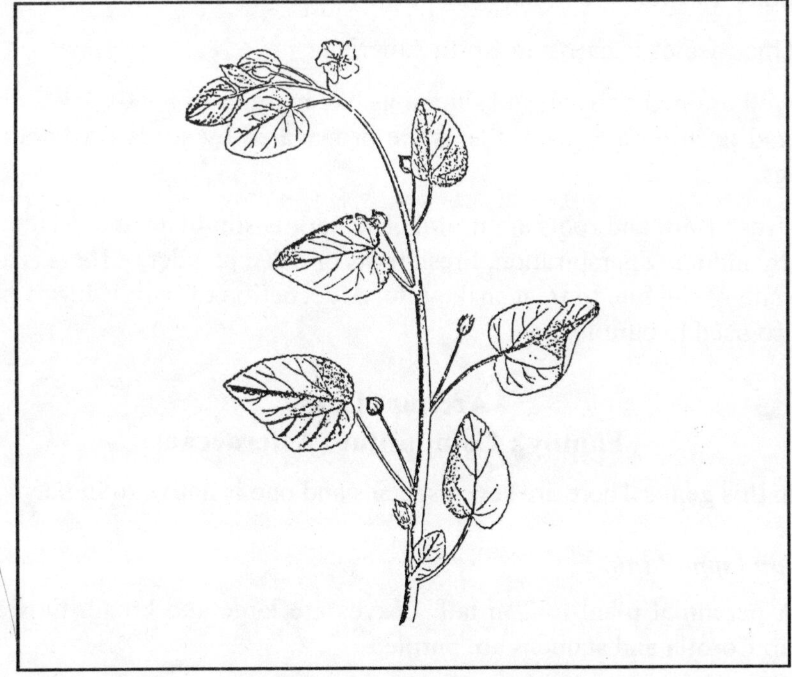

Fig. 17. Kungyi, Country-mallow (Sida cordifolia Linn. Malvaceae.

Aralia Linn.
Family : Araliaceae

This genus contains forty five species.

Aralia recemosa Linn. :

A herbaceous perennial 3 m tall, with a single leaf stalk. Leaf is divided in three parts and each part contain five leaflets. Flowers are greenish white. Berry is dark purplish in colour.

The plant is native to North America.

Rich well drained loamy soil and shady situation are ideal for these plants. Propagation is by seeds or division. Commenly known as American Spikenard.

Uses : Roots are medicinal. Tea made from the roots is used to treat backache. Rhizomes are effective in cases of skin diseases, asthma, blood poisoning, colds, coughs, gout and also to increase perspiration. Poultice of grounded root is used to treat sores and inflamations.

Aralia spinosa Linn. :

It is a tree of 10 m tall, very spiny. Leaflets are toothed, spines are on the leaf stalks. Flowers are minute and white in colour.

This species is native to North America.

Well drained rich soil and situations like margins of ponds, lakes or moist sheltered position are good. Plants are propagated by seeds, suckers or root cuttings.

Uses : Bark and roots are useful. The bark is stimulant and used to break fever by inducing perspiration. Fresh roots or dried powder of roots is applied to the site of the bite to treat snake-bite. A decoction of bark is blood purifier and also used to cure fever.

Arctium Linn.
Family : Compositae (Asteraceae)

In this genus There are seven species and one is native to India.

Arctium lappa Linn. :

A perennial plant to 2 m tall. Leaves are large and broad. Flowers are tubular. Corolla and stamens are purple.

It is native to Western Himalaya from Kashmir to Simla.

Uses : The roots of first year's growth, mature seeds and leaves are useful. The roots of the plants are one of the sudorifics and is used against gout, rheumatism, skin complaints, syphilitic diseases and cancer. Poultice made from leaves are applied to skin burns and ulcers. Juice from squeezed leaves are used to treat eye irritations. To increase urination, powdered seeds are used. The roots are also used as a diuretic and diaphoretic.

Arctium minus (Hill) Bernh. :

This is a perennial plant with large broad leaves. Tubular flowers are borne in many flowered heads. Flowers are white and pink in colour. Commenly known was common Burdock.

Uses : Leaves and roots are used for medicinal purpose. Tincture of the fresh root is administered for rheumatism and stomach ache. A tea made from leaves and roots is used for lung diseases, coughs and asthma. This tea also stimulates menstruation.

Arctostaphylos Adans.
Family : Ericaceae

Arctostaphylos uva-ursi (Linn) Spreng. :

A hardy creeping evergreen shrub. The bark is reddish in colour.

The leaves are 2.5 cm long, bright green. Flowers appear in cluster, white with red markings. Berries are red in colour. This species is native to Scotland.

Uses : Dried leaves are sold for tea in health food outlets. The leaf preparation is used in the diseases of kidneys, bladder and urethra and as a strong stimulant producing uterine contractions. It is administered in urethritis, cystitis and diarrhoea. Infusion of plant is drunk to control urinary troubles, sprains and stomach pains. The plant is also useful to treat haemorrhage and bronchitis.

Areca Linn.
Palmae

Areca catechu Linn. :

This is a tall slender palm. The inflorescence arises below the oldest living leaves, nonoeceous, with the female flowers at the base of the twigs and male above.

This species is native to Malaysia and largely cultivated in Tropical Asia.

In India, this plant is commercially cultivated in W. Coast, West Bengal and Assam. It is commenly known as Betel Nut Palm.

Uses : The seed kernel of this plant is medicinal, which constitute alkaloids like arecoline, arecaine; also arecolidine, guvacine, isoguvacine, guvacoline and arecaidine.

Seeds when chewed, acts as a stimulus upon the digestive organs and as preventive of dysentery. It is used in veterinary practices as cathartic and anthelmintic. Arecoline was formrly used to treat roundworm, tapeworm and as antidote for poisoning but it is not in practice now-a-days because of systemic toxicity.

Argemone Linn.
Family : Papaveraceae

Argemone mexicana Linn. :

This is a hardy annual of 65 cm tall herb, native to Mexico. Leaves are prickly, commonly found as weed in different parts of India.

Sandy soil and well drained sunny situation are ideal for these plants. These are propagated by seeds. Commenly known as Mexican Poppy.

Uses : The yellow juice of the stem is used twice a day for one week for healing of wounds. It is also used in the treatment of eye diseases. The seeds yield a semi-dying oil used as an illuminant.

Arisaema Mart.
Family : Araceae

This is a large genus of one hundred £&y species. These are rhizomatous herbs with deciduous spathes.

Arisaema intermedium Bl. :

This is a hardy rhizomatous rooted perennial Commenly known as Indian Turnip. Flowers are arum-like in shape. Commonly found at Ghewan, Utter 'Kashi.

Compost mixture consisting of loamy soil, leaf mould, sand and cowdung manure is ideal. Adequate watering is needed. Top dressing with decayed manure after the new growth begin and liquid manuring to well grown plants are beneficial. Propagated by division of rhizomes.

Uses : The decoction of crushed leaves is good for treatment of fever. The paste of roots is applied on the ulcer to remove the puss.

Arisaema triphyllum (Linn.) Schott. :

A small perennial plant of 65 cm height. The plants are with two stalks of three leaves each. The third stalk topped with a green tubular pulpit, and a covering hood turned down at the pulpit. Berries are scarlet.

The species is native to North America.

It grows in damp woodlands, swamps and bogs.

Uses : Corm, and plants are medicinal. Corms are edible, and these are used to cure asthma, rheumatism, stomach gas, coughs, tuberculosis. Plant is regarded as expectorant. Roots are insecticidal.

Aristolochia Linn.
Family : Aristolochiaceae

A genus Aristolochia consists of three hundred fifty five species of which fifteen are native to India.

Aristolochia indica Linn. :

A hardy climbing plant

Distributed throughout India and commonly found in Eastern and Southern India. These plants require good ordinary and well drained soil and sunny position for cultivation. Propagated by cuttings of ripe shoots or by division. Commenly known as Indian Birth wort.

Uses : Dried stems and roots are used for medicinal purpose. Drug prepared from this plant in small dose, is a tonic and stimulant and used to treat indigestion, fever and irregular menstruation.

Aristolochia serpentoria Linn. :

An erect perennial herbaceous plant of 65 cm height. Leaves are heart shaped and tip pointed. Flowers arise from the base of the stem, brownish in colour.

It is native to America. Commenly known Birth wort.

Uses : The roots are used to treat malaria, typhus fever, small pox and pneumonia. Poultice of roots are used on open wounds and skin ulcers.

Aristolochia bracteoleta Lank.

These species are very useful for treating worms and ulcers.

Arnebia Forsk.
Family : Boraginaceae

Arnebia benthamii DC. ex. Johnston. :

A pubescent herb with purplish root stock, occurring on the open rocky grounds between 3800-4200 m elevations in hill regions.

A native to Himalaya. Ordinary rich soil and sunny situation are suitable. Propagated by seeds, suckers, layering or cuttings.

The roots of these plants are used as hair tonic.

Arnica Linn.
Family : Compositae (Asteraceae)

Arnica montana Linn. :

This is a hardy herbaceous perennial produce yellow flowers. Plant attain a height of 40 cm.

This species is native to Europe. Plants are propagated by division of roots. Ordinary rich soil and sunny situation are preferable. Commenly known as Mountain Tobacco or Arnica.

Arnica soroia Greene :

The plant is occurring in Western Canada and United States.

The flowers and roots of these plants have medicinal properties and contain arnicin, amisterol, anthoxanine and Tannin.

It is used in homoeopathy medicine.

Uses : The leaves of these plants have a sternutatory property. It is used as a diuretic, tonic and febrifuge. This is also used for some neuroses, paralysis, rheumatism, gout, asthma and pulmonary catarrh. Arnica is administered externally as liniment on bruises, sprains and swellings.

Artemisia Linn.
Family : Asteraceae (Compositae)

This is a large genus of about four hundred species and about thirty four are native to Jndia. These are herbs or low shrubs, usually possess strongly scented aves. The leaves are often dissected and alternate.

Artemisia annua Linn. :

This is an annual herb, glabrous and attain a height upto 120 cm. It is native to China, Turkey, Iran, Afganistan. This species is naturalised in USA.

Well drained organic rich light loamy soil is suitable for its cultivation. They prefer cold winter and moderate summer. This species can be grown in subtropical regions only during winter season and can be cultivated in the hills of H.P., U.P. and Kashmir. Propagation is done by seeds. Seeds are sown in nursery beds during March-April in hills. Seedlings are transplanted during April-May in a well prepared fertile land. The crop is harvested during July at the stage of flower initiation. Plants are cut 40-50 cm above the ground level. Second harvest can be obtained in September. These are dried in shade and after drying leaves and flower buds are separated from the stalk. One hectare of land may yield 5-6 tonnes of dried plant matter.

Uses : The plant contain artemisimin essential oil and flavonoids. The extract of herb is effective in curing malaria. A decoction of leaves in water is administered for fever.

Artemisia cina Berg. :

The plant contain santonin, cineole, thujone, camphere and lemisin. The leaves and flower heads are used in medicine.

Uses : The plant is used as an emmenagogue and administered in a white w'ine infusion. Leaves are regarded as tonic, stimulant and antispasmodic. The powdered root of this plant is used for treatment of epilepsy. It is also known as anthelmintic.

Artemisia dracunculus L. "Tarragoa" :

It is a perennial herb. It is native to Afganistan, Western Asia and Russia. This plant is commercially cultivated in France, Spain and Temperate zones of USA.

The herb yield 0.4% volatile oil. Methyl chavicol is the chief constituent of the oil and other constituents are phellandrene and ocimene.

Uses : The leaves are aromatic and possess aperient, stomachic, stimulant and febrifuge properties.

Artemisia maritima Linn. :

This is an aromatic perennial shrub upto 1 m tall. Leaves are 5 cm long, whitish in colour, leaves are divided into numerous fine, linear segments. Upper leaves are simple, undivided, linear. Flowers arise in short spikes

This species is native to Western Asia. In India, this species occurs in Kashmir and Kumaon at an elevation of 2200 to 3500 m.

The plant contain santonin, cineole, thujone, camphere and lemisin.

Uses : Dried leaves and immature flower heads constitute the drug. The plant is used as an emmenagogue. Santion has been widely known as anthelmintic. Leaves are regarded as tonic, stimulant and antispasmodic. It is also used in fever and dropsy. The powdered root of this plant is used for treatment of epilepsy.

***Artemisia paliens* Wall.:**

This is an aromatic annual herb, erect in habit, upto 60 cm tall. Leaves are much divided. The inflorescence is a capitulum which is shortly peduncled to sessile, axillary or forming lax raceme. Flowers are yellow, small and inconspicuous. This species does not occur wild.

It is commercially cultivated in India, in the states like Karnataka, Andhra Pradesh, Tamil Nadu, Kerala and Maharashtra.

These plants require well drained rich sandy loam soil and bright winter sunshine. Propagated from seeds which lose their viability rapidly. Seeds are sown in November in a nursery bed. Seedlings of 30 to 40 days old are transplanted at a distance of 10 cm x 15 cm. Before transplanting in a v/ell prepared land about 6 tonnes of FYM is incorporated with the soil. A fertilizer mixture of 120 kg N, and 40 kg each of P_2O_5 and K_0O per hectare is required. The plant bloom within 110 to 115 days time. At flowering stage the crop is cut by sickles 10 cm above ground level during bright sunny days. From a well cultivated field around 15 tonnes of fresh herbage per hectare can be obtained. The herbage is shade died for 2 to 3 days, which produces 15 kg oil per hectare on distillation.

The oil of this herb contain hydrocarbons, esters and oxygenated compound.

Uses : The oil possess antibacterial and antifungal properties. It is also used for flavouring food items, beverages and tobacco products.

***Artemisia tridentata* Nutt.:**

It is a much branched shrubby plant grows upto 3 m height, covered with silvery fuzz. The foliage is strongly fragrant. The flowers appear in densely packed heads, yellow or white in colour.

This species is native to America. Commenly known as Bigsagebrush.

Uses : Leaves and branches have medicinal properties. Branches are burned to purify the air in a room. The leaves are chewed to ease stomach gas. The tea made from leaves are used to treat sore throat, stomachache, headache, vomiting and it is also used as an antidote for poisoning. To treat wounds and cuts squeezed leaves are applied. Boiled and steeped leaves are applied as hot poultice in treating rheumatism, to relieve colic and placed on abdomen to induce menstruation.

Asarum Linn.
Family : Aristolochiaceae

Asarum canadens Linn. :

This is low growing perennial, stemles's plant with a spicy odour. Leaves are heart shaped and radical. Flowers are solitary, terminal, purple and bell shaped. Commenly known as Indian Ginger.

Cool moist shaded place and calcarious soil are suitable for these plants.

Uses : The leaves and roots constitute the drug. The root is said to be stimulant and used to treat heart palpitations, nervous problems and fevers. The powdered dried root is inhaled as snuff to relieve aching head and eyes. A tea prepared from roots is used to relieve stomach gas.

Asclepias Linn.
Family : Asclepiadaceae

A genus Asclepias consists of one hundred and twenty six species, mostly American. These are perennial herbs. Flowers appear in terminal or axillary umbels.

Asclepias speciosa Torr. :

It is a perennial plant upto 3 m in height. The entire plant is covered with a short fuzz. Plant excrete milky latex. Leaves are heart shaped. Flowers form a head, are purple lilac in colour and fragrant.

It is native to north-west America. Commenly known as showy Mikweed.

Sunny situation and rich moist soil is good for its cultivation. This species is propagated by seeds or by division.

Uses : Entire plant is useful. Infusion of the plant is used to treat sore breasts of women. A tea prepared by boiling roots is useful for treating coughs, tuberculosis and measles. Milky latex is antiseptic and used for cuts, sores, treating ringworm, removing corns and calluses.

***Asclepias syriaca* Linn. :**

It is a perennial plant upto 1.8 m tall. Leaves are arranged opposite or in whorls, oval shaped, fuzzy below upto 30 cm long. Flowers are greenish purple in colour and fragrant.

The species is native to North America. Commenly known as common Milk Weed.

Rich and light soil, sunny, position and moist borders are suitable. Plants are propagated by division of roots or sowing of seeds.

Uses : The decoction of root is used for curing asthma, coughs, indigestion and gonorrhoea. This decoction is laxative, increase urine flow and induce vomiting. The milky latex is useful for treatment of warts and moles.

***Asclepias incarnata* Linn. :**

Plants grow upto 70 cm tall and produce red flower.

It is distributed in north America.

This species is emetic and cathartic.

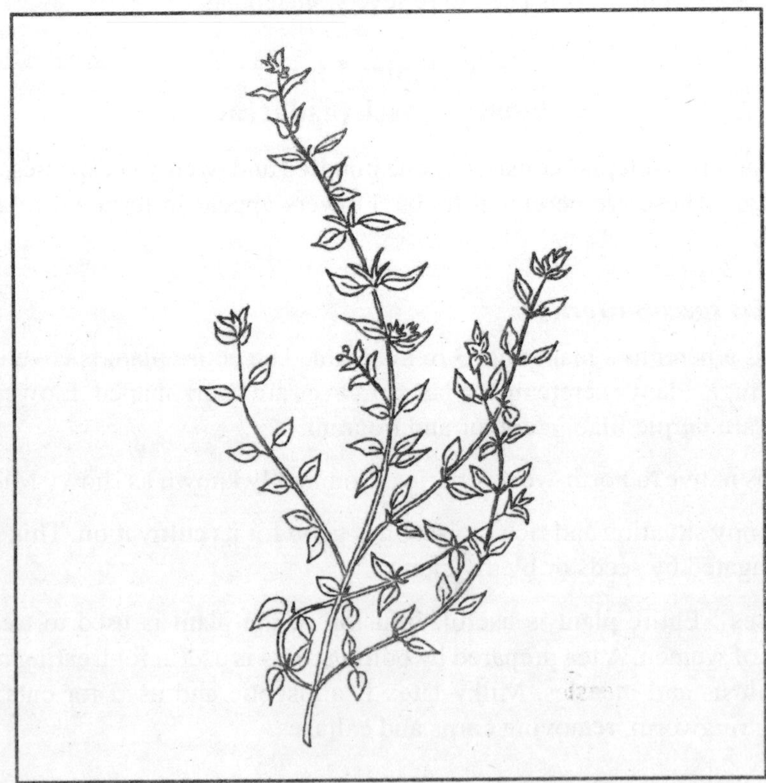

Fig. 18. Swertia chirata

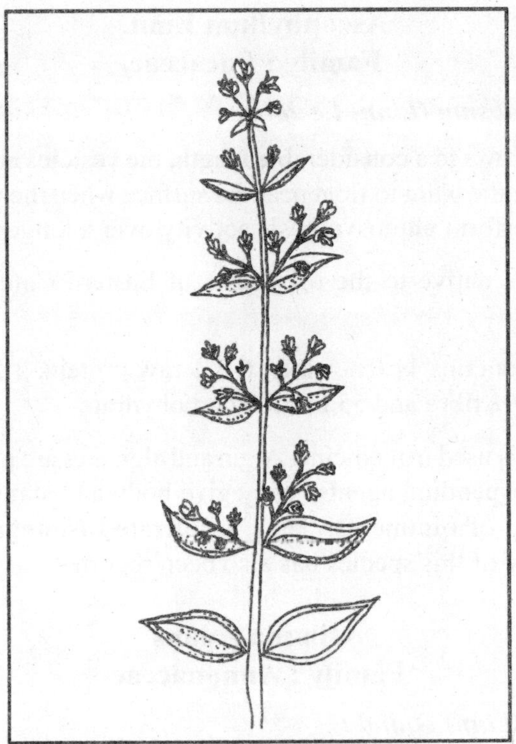

Fig. 19. Thymus vulgaris

Asclepias incarnata Linn. :

Plant attain a height of 70 cm and produce orange coloured flowers. This species occurs in north America.

Uses : It has specific actions on the lungs.

Asclepiodora A. Gray.
Family : Asclepiadaceae.

Asclepiodora viridis (Walt) Gray. :

This is a perennial plant upto 80 cm tall with milky sap. Flowers are 15 cm in diameter, green or purple in colour. Plants grow in higher elevation in dry forests.

Uses : Roots of these plants constitute the drug. A tea made from roots is used to treat asthama and shortness of breath. A decoction of root in cold water is given as drink to relieve fever. Powdered root is rubbed over the heart area to relieve palpitation. To treat neck and rib pains a poultice of powdered root is used.

Ascophyllum Linn.
Family : Fucaceae

Ascophyllum nodosum (Linn) Le Jolis. :

This often grows to a considerable length, the vesicles are normally full of gas, which enable the plant to float near the surface when the tide comes in, this way the plants perform photosynthesis activity over a longer period of time.

This algae is native to the off coasts of Eastern Canada, England and Norway.

This algae contains 11.0% water, 6-7% raw protein, 3.5 to 4% fat, 18 to 20% ash, 6 to 6.5% fibre and 55 to 62% carbohydrate.

Uses : Algin is used in medicine. Algin and alginates are protective colloids, emollients and suspending agents. They give body and stability to emulsions and spreadability of ointments. They accelerate disintegration of tablets. Antibiotic activity of this species has also been reported.

Asimina Adans
Family : Annonaceae

Asimina triloba (Linn) Dunal :

This is a strong hardy deciduous tree of 12 m tall. Leaves are large upto 40 cm long and oblong shaped. Flowers are 6 cm across, dark purple in colour. Fruits are up to 18 cm long and 6 cm wide, fleshy, edible yellow in colour changing to dark brown. Seeds are flattened.

The species is native to America.

Rich moist alluvial soil and sunny position is ideal for cultivation. Plants are propagated by seeds.

Uses : Ripe fruits and seeds of these plants constitute the drug. To treat intestinal worms fruit juice is useful. Seeds induce vomiting. To destroy head lice powdered seeds are used.

Asparagus Linn.
Family : Liliaceae

This is a large genus of about three hundred species of which .twenty are native. to India. Plants are scandant or erect under shrubs, rarely herbs. Root stocks are stout, creeping, sometimes bearing tubers. Leaves are minute.

Asparagus recemosus Willd. :

This is a hardy climbing perennial. The plant is adapted to the temperate regions and in hills in tropics and subtropics; and can also be grown in plains. Well drained, rich, friable, soil with high fertility and plentiful of organic matter are ideal. Plants are propagated from seeds.

In plains, seeds are sown during July-November, while in hills seeds are sown in spring and summer and thereafter transplanted in next spring.

About 10 kg of seeds are required per hectare. Seeds are sown in raised nursery beds which remains in the nursery for 2-4 months. Seedlings are transplanted at a spacing of 45 cm x 45 cm. It requires 10 tonnes of FYM per hectare. Every year top dressing with fertilizer mixture is done for which 500 kg of mixed fertilizer per hectare is recommended. Irrigation is required at weekly intervals. Around 4000 kg of shoots can be harvested per hectare. Commenly known as garden as paragns.

Uses : The root preparation of asparagus is used as diuretic. Shoots of asparagus have a marked sedative effect on heart contraction. The young shoots are eaten.

Asparagus racemosus Willd. :

This is a perennial foliage plant It is occurring at Kaream in the Chhindwara district of M.P. They require rich sandy loam soil and sunny position for proper growth. Plants are propagated by seeds.

Uses : The dried roots about 700 gm are burnt and fumes are inhaled under a blanket for curing fever.

Asplenium Linn.
Family : Asplaniceae

Asplanium lunulatum, SW. :

This is a hardy fern with fibrous root system. It is occurring at Mt. Abu of Rajasthan.

Compost consisting of equal parts of well decayed manure, loamy soil, leaf mould, sand and old mortar rubble is good for these plants. Moist shady situation is ideal. These are propagated by spores.

Uses : The decoction of the leaves of this plant is used in jaundice and malaria fever.

Aster Tourn ex. Linn.
Family : Asteraceae (Compositae)

Aster thomsonii C.B. Clarke. :

A hardy herbaceous perennial plant upto 60 cm tall. The plant is hairy and the stem bends at the joints. Flowers are pale blue in colour and achenes pubescent.

The species is native to Himalaya.

Plants grow in temperate climate. Organic rich soil and sunny position are ideal. Only overcrowded plants are lifted, divided and replanted. Plants are propagated by seeds, cutting of young shoots or by division.

Uses : Flowers of these plants are used in Tibetan system of medicine.

Astragalus Linn. :
Family : Leguminasae (Fabaceae)

Astragalus gummifer Labill. "Tragacanth" :

This is a perennial shrub of 70 cm tall, bushy, branches are smooth.

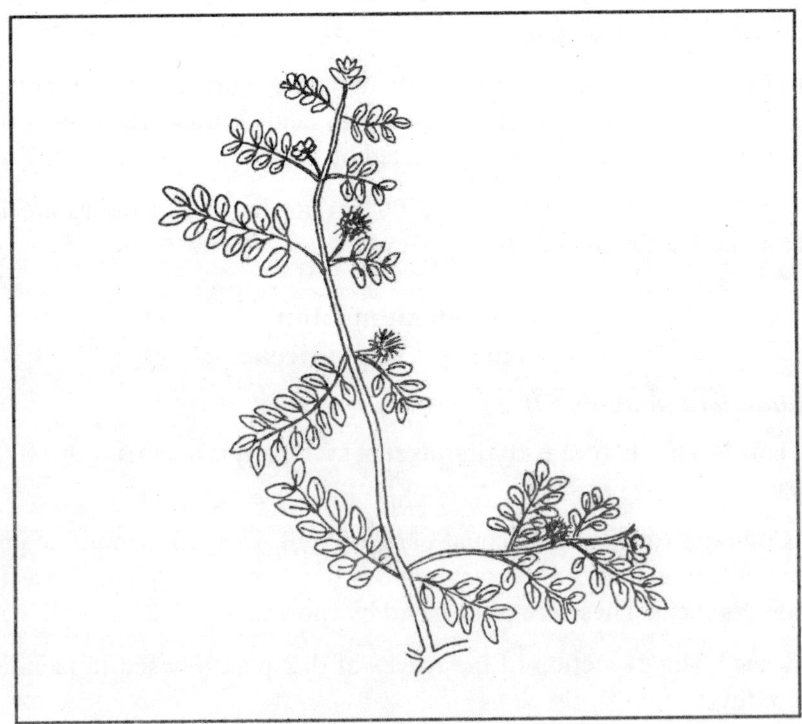

Fig. 20. Tribulus terrestris

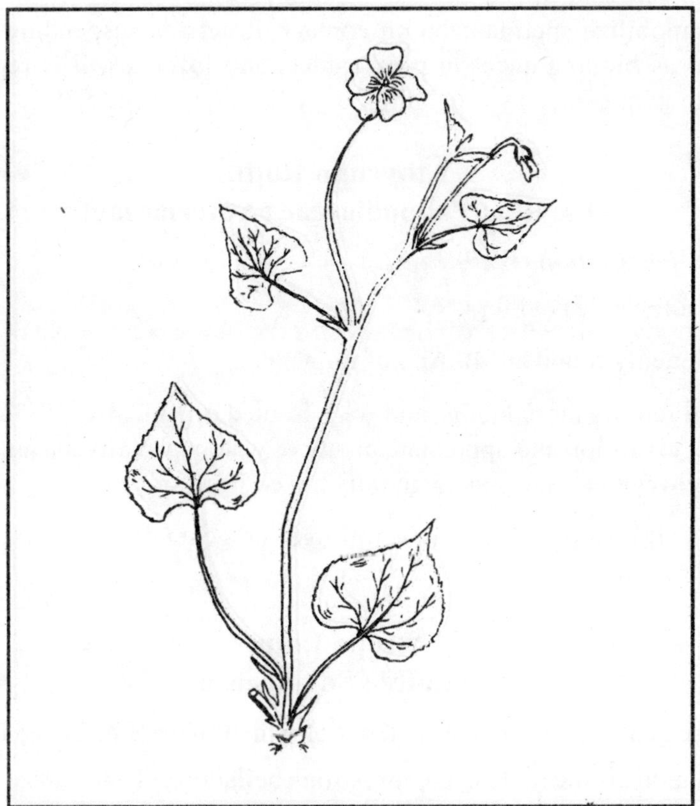

Fig. 23. Viola canadensis

Leaves are pinnately compound glabrous with 4 to 6 pairs of leaflets. Leaves shed in dry season leaving the sharp rachis. Flowers 'are white, calyx is tubular and hairy. The plants bear large tap root, the woody parts and the taproot excrete gummy substance on injury.

It is native to Greece, Corsica, siatic Turkey, Iran, Lebanon and Syria in dry regions between 1200 to 4000 m altitude.

It is a wild crop and the gum is collected from injured plant. The best gum is obtained from the roots. The plant start yielding gum for 2nd year onwards upto seven years. The root is punctured with slender pick, the exuding gum is germiform.

Uses : Gum excretes from this plant contain bassorin, tragacanthin and small amount of starch.

Gum is having demulcent and spermicidal properties. It is used to treat coughs and diarrhoea. It is also employed in spermicidal jellies and creams, which immobilize spermatozoa on contact. It acts as suspending ag^nt and employed as binding agent in pills, tablets and lozenges. It is reported that tragacanth gum exerts a strong inhibitory action on cancer cells.

Athyrium Roth.
Family : Polipodiaceae (Athyriaceae)

Athyrium pectinatum (Wall) Presl. :

It is a fancy leaved fern.

Commenly found at Mt. Abu of Rajasthan.

They require more humus and well drained organically rich loam. Plants like moist condition and appreciate plenty of water specially during periods of hot or dry weather. Plants are generally raised from spores.

Uses : It is said to be strong vermifuge and generally given to children by the tribal people.

Atropa Linn.
Family : Solanaceae

It is a genus of six species and out of which two are native to India.

Belladona is the trade name for Atropa belladonna Linn and A. acuminata Royle.

Atropa belladona is a native of Central and Southern Europe. It is cultivated in different parts of the world as well as in USA, UK, Germany. Poland, Russia, Czechoslovakia and Hungary. In India, A. belladonna and A. acuminata are grown in Kashmir at altitudes of 2000 to 3500 m. Previously most of the Belladonna was collected from forests of Jammu & Kashmir for exporting crude drugs.

A. belladonna Linn. is a spreading perennial herb reaches a height of about 1.4 m with a long, cylindrical, branched, fleshy creeping root. Roots are yellow brown externally and white within, the leaves were alternate, usually on the upper branches each leaf is accompanied by smaller stipule-like leaf. Leaves are ovate or oblong-ovate, 8 to 20 cm long and 5 to 10 cm wide. Flowers are large bell shaped, borne singly on the leaf axile, dull purple within and yellowish on the outside. Fruits are globular to 1.25 cm in diameter. Seeds are kidney shaped, many in number and embedded in the pulp.

All plant parts contain alkaloids chiefly hyocyamine and little atropine and hyoscine. The alkaloid content varies from 0.14 to 0.7%. Maximum total alkaloid content is 0.9% in seeds, 0.8% in roots and 0.5% in leaves. For the drug trade 0.40% hyoscyamine USP 0.40% BP in leaves and 0.45% USP, 0.5% BP in roots are essential. Fresh leaves, seed and fruit-are very high in alkaloid content. Before' the appearance of flower, alkaloid content in root is highest.

These plants grow in a temperate climate as a perennial plant in an open situation. In subtropical climate they may, however, be grown for a short duration in winter where yield is very poor.

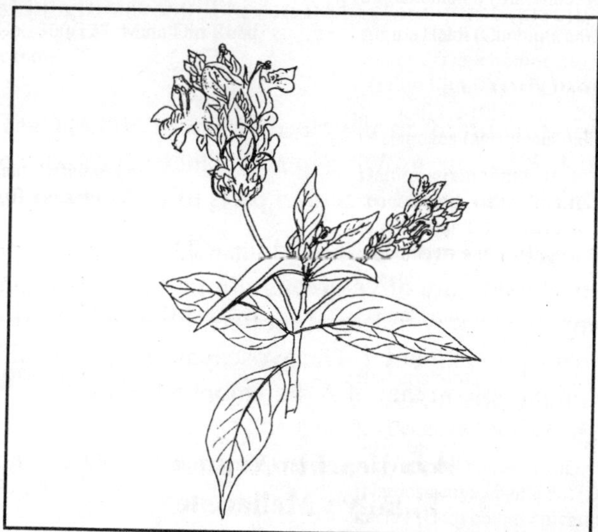

Fig. 22. Adhatoda zeylanica

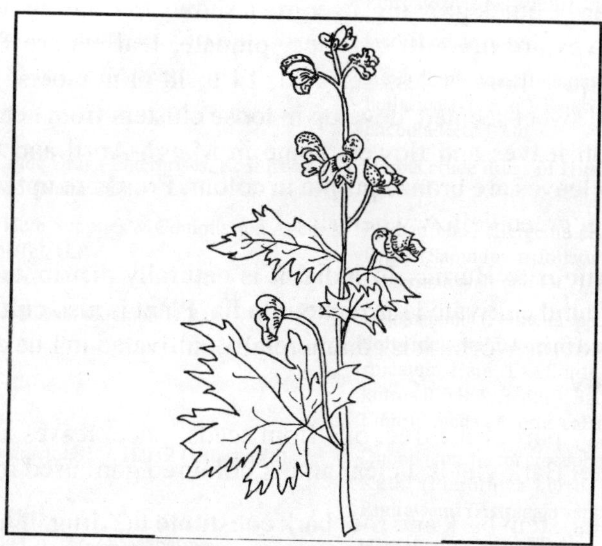

Fig. 23. Aconitum falconeri

Uses : The plant is used for number of complaints, but it is an extremely poisonous plant. Root preparation of these plants are used for the external treatments of gout and rheumatism, ff the drug is administered, carefullv measured in subtoxic doses. A tincture and several extracts are prepared from the leaves. The drug has its pronounced effects in nervous complaints such as convulsions tetanus, hysteria, nervous trempling and delirium tremors. It is also employed in opthalmology, bradycardia, psychiatry, whooping cough, asthma, gastric ulcers, complaints of stones in kidneys or gall bladder.

Atropa acuminata Royle :

It is occurring naturally in the North-Western Himalayas, in Himachal Pradesh, Jammu & Kashmir, growing in abundance at altitudes between 2500 m to 4000 m. This is also found in certain parts of Europe and North Persia.

The plant closely resembles A. belladonna. This perennial plant reaches a height upto 1.8 m. Leaves are olive-green, elliptic-lanceolate, pointed at both ends, ft has a large tap root with lateral rootlets. Flowers are brownish yellow and fruits are purple pink, globular. The hyoscayamine content of the leaves of this species is comparable to that of A. belladonna leaves.

Azadirachta A. Juss.
Family : Meliaceae

Azadirachta indica A. Juss :

This is a hardy, quick growing, evergreen showy tree. Grows to a height of 14 meters. Leaves are upto 40 cm long, pinnate, leaflets are 8 cm, lance-acuminate, oblique, more or less serrated, 14 to 18 in numbers. Flowers are small, white and sweet scented, develop in loose clusters from near the base of the leaves. Fresh leaves and flowers come in March-April and then the old leaves fall; new leaves are bronze purple in colour. Fruits are upto 1.8 cm long, rounded, oblong, green yellow when ripe.

It is indigenous to Burma. In India, it is naturally distributed in Deccan peninsula and found cultivated throughout India. Plant is also cultivated in the tropical regions of the world. It is commercially cultivated in Lucknow, Ajmer, Bihar and Bellary.

Uses : Every part of the tree, oils from seeds, juice, leaves and barks are used in medicine. Bark yields a clear amber coloured gum used in medicines.

Dried leaves, stem bark and root bark constitute the drug. It has antibiotic activity. The leaves are bitter, used in skin diseases and boils. The bark preparation is a bitter tonic and astringent and used to treat fever and skin diseases. Seeds yield an acrid, bitter oil of deep yellow colour and has a disagreeable odour. Seeds yield 30-409r oil. It is used as an anthelmintic and antiseptic and is much in demand. Young shoots and green twigs are crushed and used as tooth brushes. This can cure toothache, bad breath and gum diseases. Even the breeze from leaves is considered healthy. Neem protects the mouth from various infections. It is also useful for curing eye diseases. A decoction made from fresh leaves can cure fevers and remove phelgm from bronchial tubes. The sap can cure leprosy. Ointments made from neem are considered good for ulcers and wounds. Leaves are boiled in water and the same is used for bathing. Smoke from burnt leaves is said to repel mosquitoes. It is as heavy as teak and can be seasoned well. Wood is durable and used for various purposes like timber, house building, cart building, furniture, agricultural implements, ship building etc. Neem oil cake is a good manure.

Bacopa Aubl
Family : Schrophulariaceae

Bacopa monnieri (Linn.) Pennel :

It is a glabrous, creeping succulent herb, leaves oblong, spatulate and obtuse. Flowers are upto 1 cm in diam., pale bluish, pimple veined, fading to white, arises from the axils of leaves and are borne on short pedicels. Roots appear from the nodes of the stem.

It is found in the moist places of Nalagarh, Una, Hamirpur, Kangra, Mandi and Oochghat of H.P. below 1400 m and also occur commonly in moist and wet places in different parts of India.

Uses : The entire plant is used as drug, and the plant contain alkaloid bramhine. The powdered leaves of about 5 gm with 2-3 black pepper are given in a single dose in bone fracture. It is used as a nerve tonic, constipation, diuretic. Also used as a remedy for bronchitis and rheumatism.

Balanites Del.
Family : Balanitaceae

Balanites aegyptiaca Linn. Del. :

This is a spiny and dwarf tree. This species is native to India and distributed in Bihar, Punjab, Delhi, Gujarat, Sikkim and Deccan Peninsula.

Uses : The seeds yield oil, which is used to treat cuts and wounds. Fruits are used for washing clothes and also for making garlands. In the modern medicine so far no effective remedy is there for the herpes zoster disease. Root bark of this species, ground into fine powder, then made to a paste with water

Fig. 24. Hasminum pubescens

Fig. 25. Jasminum sambac

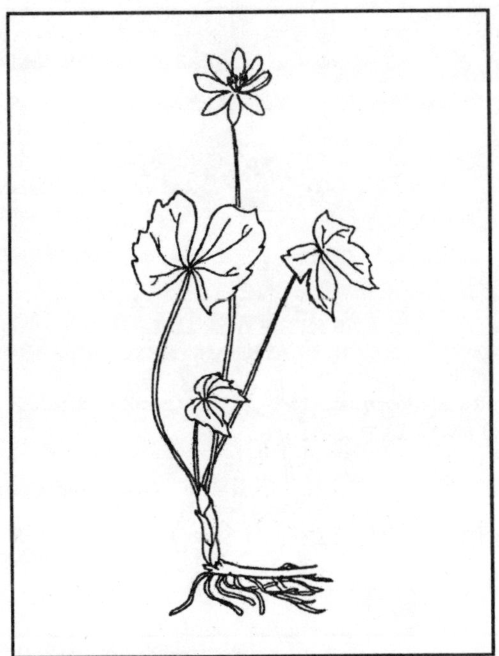

Fig. 26. Jeffersonia diphylla

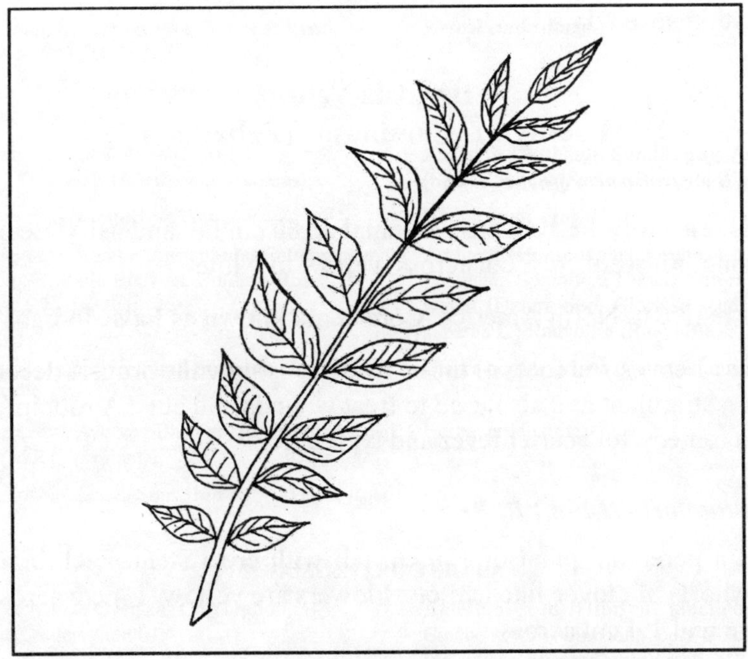

Fig. 27. Juglans nigra

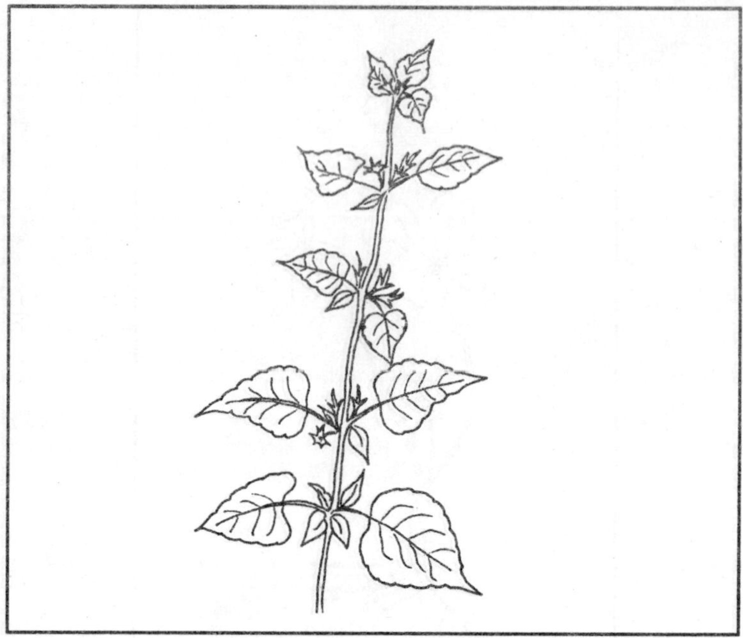

Fig. 28. Melissa officinalis

and applied to the infected area twice a day and healing occurs in case of herpes zoster within seven days.

Baptisia Vent.
Family : Leguminasae (Fabaceae)

Baptisia leucophaea Nutt. :

This is a hardy herbaceous perennial up 80 cm fall and hairy. Leaflets are 8.5 cm long. Flowers are 5 cm across, yellow ui white.

It is native to North America. Commenly known as False Indigo.

Uses : Leaves ?nd roots of these plants constitute the drug. A decoction of leaves is a stimulant and also used to treat wounds and cuts. A root infusion is used as a remedy for scarlet fever and typhoid.

Baptisia tinctoria (Linn.) R. Br. :

It is a perennial plant of 130 cm tall with erect stem. Each branch has several whorls of clover-like leaflets. Flowers are yellow, 1.4 cm across. Pods are brown and 1.4 cm across.

It is native to North America.

Roots of this species constitute the drug. Roots are astringent, antispetic and laxative. Roots are used to treat sore throat, fever, scarlet fever and typhus. A poultice made from the root is used for ulcers and wounds.

Barringtonia Forst.
Family : Euphorbiaceae

Under the genus there are about one hundred species of Palaeotropical region, species are found in India. Hence are smalfand medium sized trees. Flowers in elongate, terminal or lateral racemes or interrupted spikes.

Barringtonia acutangula Linn. Gaertn. :

The tender leaves of this are eaten. The powdered seeds are given to children as an expectorant ana emetic.

The leaves and roots of this species are used as bitter tonic. The bark contain tannin. In case of cough, asthma and diarrhoea the fruit of B. racemosa Roxb. is effective, while powder of the fruit is used in skin diseases. The seeds of this species are used in colic and ophthalmic condition. B. butonica Forst. B. careya F. Muell and B, edulis fruits are edible. The young leaves of B. fusifonnis, B. insignis Miq, B. racemosa and B. spicata Bl are cooked and eaten. Commenly known as Indian Oak.

Bauhinia Linn.
Leguminasae (Caesalpiniaceae)

Under this genus there are about one hundred species and nine are native to India. A small trees, shrubs and climbers are available with showy flowers in this genus.

Bauhinia vahlii W. & A. :

This is large extensive woody climber with many branches. Leaves are coarse and thick upto 30 cm long. Flowers are white-yellowish, appear in many flowered terminal clusters, found in different parts of fndia and also at Karean in the Chhindwara district of Madhya Pradesh.

Uses : The root decoction about 60 ml is taken three times a day for four davs in fever.

Bauhinia variegata Linn. :

This is a small sized tree with dark brown and more or less smooth bark, upto 8 m tall. Leaves 8 cm across. Flowers grow in clusters of 5-7 in a short

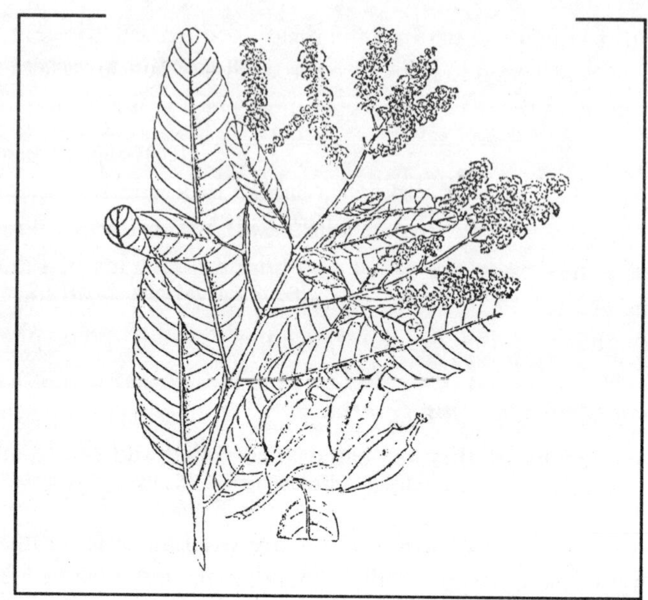

Fig. 29. Arjuna (Terminalia arjuna (Roxb.) W & A. Combretaceae

raceme or corymb, fragrant, beautifully coloured with various shades of pink, purple and white. Pods are 25 cm long, seeds nearly circular, flat and brown.

It is commonly grown in gardens and also occur at Chimtipur of Chhindwara district of M.P. It is' native to India and China.

Uses : The bark decoction about 30 ml is administered for a fortnight twice a day in tubercular lymphadenetes. The flower powder is used for haemorrhage.

The leaves and young shoots of B. malabarica Roxb. are acrid and eaten. It is used for flavouring meat and fish; and young shoots are used as vegetable. The yound leaves of B. tomentosa are sour and are used to prepare the acid savor. The buds and leaves of B. racemosa Lamk are edible. The pods of B. reticulata Roxb. are eaten and a drink is prepared from seeds. The plant of B. retusa Roxb. and B. purpurea L. yield gum.The bark of B. purpurea L. contain tannin.

Begonia (Tourn.) Linn.
Family : Begoniaceae

Under this genus there are about one thousand species and around forty-five species are native to India. These are succulent herbs or undershrubs. Stems sometimes reduced to rhizomes or tubers. Leaves are often attractively coloured.

Uses : The leaves B. barbata, B. cucullata Willd., B. dominicalis DC., B. evansiana Andr. and B. picta are eaten. The leaves of B. jirtella are used as a flavour. The stems of B. rex. Putz. are used as pot herb.

Bellis Linn.
Family Compositae (Asteraceae)

Bellis perennis Linn. :

This is a hardy herbaceous perennial native to Britain. Ordinary sand soil and sunny situation are preferred for these plants. Propagated by division of old plants or bv seeds. It generally multiplies and hibernates in nature by short rhizomes.

Berberis (Tourn.) Linn.
Family Berberidaceae

A large genus of five hundred species, and seventy species are native to India. These are erect shrubs with yellow flowers.

Berberis asiatica Roxb. ex. DC. :

It is an evergreen shrub of 4 m tall.

The species is native to Himalava, commonly occurring in Oat arid Rhododendron forests at Kaliz between 1500-3000 m of Utter Kashi. Also found in the Chhindwara district of M.P. at Kaream. Commenly known as Indian Berberry.

Uses : The extract of root is used in the eye infection by local people. The root decoction is used in body inflammation and stomachache.

Berberis cristata DC. :

This is woody shrub and thorny. Leaves are fascicled in axils of branches or simple spines. Flowers are yellow. Plants are distributed in Nilgiri Hills. Commenly known as Daru Haldi.

Uses : The bark, roots and stem extracts of these plants is useful against eve disease, fever, ulcers etc. The chief constituent of this drug is the alkaloid berberine.

Berberis vulgaris Linn. :

It is a perennial shrub, may grow upto a height of 4 m. Colour of inner wood of the stem is yellow. Leaves are small, upper surface green and lower surface gray, leaves on young sprouts are spin. From each nodes sharp spines appear. Flowers are yellow in colour. Fruits are scarlet.

It is native to Britain. Commenly known as European Barberry.

Uses : The stem and loot bark, and root of these plants have medicinal value. The fruit is rich in vitamin C. A wine is prepared from berberry fruits and used in malignant, choleraic and pestilential fever, diarrhoea and menstrual fluxes. Squeezed blooms are applied to clean old ulcers. The concentrated juice of berries is used for gums and teeth trouble. 1 he root and wood which are bitter in taste is used as purgative tonics. The bark has been used to treat dysentery and indigestion. The plant preparation is also used for jaundice, dropsy and congestion of liver and spleen.

Bergenia Moench.
Family : Saxifragaceae

Bergenia stracheyi (Hook f. et. Thoms.) Engl. :

This is a hardy evergreen perennial plant upto 90 cm tall. Flowers are white.

The species is native to Himalaya occurring commonly in Mounda of Utter Kashi.

Uses : Rich and fairly deep heavy soil and semi shady location are ideal. Propagated by seeds or division of old plants.

The leaves and roots of the plant is given to the patient of kidnev stone.

Betonica Linn.
Family Labiatae

Betonioca officinalis Linn. Stachys officianlis) :

This is a hardy perennial plant of 1.5 m tall, produce purple flowers. The is native to Europe and Asia Minor. Commenly known on word Betony.

Uses : The dried powdered leaves are used as sternutatories. It is also useful in ophthalmology, odontalgia, persistent migraine and chronic bronchitis. The root is emetic and purgative.

Betula Linn.
Family : Betulaceae

Under this genus, there are about fifty species.

Fig. 30. Asgandh, Punir, Ashvaganda
(Withania somnifera Dunal. Solanaceae)

Betula lenta Linn. :

This is a hardy deciduous tall tree upto 30 m height. Bark is reddish brown and dark. Leaves are upto 15.5 cm long, with long hairs in the underside of the veins.

The species is native to North-East America. Commenly known as Cherry bark birch.

Uses : Bark, twigs and leaves of these plants constitute the drug. Oil is extracted from bark or twigs which has medicinal properties. An infusion of twigs and bark is used to treat gout and rheumatism, and it is said to purify blood and stimulate urine flow. Dusting powder prepared from dried leaves and twigs soothe chafed skin.

Betula utilis D. Don. :

This is a hardy perennial plant. Commonly found in Chyangseel of Utterkashi above 3500 m. Ordinary rich soil and sheltered position are suitable for the growth of this species. Plants are propagated by seeds and layering.

The bark of the tree is used in bone fracture.

Blumea DC.
Family Compositae (Asterdeae)

Blumea fistulosa (Roxb.) Kurz. :

This is occurs in peninsular India and also found at Chimtipur of Chhindwara district of M.P. Bracts aie reddish in colour.

The root decoction is used as diuretic.

Boehmeria Jacq.
Family : Urticaceae

Boehmeria platyphylla D. Don. :

This is an undershrub with rough broadly ovate, wrinkled, cordate, toothed, acuminate leaves. Male and female flowers are borne on spikes to 30 cm in length during August-September.

It is native to Asia and Africa, occurring in different places of H.P. in shady ravines and slopes, below 1500 m.

Uses : The paste of leaves and 4-6 fruits of blackpepper is applied on dry eczema.

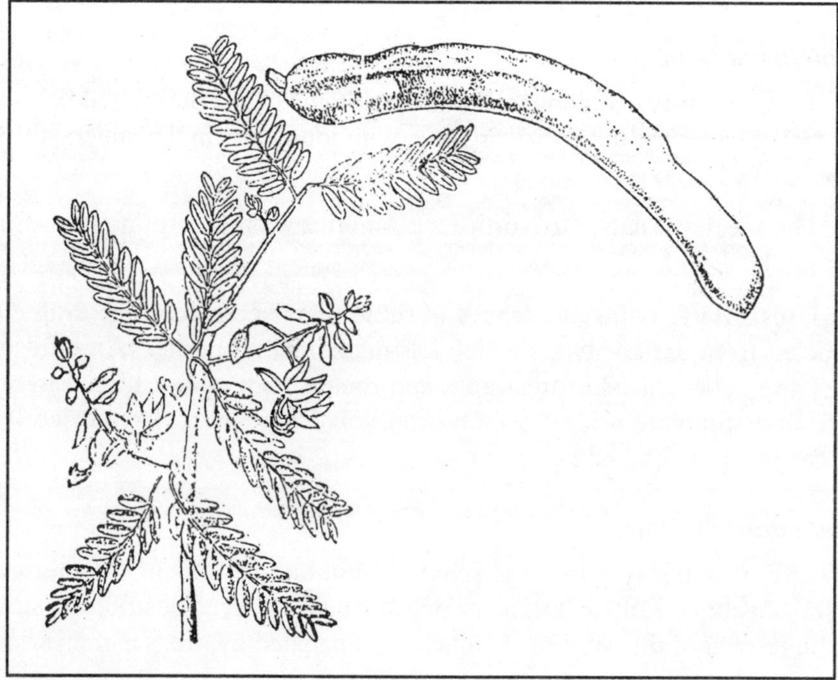

Fig. 31. Imli, Tamarind tree (Tamarindus indica Linn. Caesalpiniaceae)

Boerhavia Linn.
Family : Nyctaginaceae

Boerhavia diffusa Linn. :

This is a deep rooted perennial spreading herb. Leaves appear two in one node, one smaller than the other, upper surface green and lower surface whitish. Base of the leaves cordate. Flowers arises in short clusters on long axillary stalks, very small in size and reddish in colour. Fruits are glandular with fine ridges.

The plant is found throughout India.

Uses : All parts of the plarvt constitute the drug which contain alkaloids. It is diuretic and laxative. It is also used to treat asthma, dropsy, jaundice, intestinal inflammation and gonorrhoea. Tender shoots are eaten as potherb. Root powder preparation is used in eye diseases. The plain juice is antidote to rat-poisoning.

Bombax Linn.
Family : Bombacaceae

Bombax ceiba Linn. :

This is a common and attractive tree.

Distributed in different parts of India and also occurring at Chimtipur of Chhindwara district of M.P. Commenly known as Red Silk cotton tree.

Uses : The resin powder is used in diarrhoea. The wood is used in the manufacture of matches. The cotton from the fruits of this plant is used commercially for stuffing pillows and mattresses.

Borago Linn.
Family : Boraginaceae

Borago officinalis Linn. :

This is a hardy annual plant upto 80 cm tall, with blue flowers. This species is cultivated forjbee feeding. It has a typical bee-flower. Flowers are blue and pendulous.

The species is native to Europe and Asia. Commenly knowns common Borage or Bee Breed.

Uses : This is recommended for various skin diseases, eruptive fevers and phlegmasia. It is also used for flavouring claret-cup and as a bee food.

Brassica Linn.
Family Cruciferae

A genus Bmssica consists of one hundred and fifty species. These are annual and biennial herbs and several specie^ are commercially cultivated as oilseed crops.

Brassica juncea Coss. :

This is an annual crop, leaves are narrow. This is a self fertile species. The seeds are smaller, rugose and reddish brown in colour. There are two types in this species, one type is tall and late; while another one is short and early in which rough-leaved and smooth-leaved types are available.

It is distributed in India and China. This species is commercially cultivated in Africa and Russia and in India this crop is commonly cultivated in West Bengal, Bihar and U.P. Commenly known as Indian Mustord or Rai.

Uses : The oil content of the seeds is usually 35-40% and the volatile oil content varies from 2.8 to 30%. It is reported to contain ally I isothiocyanate and related compounds, among which crotonyl isothiocyanate has been identified.

It possesses rubefacient, vacicant, emetic and preservative properties.

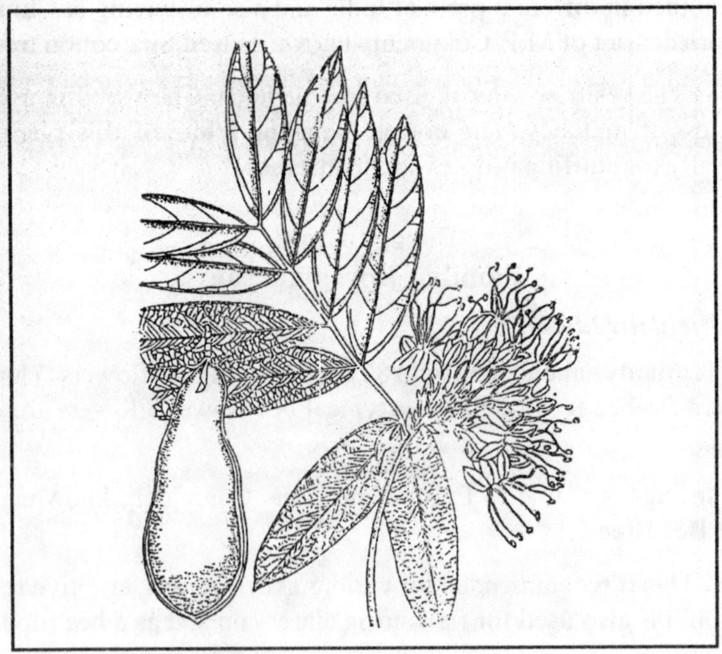

Fig. 32. Ashok (Sarara asoca (Roxb.) De Wilde Caesalpiniaceae.

Brassies nigra (Linn.) Koch. :

This is quite distinct from other Brassicas and is highly self sterile. The fruits at maturity are closely oppressed to the inflorescence axis. It is native to Europe and commercially cultivated in different parts of the world. The species is grown as a cold season crop in Tamil Nadu, Uttar Pradesh and Punjab. Commenly known as Block Mustand.

Uses : The seeds yield the condiment and also constitute the drug. The seeds contain 25 to 35% fixed oil and yield volatile oil which contain 0.8 to 1.4% allyl isothiocyanate. Apart from this sinigrin a glucoside and myrosin an enzyme are also available.

This is used in cases of pleurisy and pneumonia. It has emetic rubefacient and vesicant properties. Black mustard are used in pickels and curries. It is also used for making soap and for burning and as a lubricant.

Brassica rapa Linn. :

This is a hardy annual, slender and erect upto l.5 m tall. The leaves are lobed upto 18 cm long. Flowers are yellow in colour, 1.5 cm in diameter. Commenly known as Rape.

Uses : Seeds are useful which are digestive, laxative and induce vomiting. A pluster made from flour of seeds with water, wrapped in brown bag and applied to the chests against flux, fever and colds.

Bryonia Linn.
Family : Cucurbitaceae

Bryonia alba Linn. :

This is a scandent herb with simple tendrils. The species is native to Europe and Persia. Commenly known as While Bryony or Wild Hops.

Uses : It causes vomiting and excretion. This is prescribed in cases of dropsy, chronic rheumatism, pneumonia, particularly in dysentery.. Bryonia is used in popular medicine. Fresh puip is used as a resolvent poultice externally.

Buchanania Spreng.
Family : Anacardiaceae

Buchanania lanzan Spreng. :

This plant is native to India and Burma, commonly occurring at Rater of Chhindwara district of M.P. and also in drier parts of India.

The resin powder is used in diarrhoea. Fruits and seeds of these plants are edible. Oil obtained from kernels is used as a substitute for almond oil.

Bupleurum (Tourn.) Linn.
Family : Umbelliferae

Bupleurum candollii Wall. ex. DC.

This is an evergreen shrub. Leaves are glaucous. Flowers are yellow. The species is native to Europe and commonly found in Damtee of Utterkashi, above 3000 m elevation.

Uses : The paste of dried seeds applied in wounds and sores. This paste is also used in the laceration of toes.

Butea Koen. ex. Roxb.
Family Leguminaosae (Papilionaceae) Fabaceae

Under this genus there are about thirty species and three species are native to India. Trees and large woody climbers are available in this genus.

Butea monosperma (Lamk) Taub. B. frondosa Koen ex. Roxb. :

This is a medium sized tree, deciduous, untidy in growth and ragged in shape, with twisted and gnarled trunk. Bark is fibrous and light brown in colour. Leaves are rough in texture, pinnately trifoliate, leaflets are coriaceous, ovate, 10 to 15 cm long and broad, and fall off in cold weather. The tree is gorgeous in bloom during February-May when it is leafless. Flowers are bright flaming scarlet orange with black calyces, each shaped like the beak of a parrot and borne closely packed. Petals are 5 to 7 cm long, covered with grey pubescens. Ripe pods are light and found scattered far and wide by hot winds in June.

It is native to Indo-Malayan region. Tree is found throughout India mostly in the deciduous forests of Central and Western India. It grows wild in West Bengal, Bihar and also cultivated in gardens and roadside. Commenly known flame of the forest.

Uses : The flowers and leaves of B. monospenna are astringent, depurative, diuretic and aphrodisiac. These are used against boils and pimples, and are also prescribed to take internally in flatulent colic, worms and piles. Red coloured gum, root bark and seeds of the tree also possess medicinal properties. Gum is used for the treatment of diarrhoea which contains tannins. The flowers and seeds are mixed in a decoction and used as wormicide against tapeworms and ring worms. When several leaves are stitched together, it serves as dinning plates, and the leaves are also used in beedi factories. Lack-insects can be reared on the twigs. Bark and roots yield fibre for making ropes. Flowers yield a yellow die and are used in textiles. Bark is used for tanning.

Caesalpinia Linn.
Family : Leguminosae (Caesalpiniaceae)

Under this genus around one hundred species are available and nine species are native to India. Trees, shrubs or woody prickly climbers are found in this genus. Flowers are attractive yellow or red, appear in axillary or terminal, racemes or panicles.

Uses : The pods and bark of C. coriaria VVilld. are said to be antiperiodic, and the decoction of pod is used in the treatment of bleeding piles. The roots of C. digina Rottl. is prescribed for phthisis, scrofula and diabetes. The roots of C. nuga Ait possess diuretic and tonic properties and are useful for the treatment of stone in the bladder. A decoction of C. sappan L is used to treat mild cases of dysentery and is also effective in certain skin diseases. The bruised leaves of C. sepiaria Roxb. are applied on burns.

Calendula Linn.
Family : Compositae (= Asteraceae)

Calendula officinalis Linn. :

This is a hardy annual of 40 cm tall, with bright orange-yellow flowers. It is native to South Europe and Australia.

Ordinary rich soil, sunny or semishady situation are ideal for these plants. Multiplication of plants is through seeds. Commenly called pot more gold or scotch marigold.

Uses : The plant is reported to be a stimulant, sudorific and febrifuge.

Callistemon R. Br.
Family : Myrtaceae

This genus have about twenty five species available. Those are mostly trees and shrubs.

The leaves and twigs of Callistemon lanceolatus Sweet "Red Bottlebrush" contains 0.06-0.22% essential oil rich in cineole. C. citrinus (Curtis) Skeels, C. linearis DC. and C. rigidus R.Br. are also cultivated in India.

Callotropis (R.Br.) G. Don
Leguminaceae (Asclepiadaceae)

Callotropis gigantea (L) R.Br. :

This species is native to India and commonly occurring at Chimtipur of Chhindwara district of M.P. Bihar and Jharkhand. Commenly called Bow String Itemp or Akban.

Uses : The powdered flower is used in cold and cough. The root, leaf and bark paste are used in elephantiasis.

Caltha Linn.
Family : Fanunculaceae

Caltha palustris Linn. :

It is hardy perennial plant 3nd considered as bog or marsh plants, sometimes attain a height of 90 en:. Leaves are shiny and flowers in June-July, spathe is white or yellow, Arum like.

The plant is native to Britain.

They require rich soil and can be planted in damp borders, or banks of ponds, streams or lakes. The roots are planted in submerged soil or moist soil. Plants are propaga.ted by division oof roots. Commenly called marsh marigold or water cowslep or king cup.

Uses : The leaves of these plants are vesvant and very bitter. It is used as febrifuges.

Camphorosma Linn.
Family : Chenopodiaceae

Camphorosma monspeliacum Linn. :

It is native to Mediterranean region and Central Asia. It is rarely used now-a-days in medicine.

Uses : In the ancient medicine this plant had a place for vulnerary, aperitif, cephalic, sudorific and emmenagogue. It has been used for dropsy and asthma.

Canna Linn.
Cannaceae (Zingiberaceae)

Canna edulis Ker-Gawl. :

A herbaceous perennial plant with rhixomatous roots. The plant attain a height of 3 m and produce brighc red flowers. It is native to West Indies and South America. Another species C. indica is commonly known as "Indian Shot" of 1.25 m tall with yel'ow and red flowers and is native to West India.

They grow well in warm humid climate with well drained organic rich soil. Plants prefer sunny situation and are propagated by division of rhizomes. It needs replanting after even three years and liberal manuring in each year. Plants grow well with adequate watering during dry seasons.

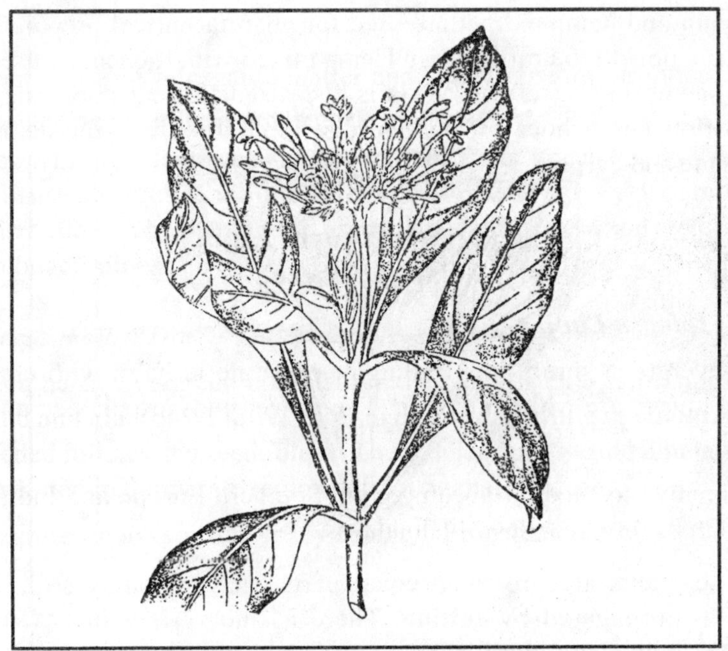

Fig. 33. Chota-chand (Rauvolfia serpetina (Linn.) Benth. ex Kurz. Apocynaceae.

Uses : The rhizomes of C. ednlis provide a nutritive and fortifying faecula. These two species are used to make emollient poultices for abscesses and tumors.

Cannabis (Tourn.) Linn.
Cannabinaceae (Urticaceae)

Cannabis sativa Linn. C. gigantea E. :

A tall annual plant attain an ultimate height of 6 m. The leaves are long narrow, margins toothed, with several leaflets, and arranged on the stem alternately. Male and female flowers borne on separate plants. By drying young leaves and flowering shoots bhang is obtained. The dried flowering-tops of the female plants consists of ganja, which become coated with a resinous exudation from glandular hairs. The plant is native to Central Asia and Himalaya. It is grown throughout India.

Ordinary rich soil and sunny situation are suitable for these species. Plants are propagated by seeds. Commenly known as Marijuma or Corgotobacco.

Uses : Mature leaves and flowering top of the female plants are used medically. It contains a resinous matter, cannabin, Hashish is the best known product of India Hemp, it has remarkable hallucinatory properties. The tincture of this plant and hemp extract are used for pharmaceutical preparation. This tincture is a poison to brain. Indian Hemp mixed with tobacco is used to treat asthma, haemorrages, tetanus. This is also used to treat depressia, bladder inflammation, gonorrhoea and nervous disorder. Fibres from the stem, used for ropes, twine and carpets.

Capparis (Tourn) Linn.
Capparidaceae

Capparis spinosa Linn. :

An evergreen shrub, low trailing or prostrate in habit with close heavy foliage. Flowers are white in colour. The flower buds are pickled and sold as capers.

It is native to Mediterranean region, Southern Europe and India. Mostly Nestorn Ghats, Inner Valley of Himalayas.

These plants are grown in equal parts each of loamy soil, sand and leafmould. Propagated by cutting. There is one variety in it "Rupestns". Commenly known as capor Plant.

It is mostly valued for its flower buds which are pickled and sold as "Capers". The flowers contain a glycoside and rutin. Flower buds contain 6% pentosans, rutic acid, pectic acid, a volatile emetic constituent and saponin. The seeds yield 35-40% pale yellow oil.

Uses : The caper buds as well as the fruits are considered useful in scurvy. The bruised leaves are used as a poultice in gout. The bark and root of this plant is slightly bitter and tart. It is aperitive, diuretic, resolvent and tonic. It facilitates digestion and stimulates appetite. It is used in medicine as a refrigerant and antiscorbutic. For treatment of rheumatism, paralysis, toothache and affection of liver and spleen and tubercular glands, it is used. Capers are used for flavouring pickles, sauce, salads and other cooked food.

Capsella Medic.
Cruciferae

Capsella bursa-pastoris (L.) Medic. :

An annual plant of 65 cm tall, arises from the basal rosette of leaves. Leaves are variable in shape and degree of division in various situations. Flowers are small and white. Pods are heart-shaped, triangular. It is native to temperate and subtropical regions. These plants grow in all types of soil in open sunny situation.

Uses : Entire plant is medicinal. It is rich in vitamin C. A decoction is used to treat bloody urine, diarrhoea and haemorrhoids. The juice of the plant is used to clear phlegm from chest, to stop nose bleeding, to increase urine flow and menstrual flow.

Capsicum Linn.
Solanaceae

This genus consists of fifty five species of annual or perennial herbs.

Capsicum frutescens Linn. C. minimum Roxb. :

A perennial plant characterised by the small size of the pod and high pungency and these are known as "Bird Chilli" and "Tobasco Chilli". Plants attain a height of 0.8 to 1.4 m, bear extremely pungent conical fruits of 15-30 mm in size. These plants degenerate under cultivation in one or two years. It is native to Tropical America and is cultivated in different parts of the world. Commenly called Red Pepper.

Dried fruits are useful. It contains capaisin, dihydrocapsaicin, nor-dihydrocapsaicin, homocapsaicin, homodihydrocapsaicin, ascorbic acid, riboflavin, thiamine, red carotinoids, carbohydrates, fat fibre, iron etc.

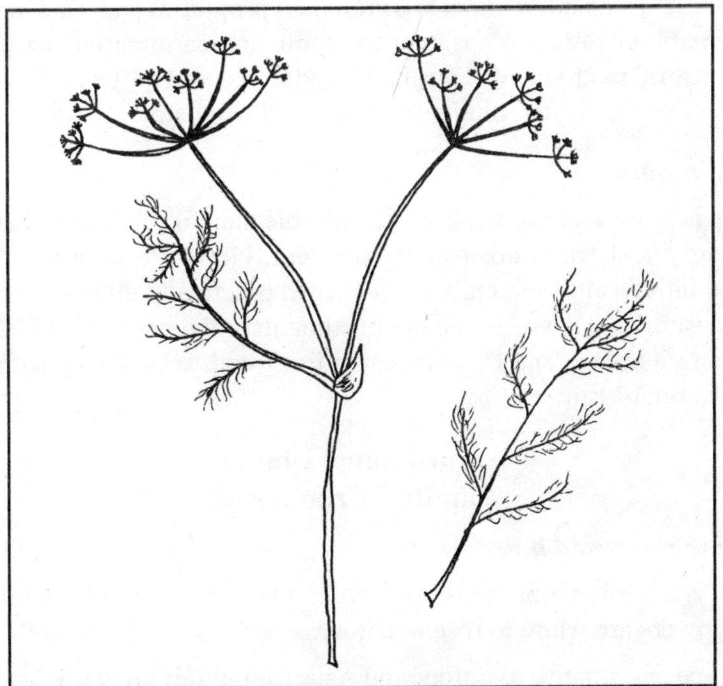

Fig. 34. Foeniculum officinalis

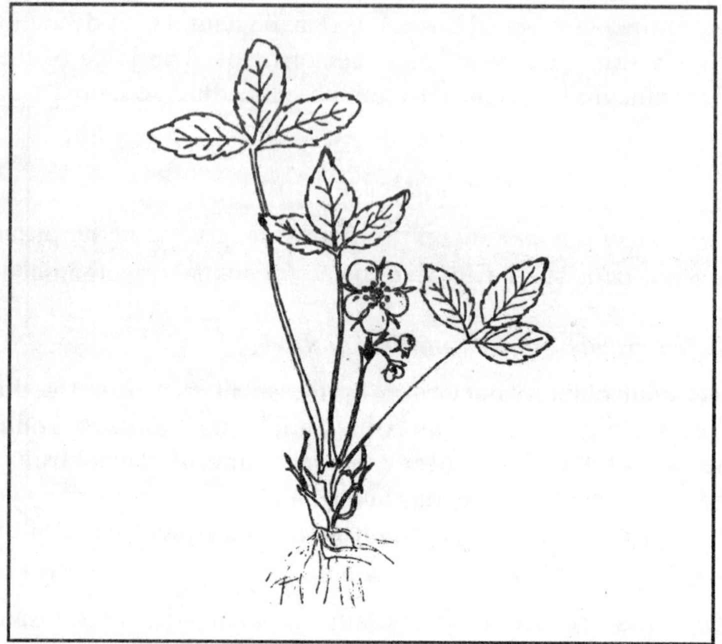

Fig. 35. Fragaria virginiana

Uses : It has counterirritant and stimulant properties and used in neuralgia and rheumatic disorders. Capsicum has tonic and carminative actions and is specially useful in atomic dyspepsia. It is administered in the form of powder, tincture, liniment, plaster, ointment, medicated wool etc.

***C. annuum* Linn. :**

This is cultivated for fruit. It is a variable annual sub-shrub. Flowers are borne singly and fruits are usually pendent. Native to Tropical America, cultivated universally, and chillies are grown practically all over India. Dried fruits are useful and possess similar constituents and properties of C. fruticans. It is rich in vitamin A and C, when eaten fresh with salads. Green chillies are reported to retard cancer.

Cardamine Linn.
Family : Cruciferae

***Cardamine pratense* Linn. :**

This is a hardy herbaceous bog garden plant, attain an ultimate height of 60 cm. Flowers are white to rose in colour.

The species is native to Europe and Asia. Commonly known as Lady Smock or Cuckon Flower.

Ordinary rich soil, moist and shady location are ideal. Plants are propagated by seeds or division of roots. It is also vegetatively propagated by adventitious roots arises on the radical leaves.

Uses : Seeds of these plants are used as stomachic, carminative and stimulants. It is also known as digestive and diuretic. The seeds are eaten as a condiment after meals to accelerate the digestive process.

Carduus Linn.
Family Compositae (= Asteraceae)

Carduus benedictus, Linn. :

This is a perennial herb native to Europe. Commonly known as holy thristle. The plant is considered as an effective remedy for pleurisy.

It is used as a febrifuge, diuretic and sudorific.

Carica Linn.
Family : Caricaceae

Carica papaya Linn. :

A herbaceous perennial, with milky latex which attain a maximum height of 5 m. Leaves are clustered around apex of the stem and branches. The leaves and unripe fruits contain milky juice in which the protein- ferment papain is available.

The plant is native to Southern Mexico and Central America. Papaya is grown in all tropical countries as well as in some subtropical areas, it is commercially cultivated in Hawaii, Colombia, Venezuela, Puerto Rico, South Africa, Queensland, The Philippines, Ceylon and India. Commonly known as Papaya.

Uses : In all parts of plants, latex is available. Latex contains two very active crystalline enzymes viz., papain and chymopapain. Apart from these, small amount of amylase, lipase, pectase and other enzymes are also reported.

The latex has been used to treat eczema, ringworm, warts, corns, calluses, infected wounds, malignant turners, pain of burns, bleeding haemorrhoids. For expelling roundworm an infusion of latex and honey is given to children. The latex in glycerine solution- is applied to the pharynx in diphtheria. Small quantity of latex mixed with sugar is given for the treatment of enlarged liver and spleen, whooping cough; and it is also regarded as emmenagogue and stomachic. Papain is known for its digestive properties to relieve dyspesia and also to cure diarrhoea.

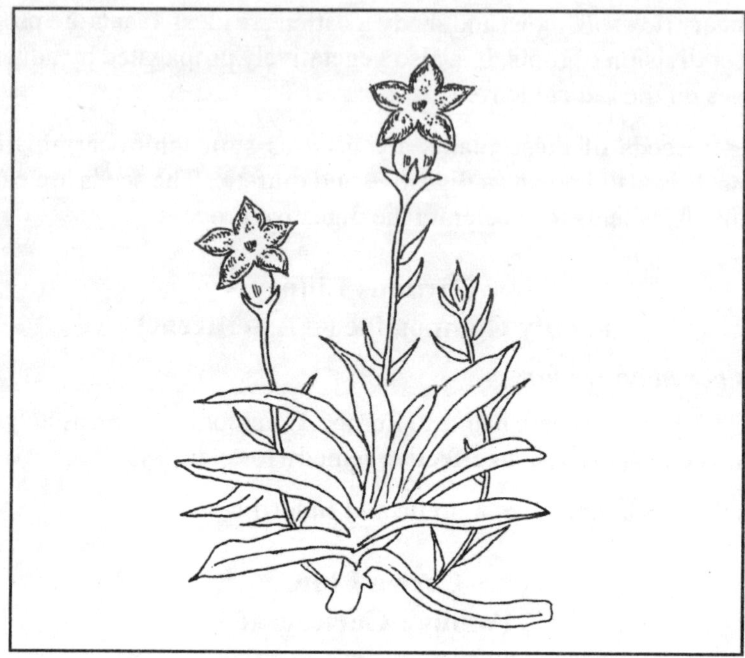

Fig. 36. Gentiana Kurroo

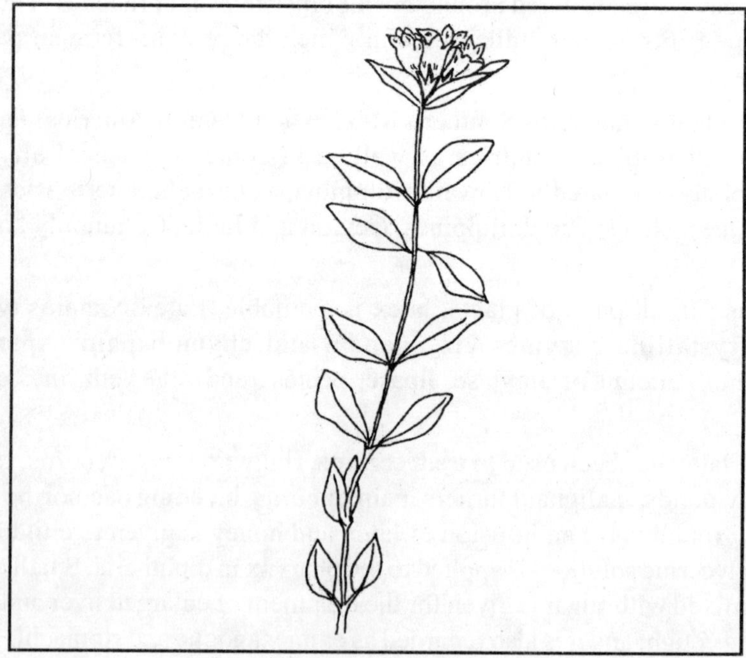

Fig. 37. Gentiana villosa

To relieve some pollen causing allergies, papain containing nasal spray is prescribed. Chymopapain is in use now-a-days for neurosurgery and orthopedics to overcome the pain of pinch nerves, but its toxic effects are also reported. In veterinary practices papain is used as anthelmintic. Apart from the medicinal use, ripe papaya is eaten as fruit and green fruit is cooked as vegetable. Young papaya leaves are also cooked and eaten. The latex is an ingredient in chewing gum and confectionaries. Papain is used in facial creams, fabrics and rubber industry. Fresh latex is valued as vermifuge, styptic and anti-chigger application.

Carlina Linn.
Family : Compositae (= Asteraceae)

Carlina vulgaris Linn. :

A hardy perennial plant of 50 crn in height with white flowers. The plant is native to Europe.

Ordinary rich soil and open dry situation are preferred for cultivation of these plants. These are propagated by seeds. Commonly called Carbina thistle.

Uses : The fresh roots and flower heads are poisonous but dried roots and flower heads soaked in honey and sugar is not poisonous. These parts of the plant give resin gum and is used for tapeworm infestation.

Carthamus Linn.
Family : Compostiae (= Asteraceae)

Carthamus tinctorius Linn. :

An annual of l2 rn tall. Leaves are alternate, spiny. Flowers are orange-yellow in colour. Seeds are white and shiny.

This species is native to Afganistan. It is cultivated throughout the greater part of India, known satt lower.

Uses : Plant in bloom, flowers and seeds are medicinal. It is laxative, stimulates menstrual flow and induce perspi.ation. Paste made from flower is applied to boils. Tea made from flower is used.to treat fever and skin eruptions. The plant is used to dry up the skin symptoms of measles. Its flowers are used in dyeing, powdered and mixed with talc, they form rouge. The dye is used for colouring food and cloth. Seeds yield an edible oil which are used for soup, paints, varnishes and as an illuminant.

Carum Linn.
Family : Umbelliferae

Carum carvi Linn (= Apium carvi Linn. or Seseli carvi :

A hardy thick rooted herbaceous biennial plant of 70 cm tall. It has a fleshy root and a slender branched stem which attains a height of 60 cm. The leaves are compound, pinnate and divided into very narrow segments. Flowers are small white and borne in flat compound umbels. The dried fruit or seed is brown in colour, has a pleasant odour. Seeds are hard and sharp. The species is cultivated in cool climates.

It is native to North and Central, Europe and extensively cultivated in Holland, Russia, Poland, Bulgaria, Rumania, Syria and Morocco. It grows wild in India in the North Himalayan region.

Plants are grown in the plains as cold season crop and in the hills of ^"Kashmir and Kumaon as a summer crop. In India, the fruits are collected before ripening. The plants are dried and fruits threshed out, cleaned and stored in bags. Yield of this crop in well prepared rich soil is upto 2000 Kg/ hectare.

Uses : Fruits are useful for essential oil. It contains carvone, d-limonene, carveol, diacetyl furfural, methyl alcohol, acetic aldehyde, thiamine, nboflavin, niacin and ascorbic acid.

It is widely used as a spice for culinary purposes and for flavouring biscuits, cakes, cheese and bread. Oil is used in the manufacture of "Kummel". This is a strong carminative, stimulant and aromatic. It is used in the form of poultices for swellings in the breast and the testicles. It is a mild stomachic and occassionallv used in flatulent colic. Carvone is anthelmintic.

Cassia Linn.
Family : Caesalpiniaceae

This is a large genus of more than five hundred species and about twenty-six are native to India. This genus consists of herbs, shrubs and trees.

Cassia angustifolia Vahl. :

The leaves and pods of this species produce crude drug senna and is often referred as Indian Senna and C. acutifolia Delile is known as Alexandrinian Senna.

Indian Senna is cultivated in south-western arid tracts of Tamil Nadu in. marginal lands. Alexandrinian Senna is obtained from wild growth in Sudan.

Uses : Extracts of these plants contain senosides and these are used as laxatives.

Cassia angustifolia Vahl. is a small perennial under-shrub with pale suberect or obtusely angled erect or ascending branches. The leaves are large, compound and pinnate. Leaves are usually 6-9 jugate, oval-lanceolate, glabrous, bluish green to pale green in colour. Racemes are axillary erect many flowered and flowers are bright yellow. Seeds are obovate, cuneate and compressed.

When cultivated, it remains in field from 95 to 140 days, both as rainfed crop or irrigated crop. The flowers are not season bound and are borne in between 35 to 70 days of age, depending upon type of plants, season of cultivation or time of sowing. Flowers come in succession for a longer duration. Immediately after flowering pods appear. Pods are slightly curved, 4 to 7 cm long and upto 1.8 cm broad. On maturity pods turn to dark brown. There are 6 to 8 dark brown, ovate seeds in each pod. The plants shed leaves at the advent of winter season.

The plants grow veil in warm arid region. They can be grown in wide range of soil, except in waterlogged soil. Sandy loam soil, of pH range 6 to 8.5 is suitable for its cultivation. Once established, the crop can tolerate higher salt content. As irrigated crop, the crop is sown during February-March in

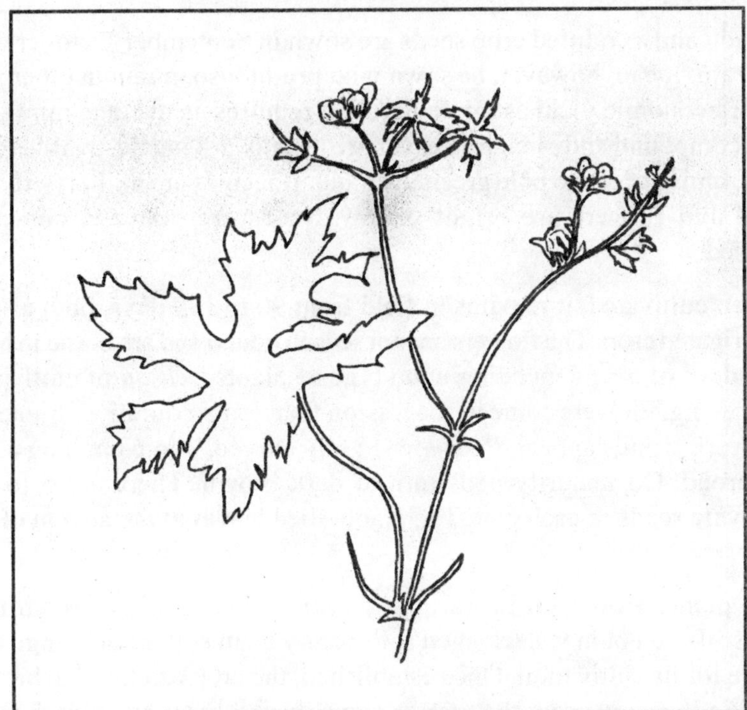

Fig. 38. Geranium nepalense

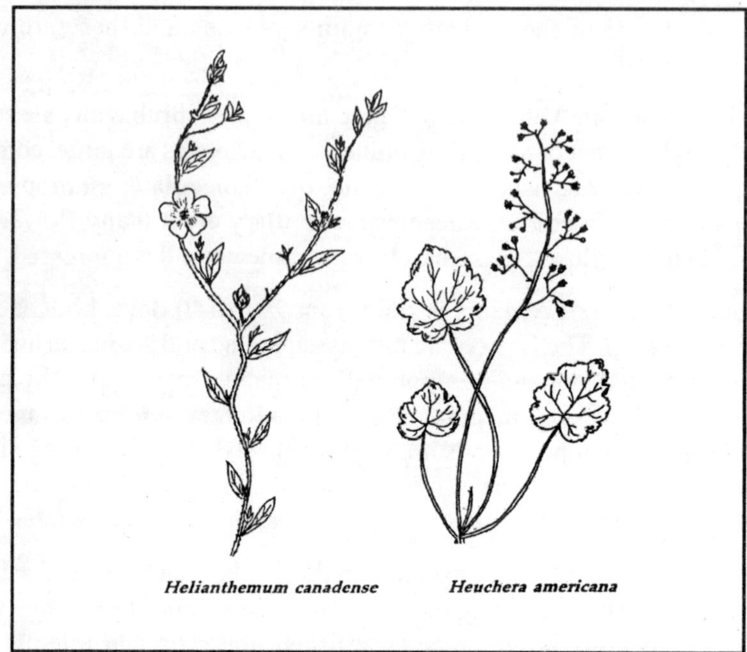

Fig. 39. Helianthemum canadense, Heuchera americana

Tamil Nadu and as rainfed crop seeds are sown in September-October with the onset of rain. It can, however, be sown with pre-monsoon rain in other parts of India. For economic yield as rainfed crop, it requires an average rainfall of 30 to 45 cm compound and pinnate. Leaves are usually 5-8 jugate, oval-lanceolate, glabrous, bluish green to pale green in colour. Racemes are axillary erect many flowered and flowers are bright yellow. Seeds are obovate, cuneate and compressed.

When cultivated, it remains in field from 90 to 130 days, both as rainfed crop or irrigated crop. The flowers are not season bound and are borne in between 35 to 70 days of age, depending upon type of plants, season of cultivation or time of sowing. Flowers come in succession for a longer duration. Immediately after flowering pods appear. Pods are slightly curved, 3 to 6 cm long and upto 1.5 cm broad. On maturity pods turn to dark brown. There are 5 to 7 dark brown, ovate seeds in each pod. The plants shed leaves at the advent of winter season.

The plants grow well in warm arid region. They can be grown in wide range of soil, except in waterlogged soil. Sandy loam soil, of pH range 6 to 8.5 is suitable for its cultivation. Once established, the crop can tolerate higher salt content. As irrigated crop, the crop is sown during February-March in Tamil

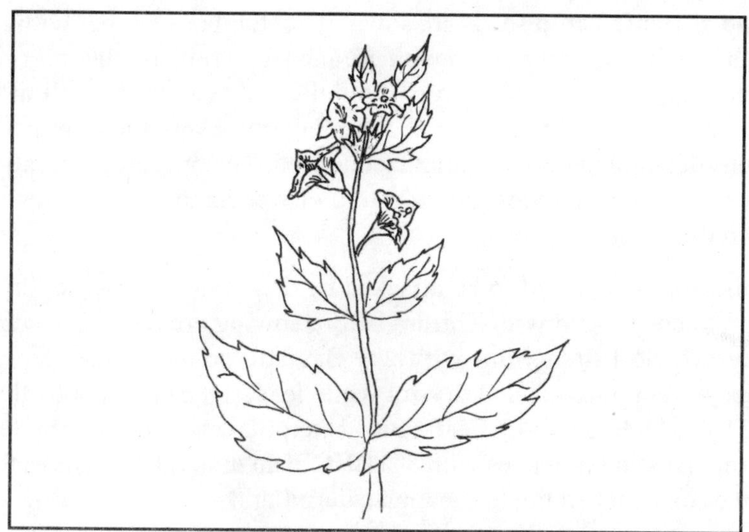

Fig. 40. Hyocyamus niger

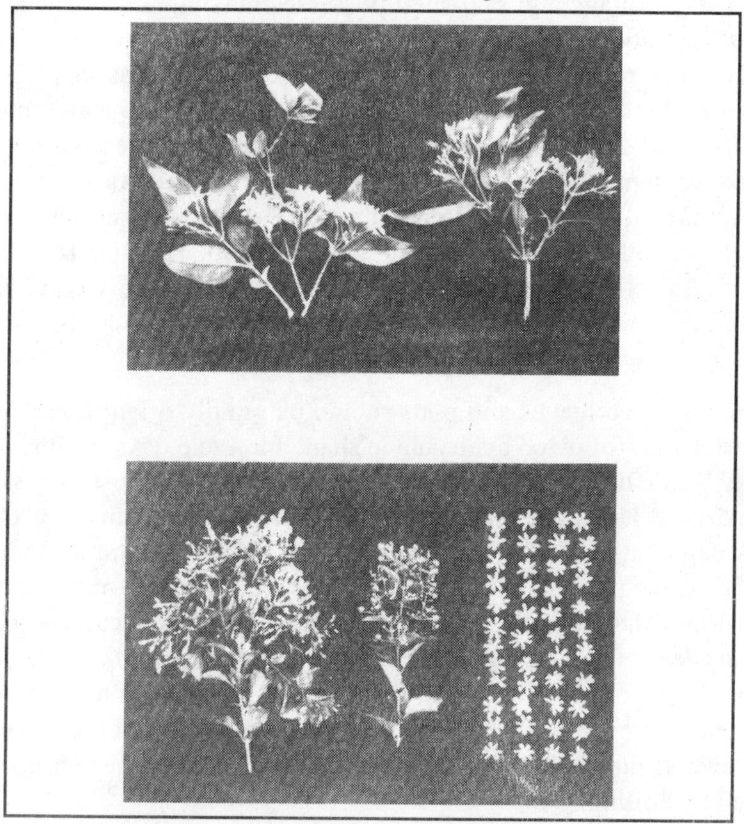

Fig. 41. Jasminum arborescens, Jasminum auriculatum

Nadu and as rainfed crop seeds are sown in September-October with the onset of rain. It can, however, be sown with pre-monsoon rain in other parts of India. For economic yield as rainfed crop, it requires an average rainfall of 25 to 40 cm require 4 to 8 light irrigation. As rainfed crop even, the crop need 2 to 4 irrigations depending on the climatic condition. Two weeding operations upto 60 days after sowing, keeps the crop free of weed and it also saves moisture loss from the soil.

This crop is reported to be attacked by damping-off disease, dry-rot and leaf spot. Seeds treated with Thirum before sowing prevents the incidence of damping-off. Soil drenching with 2% Bavistin solution also prevents the incidence of damping-off and dry-rot while leaf spot can be controlled by the spray of 0.4% Dithane M-45. Leaf-eating caterpillar cause considerable damage to the crop. Dusting of leaves with 8% BHC at an interval of 15 days for two to three times can control the leaf-eating caterpillars.

Generally, 4 stripping of leaf and pods are done within 130 days period of this crop. The foliage and pod-shell possesses maximum senosides contents and constitute about 40% by weight. Other plant parts like stem bark, stem wood, root bark, root wood, rachis and peduncles also possess varying amounts of senosides, but seeds do not possess. Fully grown mature leaves are picked and the pods of 20-25 days old are good. There is progressive decrease in percentage content of senosides in leaves and pods with age. As per experimental evidence, maximum yields are obtained under Delhi conditions when crop was picked at 60, 100 and 140 days; and at Anand 100, 120 and 160 days after sowing. The first picking contained exclusively leaves, the second had more leaves and last picking contained mainly pods. In Tamil Nadu, farmers do the first picking between 60 to 80 days of sowing.

The harvested leaves and pods should be rapidly dried under sun in the field for 8 hours, followed by drying in shade for 4 to 6 days till it reaches 8% moisture. While drying in shades the material should be constantly stirred for uniform drying. However, rapid drying through hot draft of air at 50°C is ideal. There is degradation in the senosides content if the temperature is increased over 60°C. If the produce losses the colour and turns to ash or black colour, it loses its drug value. A good crop under proper management can yield about 20 q/ha of dry leaves and 8 q/ha of pods under irrigated condition. Rainfed crop yield about 15 q of leaves/ha and 5 q of pods/ha. Senosides content of Indian Senna varies from 1.8 to 4.0%. The cathartic properties of Senna are generally not lost even if they are stored for a number of years, but these properties are destroyed by boiling.

The plant contains 0.38% B-sitosterol. The active constituents are the derivatives of anthraquinone. The leaves and pods of Cassia angustifolia are cathartic, contain senosides A, B, C, D, rhein and aloe-amine in free and compound forms. Of these senosides, A and B are present in larger amounts than others. These glucosides show high laxative properties. The leaves and leaflets contain kampferol, anthraquinone, essential oil, calcium salts, isohamnentin, rhein and emodin.

The action of glucosides is particularly on the lower bowel, where it increases peristaltic movements by its effect on the intestine wall. Senna leaf and pod is usually used in the form of infusion. Against habitual constipation these are used in the form of decoction powder. It is reported that if nursing women take this drug, their milk acquires purgative properties. The leaves are generally preferred in household remedies for manufacture of herbal tea in Central and Western European countries. It is also used as anthelmintic for intestinal worms and as a liver stimulant. For skin diseases and for removing pimples, a paste of dried leaves mixed with vinegar is used.

Different species of Cassia are well known for their medicinal value, they have cathartics, antibacterial, antibiotic and furariform activity. Several species contain anthraquinone, chrysin emodin, Rhein and tannin. To treat diabetes, haemoglobin disorders and constipation various preparations are made from these plants.

Cassia acutifolia Delile :

The leaves and pods of these species are considered an effective purgative. In high doses it causes contractions in the uterus and bladder, and in pregnant women may cause uterine haemorrages. It is used for diseases caused by worms and chronic skin diseases combined with bitter tonics, it is very effective for dyspepsia. This is also used in some brain diseases.

Cassia alata Linn.

This is a large spreading shrub with irregular, angled, glabrous branches. Flowers are bright yellow in colour. Pods are long, membranous, dehiscent with 28 or more seeds per pod.

Uses : This plant contains rhein, anthraquinone derivatives, B-sitosterol and chrysophenol. Leaves of Cassia alata homogenised with juice of lime are used for curing eczema and fungal disease. The plant contains chrysophanic acid. In Ayurvedic treatment the leaves of C. alata are recommended for use in ring worm infection and in snake bites. Decoction of leaves and flowers are used internally in constipation, bronchitis and asthma and also for washing eczematous patches.

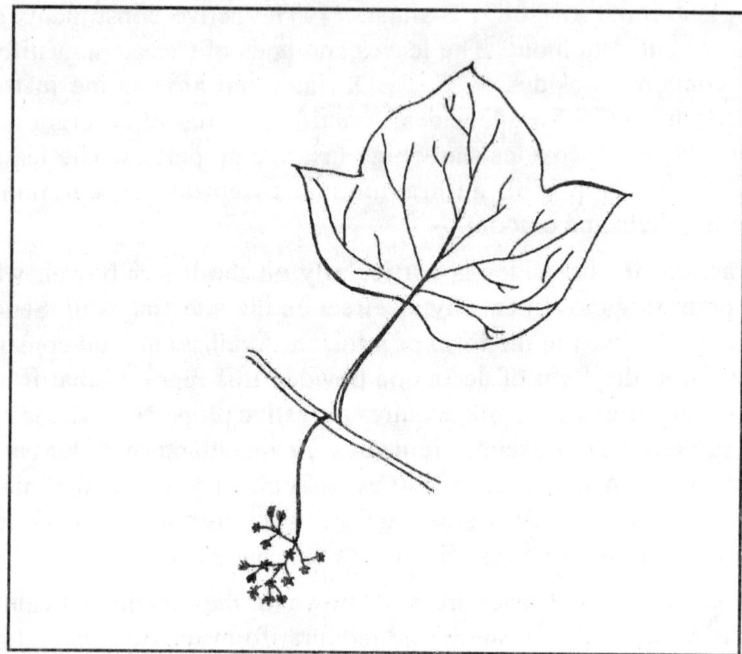

Fig. 42. Menispermum canadense

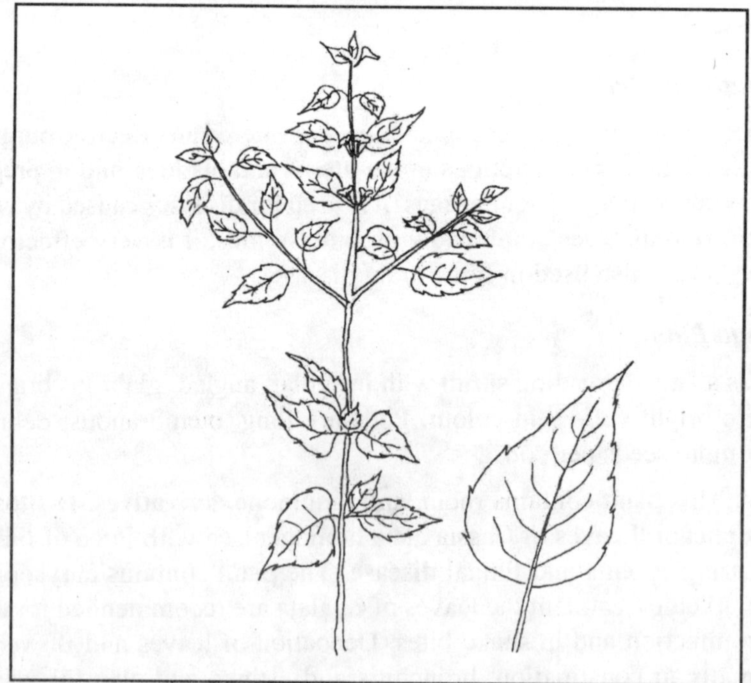

Fig. 43. Nepeta cataria

The roots of this plant are diuretic. The plant has insecticidal value and antibacterial property. The juice of the leaves is used as antiherpectic and skin diseases.

***Cassia fistula* Linn. :**

This is a medium sized tree with shining dark green large leaflets. Leaves are paripinnate, 40 to 80 cm long. Flowers appear in hanging branches, bright yellow in colour. Fruits are cylindric long stick like, black or shining dark brown. Fruit pods are 40 to 60 cm long.

This plant is found in different parts of India and common in moist or evergreen forests.

Uses : This is called a "Cassia pulp", is well known laxative. The pulp contains sugar, tartaric acid, melic acid, oxalic acid. One more acid, cathartic acid considered the purgative principles of the drug. The bark of the tree is rich in tannin.

***Cassia obtusifolia* Linn.**

An erect annual herb attain a maximum height upto 3m. Leaves are upto 15 cm long, leaflets are in 3 pairs, green, membranous. Flowers are yellow. Pods upto 30 cm long, membranous, contain around 40 seeds per pod.

It is found wild in roadsides, riverbeds, forest edges as weed in different parts of India. It was originally introduced from tropical America.

Uses : The leaves and roots are used as laxative. It contains chrysophanol, physcion, obtusion, chryso-obtusin and aurantio-obtusin. This plant is used in ayurvedic and unani system of medicine. Leaves are used in various preparations known as refrigerant, anthelmintic, antipyretic, diuretic and laxative. These are employed for the treatment of asthma, bronchitis, half headache, skin diseases, piles and itching. The fruits and seeds are anthelmintic, astringent to the bowels and recommended for treatment of skin diseases, leprosy, scabies, asthma, cough and also to cure tumours. The roots are used for tuberculous glands, ring worm and skin diseases. The seeds and roots are reported to possess antimicrobial activity.

***Cassia tora* Linn. :**

This is an annual herb upto I'Sm tall. Flowers are yellow. Pods are tapering.

It is a common weed, occurring at Chimtipur of Chhindwara district of M.P. and also in various parts of the country.

Uses : The seed paste is used in the treatment of ringworm and itch. The crushed seeds about 5 gms are taken with water in cough.

Cassiope D. Don.
Family : Ericaceae

Cassiope fastigiata (Wall) D. Don. :

A tufled, decumbent, stunted shrub upto 30 cm tall. Leaves are sessile, thick, closely oppressed. Flowers are solitary, axillary, often in clusters of 2-3, white, appearing during July-August.

It is native to alpine Himalayan region, occurring in temperate and alpine zones of H.P. between 3500 m to 4500 m elevation.

Sandy soil rich in organic matter, moist and cool situation are ideal. Propagated by seeds, layers and cuttings.

Uses : The leafy twigs are ground into a paste and applied in the fire burns.

Castanea Mill.
Family : Fagaceae

There are about twelve species of shrubs and trees under this genus.

Castanea dentata (Marsh.) Borkh (= C. americana) :

This is a hardy deciduous tree upto 38 m tall. The leaves are upto 26 cm long, margins are deeply serrated. Two or three nuts are found together upto 2.8 cm across, brown in colour, fattened on one side.

It is native to Eastern-North America.

Uses : Leaves and bark of these plants possess medicinal properties. A tea made from leaves is used as sedative, tonic and also to treat whooping cough. The bark preparation is used to treat dysentery and worms. The timber of this plant reaches maturity at 50 years of age. The wood is used as rafters in churches, cabinet work, post and rail fencing and rustic work.

Castanea sativa Mill. :

This is a hardy deciduous tree, upto 25 m tall, with edible fruits.

The species is native to S. Europe, N. Africa and Japan. It is cultivated in Himalaas.

Uses : Leaves of these plants are used for treatment of rheumatism. Tincture of these leaves are used for lumbago in Homoeopathy. Timber of this plant is most useful in young state and brittle when old. Seeds are edible.

Catalpa Scop.
Bignoniaceae

Catalpa bignonioides Walt. :

A tall tree of about 25 m height. Leaves have disagreeable odour when crushed, upto 26 cm long, sometimes with two lobes, fuzzy beneath. The flowers are in long clusters, individual flower measuring 5 cm in diameter, white with yellow and purple-brown markings. Pods are thin upto 45 cm long dark brown in colour.

It is native to America.

Uses : Leaves, tree bark and seeds have medicinal value. Leaves are applied as poultice to wounds and abrasions. The decoction of bark is a mild laxative and used to treat intestinal worms. The seeds are also used to induce vomiting and expel intestinal worms.

Catalpa spinosa (C. cordifolia) :

This is a hardy deciduous flowering tree with ornamental leaves. The plant may attain a height of 35 m, and flowers are white in colour.

Native to the South and East of the United States.

Uses : The flowers are recommended for the treatment of asthma and chronic bronchitis. The leaves of these plants have febrifugal qualities.

Catheranthus G. Don.
Family : Apocynaceae

Catheranthus roseus (L.) G. Don. (= Vinca rosea. Linn.) :

This is a perennial branched herb, 50 to 125 cm tall with flowers of white, pink and intermediate colour. This plant has become important in modern medicine. It grows throughout India mostly as escape from garden or as pot herb. It is commercially grown in a limited scale in Karnataka, Tamil Nadu and Maharashtra both for internal consumption and for export. It has medicinal importance owing to the presence of indole alkaloids raubasin and serpentine in its roots and of vincristine and vinblastine in its leaves. The dried leaves and roots are exported to USA and Western Europe.

Periwinkle is raised from seeds for which both direct field sowing and transplanting are adopted during mansoon. If the seedlings are transplanted, about 1 kg seeds per hectare, is sufficient. Freshly harvested seeds or upto one year old can be used. For optimum germination pre-soaking for a day is better. About two months old seedling are transplanted 45 cm x 30 cm spacing accommodating 80,000 plants per ha. Prospective buyers and promoters prefer that these crops should be grown with organic manure only. Hence, FYM 5,000 kg/ha, bone meal 250 kg/ha and neem cake 200 kg/ha is sufficient. Periwinkle also grows well with the application of inorganic fertilizer of 100 kg N, 120 kg P_2O_5 and 100 kg K_2O per hectare; in which full dose of phosphorus and potash is applied prior to planting as basal dose and nitrogen is applied in two equal split doses, one and two months after transplanting. During winter months the crop slows down its growth.

Leaves are harvested at 6 months, and then at an interval of 3 months. This crop yields one tonne leaves and 0.5 tonne roots per hectare under irrigated conditions.

Most of the cultivated crop in India is made of mixed genetic stock since pure lines have not been identified yet for commercial purpose. So far, over 100 different alkaloids have been isolated from various parts of this plant. These alkaloids possess antifibrillic, hypotensive, sedative and tranquilizing properties. Leaves of the pink flowering germplasm contain maximum vindoline content. Its vindoline content increases with the age of plant till ten months stage. Its leaves possess higher percentage of VLB alkaloids like vincristine and vinblastine, whereas the roots and basal stems have higher ajmalcine, vinceine, vincamine, raubasin and reserpine group of alkaloids. In addition to alkaloids, the leaves of these plants contain a volatile oil, aldehyde, sequiterpenes, furfural, sulfurous compounds, an alcohol called lochnerol, several monoterpene glycosides including adenosine and reseoside.

Uses : The leaves are emetic and diaphoretic. The plant infusion if gargled, is said to relieve sore throat, laryngitis and chest complaints. In some places, plant decoction is taken internally to arrest haemorrhages, ft is also believed that leaf decoction is antidiabetic. The squeezed leaf juice is applied to wasp stings. Leaf infusion is also taken for indigestion and dyspesia. The decoction of stems and leaves are used for regulating menstruation, for promoting abundant flow of urine in cases of scanty and reddish urine and is also beneficial for kidney. The roots are used to relieve toothache. Roots of these plants have purgative, vermifugal, depurative and haemostatic properties. It has also been reported that a decoction of few flowers with a few drops of alcohol is used as an eye wash for infants. It is said that the decoction of white, red centred flowers refresh tired and inflammed eyes. The alkaloid vincristine derived from leaves is used for treatment of acute leukaemia in children including lymphocytic leukaemia. Vincaleukoblastine is used for the treatment of Hodgkin's disease and leurocristine is employed for breast cancer. These are also used for treating malignant lymphoma. Vincal-eukoblastine has also been prepared as an ointment for treatment of psoriasis. Several combinations of different leaf alkaloids have been used in multiple drug cancer therapy with varying degree of success. Among the root alkaloids ajmalcine is said to be hypotensive and antiarrhythmic agent.

Caulophyllum Michx.
Berberidaceae (= Leonticaceae)

Caulophyllum thalictroides (L.) Michs. :

This is a hardy tuberous rooted perennial plant upto 1 m tall. The leaflets are two or three lobed. Flowers are yellowish green. Fruits are berries of blue colour. Wall of fruit early evanescent exposing the large drupe like seeds borne on accresec funicles.

The species is native to North America.

Uses : Roots of these plants are medicinal. Roots are used to relieve spasms, to increase perspiration, to induce menstrual flow, to hasten child birth, and to cure colic, rheumatism and sore throat. The tea prepared from roots are also used to treat venerial diseases and to strengthen the stomach.

Ceanothus Linn.
Family : Rhamnaceae

Ceanothus americanus Linn. :

This is a perennial deciduous shrub of l-4m tall. Leaves are .finely toothed, dark green in colour. Flowers appear on long stalks, white in colour. Root is dark red.

It is native to East America.

Uses : Leaves, seeds and roots of these plants constitute the drug. The plant is used to treat eye diseases in children, gonorrhoea and dysentery. Root is stimulant and sedative. To cure ulcerated, sore throat, a decoction of seeds and leaves are used. A decoction of roots is used as a wash for syphilis and cancer. Leaves were used as a substitute of tea.

Celastrus Linn.
Family : Celastraceae

***Celastrus scandens* Linn. :**

This is a vigorous, hardy, deciduous climber off 10 m in length. Leaves are upto 14.5 cm long and 3.5 cm wide. Flowers are minute and yellowish in colour. Fruits are like Euonymous. The seed capsules are attractive, globe-shaped, and orange in colour. Seeds are scarlet.

It is native to North America.

Uses : Root bark is useful for medicinal purpose. Ointment prepared from root bark is used to treat swellings, burns, tumors and cancers. To induce menstrual flow and perspiration, a decoction of root bark is used. The bark of the root is prescribed to take orally to increase urine flow, to treat veneral diseases, to quiet disturbed people and to induce vomiting.

Celtis (Tourn.) Linn.
Family : Ulmaceae

There are about eighty five species under this genus and ten are found in India. These are shrubs or trees and some are evergreen types.

There are several species available in South Africa.

The fruit of C. austmlis L. "European Nettle Tree", is sweet and edible and is used to treat amenorrhoea and colic.

It is native to North India and Temperate Himalya. Thewood of C. cinnamomea L. which is found in Sikkim, W. Bengal and M.P.

Uses : It is a blood purifier and used for treatment of itch and skin eruptions. Ointment made from the wood of this species is valued as medicine for headache. It is also used to treat convulsions, nervous disorder and fever. The fruits of C. aurantiaca, C. caucasia, C. integrifolia, C. korarensis, C. laevigata, C.Aeveillei, C. occidentalis L., C. reticulata Torr and C. sellomann Miq are edible.

Centaurea Linn.
Family : Compositae (= Asteraceae)

***Centaurea cyanus* Linn. :**

This is a hardy annual plant and attain a height of 1.5 m. This species produce blue, rose and white flowers.

It is distributed in Britain and found in North-West India in cultivated fields.

Fig. 44. Nelumbium nucifera

Fig. 45. Nymphoea odorata

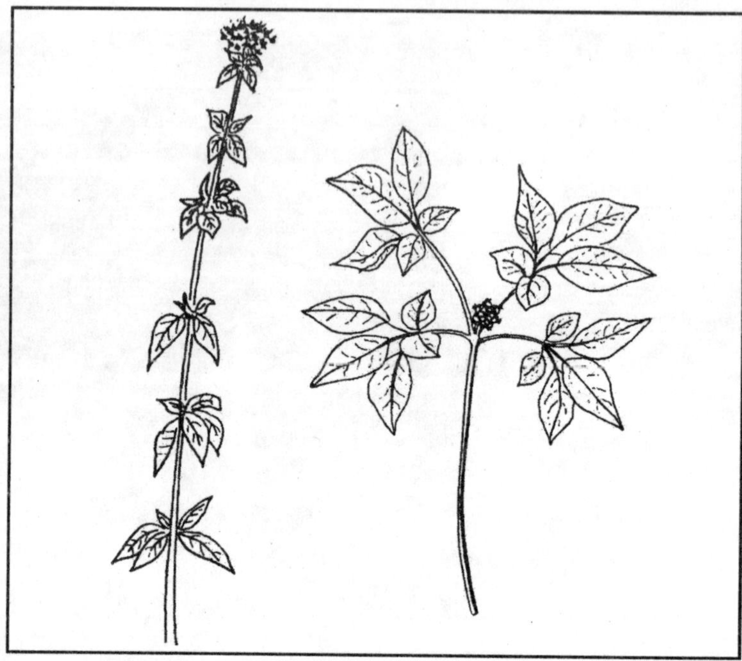

Fig. 46. Origanum vulgare (Panax quinquefolius)

Ordinary rich soil and sunny position are good for this plant. These are propagated through seeds.

Uses : Wild cornflower is regarded as a remedy for ophthalmia. Distilled water prepared from this flower is used for eye wash.

Centella Linn.
Umbelliferae (= Hydrocotylaceae)

Centella asiatica (Linn. Hydrocotyl asiatica Linn.) :

A small trailing herb and the creeping stems bear roots on their nodes. Leaves are undivided, rounded or broad, kidney shaped upto 6 cm in diameter. Flowers are minute, pinkish red in colour, 3 to 6 in clusters, umbel. Fruits are laterally compressed, small with 8 to 10 ridges.

These plants are found throughout India in marshy banks of rivers, streams, ponds and in irrigated fields. Commonly known Brichmi.

Uses : Leaves, stems, roots and seeds of these plants are medicinal. The leaves are used in brain tonic. The above ground portion of the plant contains asiaticoside and the decoction of the plant is given in the treatment of leprosy. It is also used to treat tuberculosis.

Centratherum Cass.
Compositae (= Asterateal)

Centratherum anthelmiticum (Willd.) Kuntze.
(= Vernonia anthelmintica Willd.) :

This is a tall erect herb. Leaves are upto 15 cm long, covered with minute hairs and margins toothed. Flower heads are upto 32 cm across, purplish, appear in small clusters of 40 to 45 minute flowers. Fruits are achenes, cylindric, hairy with 10 narrow ridges.

The plant is distributed throughout India mostly in waste places near habitations upto 1800 m altitude.

Uses : Seeds of these plants are medicinal. It is stimulant, antiseptic and anthelmintic. It destroys thread worms and promote urination. It is also used to treat skin diseases.

Cephaelis SW.
Rubiaceae

Under this genus there are about two hundred species of straggling herbs or shrubs.

Cephaelis ipecacuanha A. Rich. :

A small perennial herbaceous plant upto 80 cm tall. Stem is slender, cylindrical, trailing, with few or no branches. Leaves are in opposite pairs near the top of the plant, alternate below upto 15 cm long and 6 cm wide, pointed tips, margins entire, dark green above, pale green underneath, roughly hairy, short downy petioles. Flowers are white, funneliform, sessile, 2 cm wide, arises in dense clusters. The fruit is oval, purple when immature and black when ripe, with two stones containing a single white seed. Roots are slender, Spreading horizontally, mature roots moniliform, in bunches, root bark brick-red to dark-brown when mature. Roots have musty odour and a bitter taste.

This plant is native to Bolivia and Brazil. It is cultivated to a limited extent in Brazil, India, Burma and Malaya. In India it commercially cultivated in Darjeeling and parts of Assam and Sikkim and on an experimental scale in South India.

Ipecac roots contain 2.2 to 2.5% total alkaloids. The major alkaloids are emetine and cephaeline. Apart from these two alkaloids root contains psychotrine methyl ether, emetamine, ipecacuanic acid, ipeoside, erythrocephaleine, saponin, resin, fat, calcium oxalate and starch. Leaves and stems also contain emetine but these parts are not worth processing.

Uses : Roots of these plants which constitute the drug, can be harvested between 2 to 4 years. Highest alkaloid yield is obtained in 4 years old plants. Harvested roots and rhizomes are washed thoroughly, sun-dried, processed into pieces ranging from 4 to 12 cm in length and packed.

It is used as an expectorant, emetic, antiamoebic agent and vomitive. In minute doses it reduces the number and intensity of heartbeats. It is effective in the treatment of haemorrhages, amoebic dysentery, acute bronchitis, whooping cough. It also acts as a diaphoretic. In very small doses it stimulates apetite and improves digestion. There is no benefit of this drug in cases of bacillary dysentery.

Cephaelis acuminata Karst is known as Cartagena, Nicaragua or Panama ipecae. The plant is distributed in Central America and Colombia. The roots are greyish-brown or reddish-brown externally and thicker than C. ipecacuanha and higher in cephaeline content.

Cephalenthus Linn.
Rubiaceae (= Naucleaceae)

Cephalenthus occidentalis Linn. :

This is a hardy deciduous shrub may attain a maximum height of 8 m. Leaves are opposite, shiny upto 18 cm long. Flowers are white, appear in dense heads.

It is native to Mexico and India.

Uses : Bark and roots are medicinal. The bark is emetic, laxative and tonic. The tincture of bark is used to cure veneral disease, fever and coughs. Boiled roots mixed with honey is prescribed to treat pleurisy. For relieving toothache bark is chewed.

Ceratonia Linn.
Family : Leguminasae (= Fabaceae)

Ceratonia siliqua Linn. :

This is a tender evergreen tree, attain a height of 18 m. Flowers are yellow and red. The fleshy pods are edible.

It is native to Greece, Cyprus, Sicily, Italy, Spain, Middle East and California.

Uses : Seeds are the source of gum. The carob gum is a polysaccharide and contains D-galactose and D-manose.

The gum has demulcent, lubricant and absorbent properties, it also serves as a suspending agent.

Chamaelirium Willd.
Family : Liliaceae

Chamaelirium luteum Linn. :

This is a hardy herbaceous tuberous rooted .perennial, grows to a height of 130 cm.

The plant is dioeceous and native to North America.

Uses : Roots are used in medicine. Roots are used to treat intestinal worms, vitamin C deficiency, uterine disorders, pain in head, cough, poor appetite, depression and colic. It is also employed to increase urine flow, and to induce vomiting.

Chelianthes Sw.
Family : Sinopteridaceae

Chelianthes abomarginala Clarke :

This is a fern commonly occurring at Aravalli Hills. The compost mixture consisting of two parts loamy soil, one part each of leafmould and sand is ideal. They grow well in shady location. Plants are propagated by spores or by division.

Uses : The extract of the plant mixed with honey is taken after meal by the weak people suffering from tuberculosis. The young plants are used in making some tonic.

Cheiranthus Linn.
Family : Cruciferae

Cheiranthus cheiri Linn. :

This is a hardy perennial plant of 80 cm tall, flowers are of various colours.

This species is native to Europe.

Uses : The flowers of these plants are regarded as cephalic, cordial, anodyne and antispasmodic. These are used as a diuretic, emmenagogue, amenorrhoea and paralysis. The powdered seeds are administered in dysentery. The oil of the seed is applied locally for bruises, nervous a-nd rheumatic pains.

Chelidonium Linn.
Papaveraceae

Chelidonium majus Linn. :

This is a hardy perennial plant of 80 cm tall with yellow flowers.

The plant is native to temperate and subarctic Eurasia.

Uses : plant secretes milky juice from all parts when cut. It gives off a noxious odour when the plant is rubbed. This plant should be employed with great care and can be used as aperitive, diuretic, febrifuge and purgative.

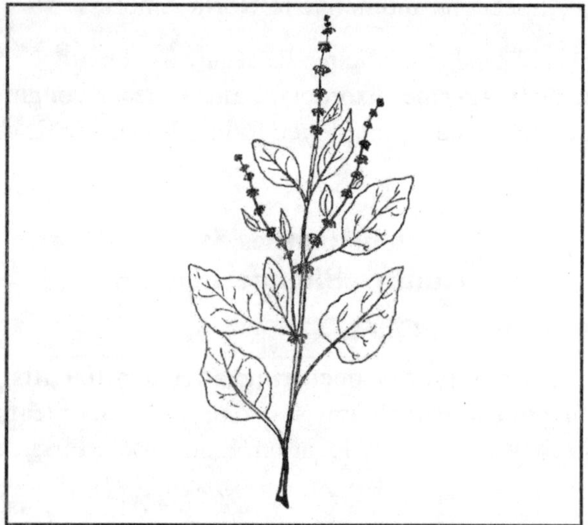

Fig. 47. Rumex obtusifolius

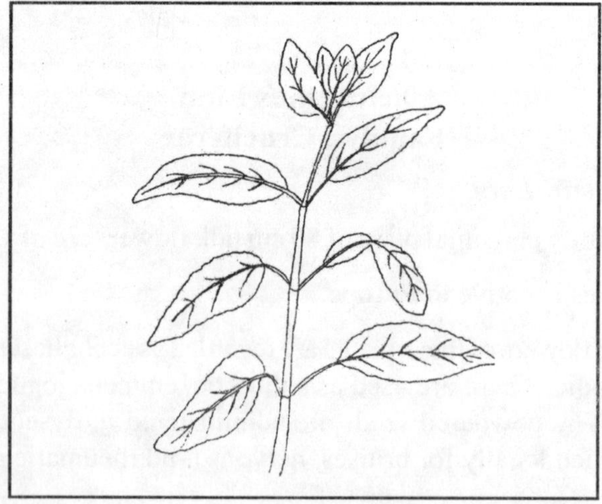

Fig. 48. Salvia officinalis

Chelone Linn.
Family : Scrophulariaceae

Chelone glabra Linn. :

This is a hardy herbaceous perennial plant upto l-fm tall. Leaves are narrow and opposite. Flowers are 3.5 cm long and white in colour.

This species is native to North America.

Uses : Entire plant is used for medicinal purposes. Powdered plant or leaves are used to treat worms and jaundice, this is also used as laxative and tonic.

Chenopodium Linn.
Family : Chenopodiaceae

Chenopodium album Linn. :

This is a hardy annual of 1.40 m height, with whitish, fuzzy surface. The lower leaf is whiter than the upper surface. Flowers are arising in spikes from the upper leaf axils and greenish in colour.

Uses : Leaves and entire plant are used. Leaves are rich in vitamin C and edible. Vegetable dishes are prepared from leaves. It is mildly laxative. A tea of the leaves and plant is used to relieve stomach pains.

Chenopodium ambrosioides Linn. :

This is a hardy annual of 1.5 m tall. Leaves are upto 14.5 cm long and alternate. Flowers appear from the axils of the leaves, greenish in colour and small sized. Leaves and young shoots are edible.

It is native to Tropical America and West Indies.

Uses : Plant tops constitute the drug. Distilled oil of this plant contains axaridol, p-cymene, 1-limonene, d-camphor and methyl salicylate.

Fruits of these plants yield oil and are used to treat intestinal worms. Plant is cooked and eaten to treat worms. A tea made from plant is used to relieve post delivery pains and also to hasten milk flow from nursing mothers. The plant is lightly narcotic and is used to treat nervous conditions, to induce menstrual flow and to reduce profuse and painful menstrua] flow. It it used as vermifuge.

Chimaphila Pursh.
Family : Pyrolaceae (= Ericaceae)

Chimaphila maculata Pursh. :

This is a hardy dwarf herbaceous perennial upto 20 cm tall. Leaves are wedge-shaped, dark shiny green in colour. Flowers are flesh coloured or pink and white to 1.45 cm long.

Uses : Leaves and fruits have medicinal properties. These parts are used as tonic, to increase urine flow, and also for ulcers, nervous disorders, skin eruptions, rheumatism and diarrhoea.

Chionanthus Linn.
Family : Oleaceae

Chionanthus virginicus Linn. :

This is a hardy deciduous flowering tree of 14 m in height. Flowers are produced on long stems in groups, white in colour. The berries are fleshy, round, purple in colour.

Uses : Flowers and barks of root and trunk of these plants have medicinal value. A poultice of crushed bark is used to cuts and bruisees. A decoction of boiled bark is used to treat skin inflammations and malaria.

To treat ulcers in the mouth and throat, diarrhoea, inflammations and sores, the leaves and flowers of this plant is used.

Chlorophytum Ker-Gawl.
Family : Liliaceae

A genus of two hundred five species and fifteen are native to India.

Chlorophytum arundinaceum Baker :

It is a rhizomatous herb. Leaves are suberect, lanceolate and many nerved. Inflorescence is an erect, dense-flowered racemes. Flowers are star like, white upto 4 cm across, sepals are acute, anthers are longer than filaments and green or yellow in colour, bracts are long. Seeds are black in colour with angular edges.

It occurs in tropical and subtropical Africa and India. In India, it is widely distributed in Tarai regions of Bihar and West Bengal and also in some parts of Orissa and M.P.

Organic rich well drained sandy loamy soil and warm humid climate is suitable for its cultivation. Plants are propagated by seeds and by division of rhizomes. Seeds remain dormant for nearly ten months. Germination of seeds take two weeks time and about 20% germination is reported. Vegetative propagation is easy and the plants obtained through fleshy root with attached stem disc are more vigorous. Plants are planted in June-July at a distance of 30 cm x 15 cm and for one hectare of land around 300 kg of fleshy roots are required. While preparing land for planting 10 tonnes of FYM/h is to be incorporated with the soil. It is a three months crop if cultivated, and crop mature in September-October when the leaves have dried. Fleshy roots are harvested through careful digging. The estimated yield is 1 tonne of fresh roots /ha and 200 kg of dry root/hectare. The dry roots are hard and bony in appearance and white in colour.

The dry roots possess less than 5% moisture. It contains carbohydrates, proteins, root fibres, saponins and minerals.

Uses : Dry roots of these plants constitute the drug and is used in powdered form. It is a well known tonic and aphrodisiac, and used to treat general debility. Leaves of this plant are also eaten as vegetable. The tubers of about 20 g is boiled with milk and is taken twice a day for a month, in cases of impotency and weakness.

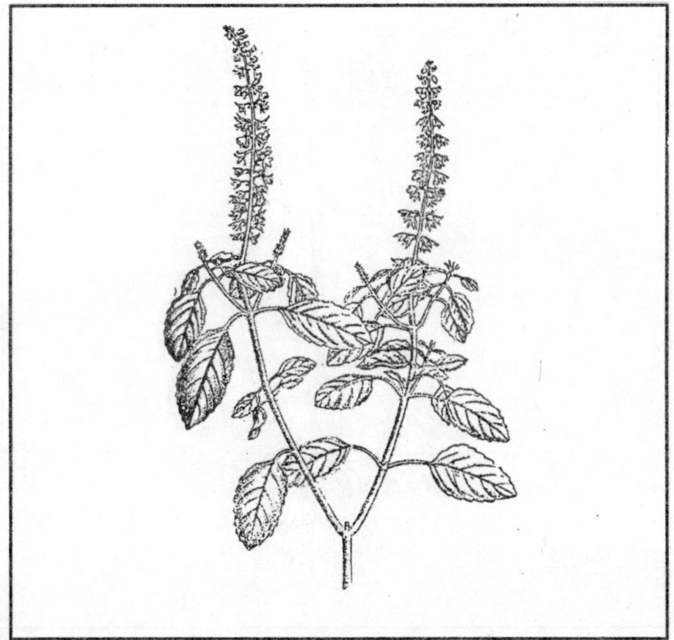

Fig. 49. Tulsi, Sacred Basil (Ocimum sanctum Linn. Lamiaceae (= Labiatae)

Choerophyllum Brongn.
Family : Umbelliferae

Choerophyllum sativum :

The fruit of this plant is regarded as a stimulant and carminative. It is aromatic, stimulates appetite and facilitates digestion,

It is used externally for swellings in the breasts and piles, and boiled in water can be applied as a poultice.

Chondrodendron Ruiz.
Family : Menispermaceae

Chondrodendron tomentosum Ruiz. :

This is a climbing woody vine. Stems are warty, bark velvety grey. Leaves are heart shaped or nearly round, upper surface slightly hairy, and lower surface woolly white. Male and female flowers borne on separate plant. Rhizomes are several feet in length and around 13 cm in thickness, dark brown in colour. Fruits are upto 4 cm long, oblong oval in shape, scarlet when unripe, ripe fruits are black in colour, fleshy and with acid flavor.

It is widely distributed in Brazil, Peru, Colombia and Panama upto 600 m altitude.

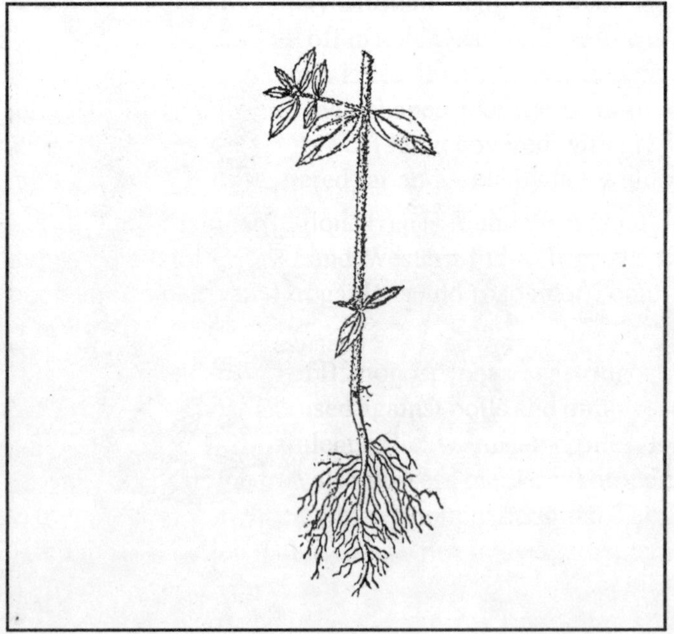

Fig. 50. Dudhi (Euphorbia hirta Linn. Euphorbiaceae).

Uses : The bark of the stem has febrifuge properties. Dried roots are used to treat rheumatism, jaundice, leucorrhoea, gonorhoea, Kidney and bladder inflammation. Tubocurarine chloride is employed to induce paralysis of skeletal muscles for the purpose of reducing convulsions, mental patients during shock therapy. It is a muscle relaxant and also acts as relaxant in neurological cases. Overdose of the drug may cause respiratory paralysis.

Chondrus Stackh.
Family : Gigartinaceae

Chondrus crispus Stackh. :

This is mostly occurs in the coasts of North Atlantic Ocean from Maine to the Carolinas. This Irish moss is about four-fifth water and one-tenth gelatinous matter with smaller percentage of nitrogen, calcium, sodium, potassium, magnesium, chlorine, bromine, iodine and sulphur. It contains sulphated polysaccharides like galactose and anhydrogalactose.

Carrageen is used in making blanc mange and other puddings, as well as in jellies. It is used as a source of agar, yields two principle colloidal substances, one more soluble in cold water than the other and giving a viscous solution, another forming a gel more readily. It has emollient and demulcent properties. This is an emulsifying and suspending agent.

Uses : It is often used in mineral oil and anticough preparations. Chondrus is employed as a base of various skin foods like face paints, pastes, creams and pomades. It is also added to soaps, and as well as used as a binding material in tooth paste.

Chrysanthemum Linn.
Family : Compositae (= Asteraceae)

Chrysanthemum parthenium (Linn.) Bernh. :

This is a herbaceous perennial, many branched, upto 95 cm tall. Flowers appear in clusters, white or yellow in colour.

It is native to Europe.

Uses : Leaves and flowers are useful. A tea made from leaves is used to cure bad colds, indigestion and diarrhoea. Dried flowers are used to induce menstrual flow and bring about abortion, to treat intestinal worms and indigestion. Leaves are chewed to relieve colic. A popular remedy against fever.

Cimicifuga Wernischek
Family : Ranunculaceae

Cimicifuga americana Michx. :

This is a hardy herbaceous perennial plant of 95 cm tall. Flowers are borne on short lateral spikes. Flowers are .small sized and white in colour. The species is native to North America.

It grows well in temperate region. Ordinary rich soil and moist shady location are ideal. Plants are propagated by seeds and division of roots.

Uses : Roots are used to treat rheumatism, lung inflammation, hysteria and skin rash. To repel vermin and to treat rattle snake bites, dried powdered roots are used. The other species C. foetida L. is used as preventive against vermin.

Cimicifuga racemosa Linn. :

This is a hardy herbaceous perennial, 2.5 m in height. Leaves are with two to five leaflets. Small white flowers appear on a slender spike. The rhizome is gnarled and twisted.

This species is native to North America.

Uses : Rhizomes and roots are used for medicinal purposes. A decoction of root is used to treat rheumatism, kidney trouble, general malaise, women's ailments, promote perspiration and for malaria. It is also used to cure ronchitis, nervous diseases, uterine disorders, snakebite, itching and yellow ever. A tea made from the root is used to treat sore throat. It is also known is emetic.

Cinchona Linn.
Family : Rubiaceae

Cinchona is an important medicinal plant which is commercially cultivated in India for more than 125 years.

This plant is native to Bolivia, Columbia, Ecuador and Peru in South Africa. About 40 species are there in the genus Cinchona, while four species are cultivated in India namely, C. calisaya Wedd, C. ledgeriana Moens, C. officinalis L. and C. succirubra Pav. ex Klotzsch. Hybrids among the various varieties are also grown as plantation crop. This crop is extensively cultivated in the Nilgiris and Annamalai of Tamil Nadu and Darjeeling Hills of West Bengal. In other parts of the world where the commercial cultivation is undertaken are Indonesia and Zaire. The bark of Cinchona tree contains 6-7% alkaloids. Among the large number of alkaloids found in Cinchona, the most important are quinine and quinidine. Cinchona species grown in India are usually large or medium-sized tree. C. celisaya is grown in Nilgiris and Sikkim which has yellow bark. In West Bengal, Assam and South India the commonest species is C. ledgeriana which is a straggling tree and bark colour in this tree is yellow. C. officinalis L.var. "Condaminea", C. officinalis L. var. "Bemplandiana", trade names are Lox Bark and Crown Bark, respectively, are grown in Nilgiris. C. succirubra is a red bark species grown in Satpura hills, Sikkim and Nilgiris. The other worth mentioning species are C. peruviana HOW, C. robusta and C. hybrida. Quinine content of these species varies from 0.5 to 1.5% and in C. hybrida it is 2 to 2.5%.

Light well drained, organic rich acidic soil of virgin forest is suitable for this crop. It is badly affected by subsoil waterlogging. In water logged soil, diseases like root rot, cankar and dieback occur. The crop grow well in a cool climate with an evenly distributed rainfall of about 4000 mm in a year at an altitude of 1200 to 2500 m.

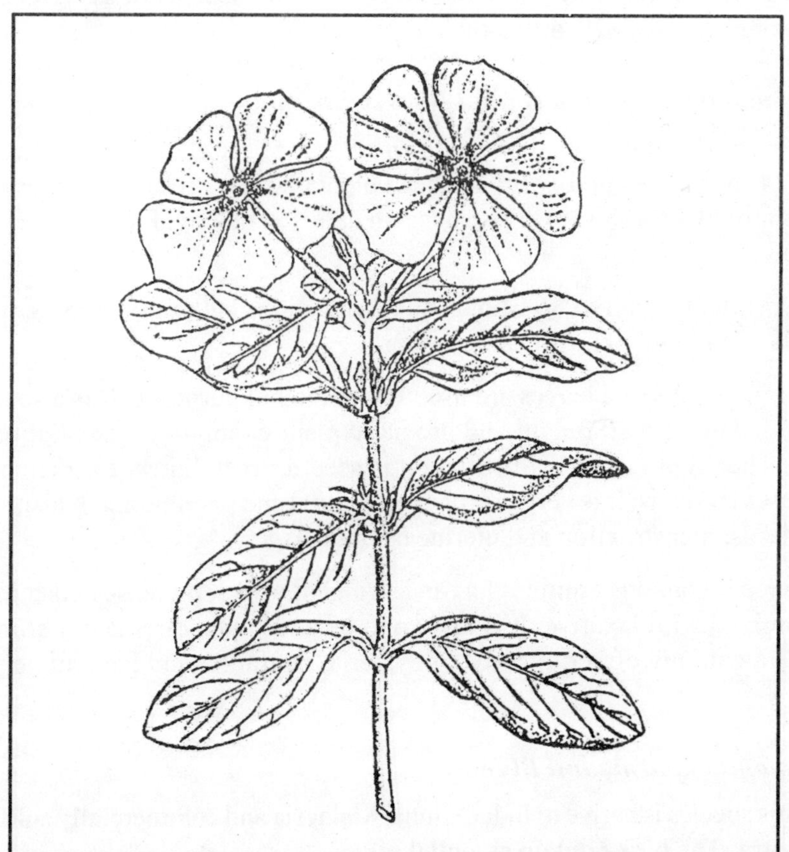

Fig. 51. Sadabahar (Catharanthus roseus G. Don. Apocynaceae).

Uses : 'The dried bark of Cinchona constitutes the drug. Its use as raw material for the manufacture of quinine was well known at one time. It is an antiseptic, acting mainly on the Infusoria. It is easily absorbed by wounds, the mucous membrane acts like an irritant. The use of quinine as antimalarial drug has gone down and the production of this alkaloid also decreased considerably. Quinine also destroys certain bacterial infections, and in certain preparations have been found useful in pneumonia, amoebic dysentery and for eye lotions. Some preparation of quinine are useful as local applications on certain rheumatic pains and as gargles. Quinine is also used as bitter principle in soft drinks in some foreign countries. Other alkaloid, quinidine is employed as antiarrhythmic agent, and used as treatment of heart disease.

Cinnamomum Schaeffer.
Family : Lauraceae

A genus of about two hundred thirty and fifty species of evergreen trees and shrubs, usually with aromatic bark.

Cinnamomum camphora (Linn.) Nees & Eberm :

This is a medium sized tree with an enlarged base and reaches a height of about 14 m. Leaves and twigs of these plants have a strong camphor smell. Leaves are smooth, shiny, whitish beneath upto 15 cm long. Flowers are yellow in colour.

This plant is native -to China and Japan; and is cultivated in Nilgiri Hills of India. Commonly known as Comphor.

Uses : Twigs and leaves are used for medicinal purposes. It is a source of camphor. This is a stimulant and aromatic plant containing essential oil and tannin. The oil obtained by distillation is used to treat diarrhoea, rheumatism and muscular pains. It is very useful in bronchitis and pneumonia. It also stimulates uterus, menstruation and uterine haemorrhages.

Since it contains tannin it has an astringent effect, increases constipation unless mixed with laxative. It is known to be antiseptic, circulatory stimulant and has a calming effect in cases of hysteria, neuralgia and general nervousness.

Cinnamomum burmannii Blume :

This species is native to India, China, Malaysia and commercially cultivated in Sumatra. The bark contain essential oil.

It is used as a carminative and flavour. Commonly known malay cinnamon.

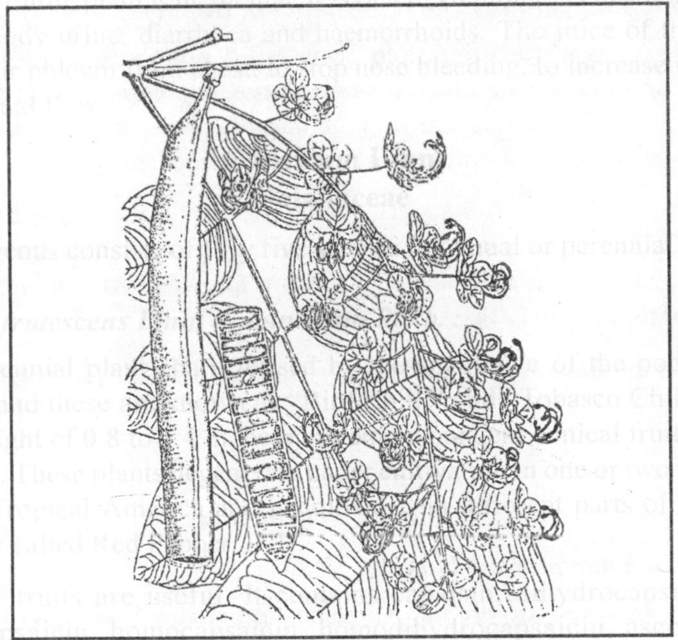

Fig. 52. Amaltas, Indian laburnum (Cassia fistula Linn. Caesalpiniaceae)

Cinnamomum burmannii Blume :

This species is native to India, China, Malaysia and commercially cultivated in Sumatra. The bark contain essential oil.

It is used as a carminative and flavour. Commonly known Malay cinnamon.

Cinnamomum loureiri Nees. :

It is commonly known as Saigon Cinnamon. This species is native to South-East Asia. The bark contain essential oil 0.8 to 1.4%, it constitute cinnamic aldehyde, phenols, pinene, phellandrine, caryophylene.

It is used as carminative and flavour. To relieve nausea, flatulence and diarrhoea it is employed.

Cinnamomum tamala Nees. & Eberm. :

This is a small tree known as Tejpata or Indian Cassia. Leaves are three nerved. Flowers are vellow white and fruit black when ripe.

It occurs in the Himalayan region of W. Bengal, Assam and Garhwal Leaves of this plant is useful. It is regarded as carminative and spice.

It is used for treatment of diarrhoea and colic pain.

Cinnamomum verum Presl. (= C. zeylanicum Blume) :

This is an evergreen tree upto 10 m tall. Commonly known as Ceylon Cinnamomum. Leaves are thick, leathery, large, ovate, pointed at tip, shining green in colour. Flowers appear in large hairy clusters, minute in size. Fruits are oblong or ovate, dark purple in colour, one seeded.

The tree is native to India and Ceylon. The bark, twigs and leaves contain essential oil. The oil contains 90% cinnamaldehyde and small quantity of cinnamic acid, cinnamyl acetate and eugenol. It is known as stomachic, germicide and carminative.

Used in thc trcatment of diarrhoea, gastric debility, flatulence, nausea and vomiting.

Cirrhopetalum Linn.
Family : Orchidaceae

Cirrhopetalum maculosum Linn. :

This is an epiphytic orchid. Rhizomes are covered with fibrous sheaths and many roots. Pseudobulbs are ovoid and shrivelled. Leaves are upto 15 cm long, linear, notched. Scape stout, to 8 cm long, umbels 2-8 flowered. Flowers are pale yellow-green.

It is native to Garhwal and Kumaon between to 1200 to 1500 m altitude.

They require high relative humidity and temperature and abundant light. Grows in baskets, shallow pans or pots. Potting media consisting of clean broken crocks, charcoal and sphagnum moss is ideal.

Uses : Roots are medicinal and is said to increase longevity.

Cirsium Mill.
Family : Compositae (= Asteraceae)

Cirsium flodmani (Rydb.) Arthur :

This is a biennial plant of 95 cm tall and bear purple flower.

The species r is native to North Temperate region.

Commonly known as Flodman's thestle.

Uses : Entire plant is medicinal. A decoction of flower is used to treat gonorrhoea. To relieve earache the juice of the roots is used. To treat dysentery, the plant is boiled in milk and the liquid is drunk. To restore falling hair, the juice of the leaves and stems are used as a scalp tonic.

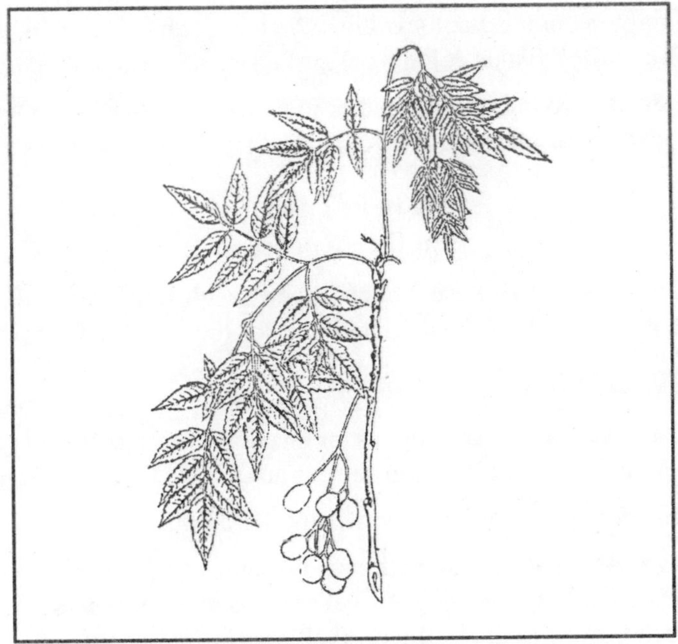

Fig. 53. Drek, Bakain (Melia azadirach Linn. Meliaceae)

Cissampelos Linn.
Family : Manispermaceae

Cissampelos pareira Linn. :

This is a shrubby climber and pubescent. Leaves are arbicular, reniform or cardate. Flowers appear in long axillary racemes. Flowers of these plants •are greenish and drupes scarlet.

It occurs at Karupan of Chhindwara district of M.P. Commonly found in the shady areas of the forests of Garhwal between 800-2800 m altitude.

Uses : The entire plant has medicinal value. The root decoction is used in diarrhoea and in case of urinary trouble. The plant possess anthelmintic, antihistaminic, antipyretic properties. It is also used as cardio tonic, and stomachic.

Citrullus Schrad. ex. Eckl. & Zeyh.
Family : Cucurbitaceae

Citrullus colocynthis (Linn.) Kuntze :

This is a tender climbing plant. Fruits are variegated dark green and white.

The species is native to India.

The fruit pulp contains colocynthin which is a combination of an unnamed alkaloid and citrullol. It also contains elaterin, -phytosterol and hentriacontane.

Uses : Fruit pulp is used in dried or powdered form. It is powerful hydragogue cathertic. In large doses, it is toxic. The fruit pulp is used as a purgative.

Citrus Linn.
Family : Rutaceae

Under this genus there are about fifteen species of trees and shrubs and eleven are native to India. Mostly cultivated for their edible fruits.

Citrus aurantium Linn. var. amera Linn. :

This is an evergreen shrub upto 8 m tall with fragrant flowers. Fruits are loose skinned, orange in colour and sour in taste, with hollow core when ripe.

The species is native to Italy, Spain and Paraguay.

It grows well in sub-tropical climate. It requires organic rich well drained soil and sunny situation. Watering at weekly interval is necessary and the plants should not be allowed to dry at the roots. Plants are propagated by seeds, cutting, budding and grafting.

Uses : Peel and flowers of this fruit yield essential oil. Flower oil contains pinene, camphene, dipentene, linalool, geraniol, nerol, methyl anthranilate, indole etc. Peel oil constitute methoxyflavones. The essential oils are flavouring agents. The variety "bergamia", the "Begaamot Orange" from which perfume is obtained.

Citrus limon Linn. :

This is a shrub of 6 m tall, with fragrant flowers. Fruits are thin skinned, pale yellow, oblong with terminal nipple, very sour in taste. It is native to India, Australia, New Zealand, Mexico, America, Portugal, Italy, Sicily, France and Spain. Commonly known as Lemon.

Uses : Essential oil is extracted from peel of this fruit. The oil contains d-limonene, terpinene, linalool, terpineol and cadinene. This oil is carminative and also employed as flavour.

Claviceps Tub
Family : Clavicipitaceae

Claviceps purpurea (Fr.) Tul :

This is a parasitic fungus exists on the heads of rve (Secale cereale L.) and belongs to the class ascomycetes called ergot.. The ascospores of the fungus infect the rye flowers through stigma. The fungus is developed from spores into mycelia, excreting honey dew filled with microscopic bodies. This is carried by insects, wind and rain from one host plant to another where they reproduce. Ergot is also found in wheat, barley and some other grasses. Heavy infestation occurs in humid cool weather. The mycelia of the fungus when comes in contact, grow in the ovaries of the flowers of the host and form fruit like bodies called sclerotia. This sclerotia is a long horn like hard structure, dark purple or nearly black in colour, which look like oversized seeds. The sclerotia takes the place of seed and prevents the development of seed. It has a fishy odour. The sclerotia fall on the ground at the end of the season. These then overwinter and germinate during next spring, emerge from the soil tiny mushroom like forms and infect the next grain crop.

Uses : There are more than fifty alkaloids have been reported from ergot. The primary constituents of ergot sclerotia are a group of indole alkaloids with a tetracyclic ring system called ergoline. Among the alkaloids ergonovine, ergotanine, ergocryptine, ergocornine, ergocristine, ergosine and ergovalide are important. Apart from the above, several amino acids and also ergosterol, fungisterol, clavicepsin, sclererythin, ergochrysin, ergoflavin, inorganic salts and complex proteins have been isolated from ergot sclerotia.

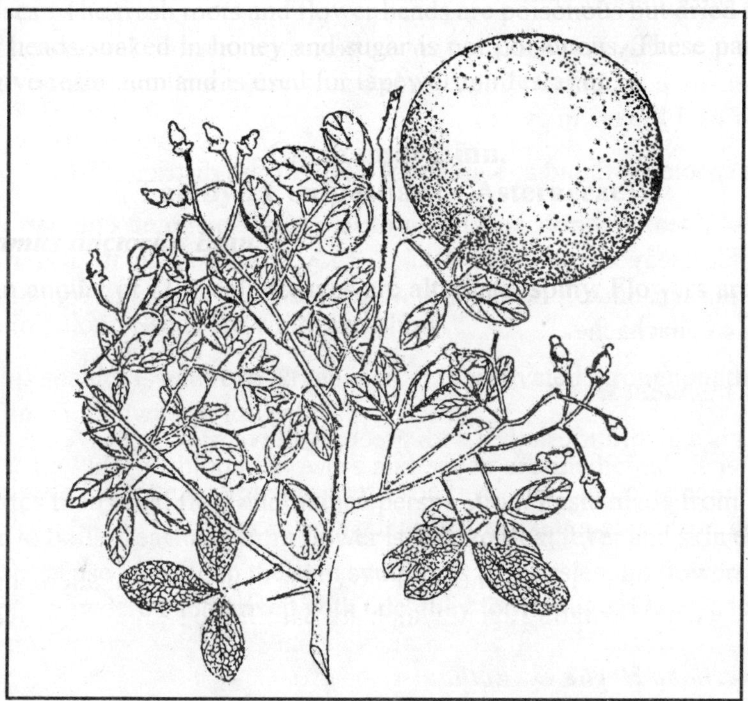

Fig. 54. Bel (Aegle marmelos Correa Rutaceae)

Physiologically active alkalods stimulate the smooth muscles of uterus, including rhythmic contractions. Ergonovine is administered to stop haemorrhage after child birth and also used against migraine. Ergonovine, when given orally or intravenously is found most effective. Ergotamine is employed with caffeine in tablet form or in rectal suppositories, in some migarine cases. Ergotamine is reported to constrict cranial blood vessels, modifying their pulsation. It has been suggested that ergotasmine must be given intravenously and it is slow in action. Ergotoxine group of alkaloids are used for controlling hypertension and peripheral vascular disorders. Dihydroergotoxine has favourable effects in some cases of senility. Bromoergocrystine, a derivative of ergocryptine is used against breast cancer and galactorrhoea. Preparations from ergot alkaloids may not be given in early stages of labour or to patients with sepsis or impaired kidney or liver functions.

Clematis (Dill.) Linn.
Family : Ranunculaceae

A genus of two hundred twenty five species of climbing undershrub. About twenty five species are available in India.

Clematis gauriana Roxb. :

This is a hardy herbaceous climber, perennial in habit. Lower sides of the petiole sensitive to contact. The petiole bends once round the support thickens and lignifies. Flowers in cymes.

The species is found at Kaream of Chhindwara district of M.P.

Uses : Ideal compost consisting of two parts loam and one part each of leaf mould, decayed manure and sand. They require adequate watering and liquid manuring annually. Pruning of weak shoots is needed. The root decoction is used in stomachache.

Clematis nepaulensis DC. :

This is a glabrous climber with greenish flowers.

Commonly found in moist shady areas of temperate zones between 1800 m to 2500 m altitude in Garhwal Himalaya.

Uses : The plant is acrid and poisonous due to presence of anemonin. The leaves are used as irritating agent which are harmful to skin.

Clematis triloba Heyne ex. Roth. :

It is occurring at Karrapani of Chhindwara district of M.P.

Uses : Raste is applied on painful boils. The leaf decoction is used in the treatment of asthma. The leaf paste is applied in leprosy and also in snake bite.

Cleome (Linn.) DC.
Capparidaceae (= Cleomaceae)

Cleome serrulata Pursh. :

This is an annual of 1.5 m tall. The flowers are purplish rose in colour. The compost consisting of equal parts of loam, leaf mould and sand, and sunny situation are ideal, propagated through seeds.

Uses : Mature leaves and flowers are used in medicine. To reduce swelling poultice of crushed leaves is applied on swelling. The flowers are used to treat anemia.

Cnicus Linn. emend. Gaertn.
Family : Compositae (= Asteraceae)

Cnicus benedictus Linn. :

This is a thistle-like annual of 70 cm tall and has spiny toothed and lobed leaves. Flowers are yellow in colour.

The species is native to Britain, Mediterranean Region and Caucasus.

They grow in ordinary soil and sunny situation. Plants are propagated by seeds.

Uses : Mature seeds, flowering tops and leaves of these plants are medicinal. An infusion of flowering tops are used as a contraceptive measure, to treat internal cancer, to expel worms, to increase sweating and urine flow; and also used to treat inflammation of respiratory system, fever, liver ailments and hysteria.

Coccoloba P. Br. mut. Linn.
Family : Polygonaceae

Coccoloba uvifera (Linn.) Jacq. :

This is a dwarf tree. Leaves are leathery, glossy and round upto 25 cm in diameter. Flowers appear in dense cluster, white in colour. Fruits are clustered like grapes, thin-fleshed and dark purple.

It is native to tropical and subtropical America.

Uses : Bark, root and fruits are useful. The bark extract is used for treatment of dysentery. Roots are used to treat diarrhoea and the fruits for fever.

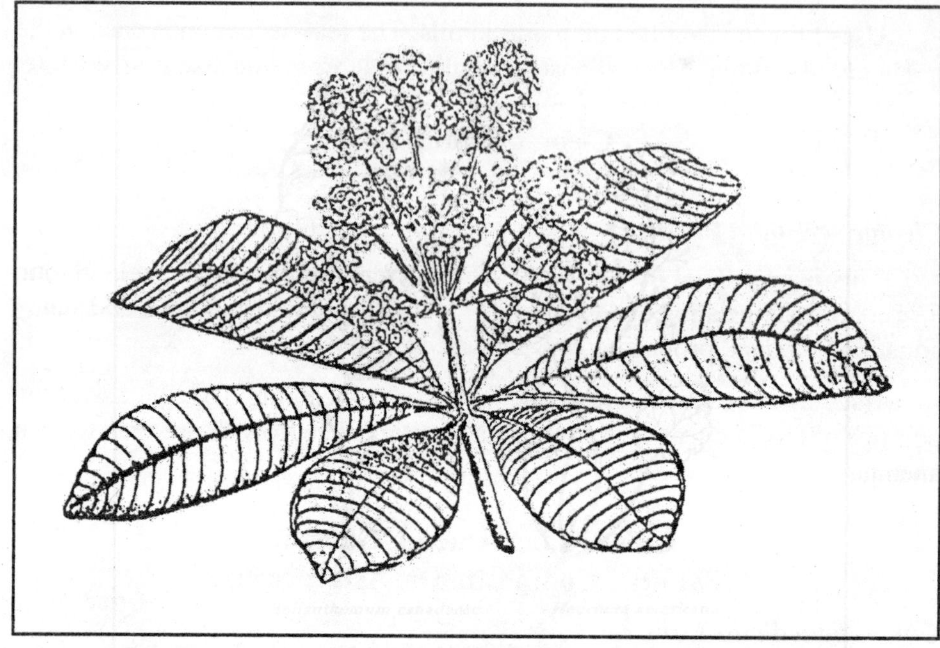

Fig. 55. Chhatian (Alstonia scholaris (Linn.) R.Br. Apocynaceae.

Cochlearia Linn.
Family : Cruciferae

Cochlearia officinalis Linn. :

It's leaves are fleshy

The species is distributed in Eastern Himalaya.

Uses : The plant is stimulant and detergent. It has pungent odour and is slightly bitter in taste. It is one of the best anti-scorbuticus.

Cochlospermum Kunth.
Family : Cochlospermaceae (= Bixaceae)

Under this genus there are about fifteen species and only one species is native to India. Which are generally shrubs or trees. Flowers are large and yellow in colour.

Cochlospermum gossypium DC. :

It is native to India.

Uses : The gum obtained from this plant is a mild demulcent used in coughs. The dried leaves and flowers of this species are used as stimulant.

Cochlospermum religiosum (Linn.) Alston :

It occurs in dry forests of India which yields an edible gum, fibre, covering the seeds and these are used as a stuffing material.

The bark infusion of about 25 ml is taken in jaundice twice a day for fifteen days.

Codonopsis Wall.
Family : Campanulaceae

A genus of thirty two species and about twenty are native to India, which are perennial, erect or climbing herbs.

Codonopsis ovata Benth. :

This is an ascending to twining herb with thick white, milky tuberous roots, often large, long and thick enough as compared to the weak stems. Plants attain a height of 40 cm. Flowers are blue

Native to Himalaya. Sporadically occurring between 2800 m - 4800 m. of Himachal Pradesh.

Uscs : The milky roots are sweet in taste, eaten raw and are considered as a good physical and sexual tonic.

Codonopsis rotundifolia Benth. ex Royle :

This is a hardy perennial plant of 50 cm tall. The root is unbranched, white cylindric, usually beset with circular wrinkles. Flowers are cup-shaped, greenish yellow with purplish streak, appearing during July-September.

The species is native to Western Himalaya. Sporadically occurring along with alpine ferns between 4000-4500 m of Himachal Pradesh.

Ordinary good soil and sunny situation are ideal. Propagated by seeds and cuttings.

Uses : The roots are considered as aphrodisiac and stimulant.

Coelogyne Linn.
Family : Orchidaceae

Coelogyne cristata Linn. :

It is an epiphytic orchid.

Commonly occurring at Thunera of Utterkashi, between 1500-3000 m and also in the hills of North-Eastern India.

Uses : The compost consisting of three parts of cut osmunda fibre and one part sphagnum moss are ideal. Plants are grown in wooden baskets or perforated shallow earthen pots. This species is propagated by division of plants. The gum of the bulb is applied on the sores.

Coffea Linn.
Family : Rubiaceae

***Coffea arabica* Linn. :**

A small evergreen tree upto 6 m height. Leaves are glossy, oblong-elliptic, opposite, pointed at the tip, upto 18 cm long. Flowers are showy, fragrant, abandant, white in colour, clustered in the leaf exils. Fruits are red when ripe, mucilaginous, round, 1.8 to 3, cm across. Seeds are oval, grey-green, enclosed in silvery testa.

It is native to Ethiopia and commercially cultivated in Tropical American countries, Brazil and India.

Uses : Caffeine is the main active principle in commercial coffee seeds which ranges from 1 to 3%. The green fruit also contain some alkaloids like theobromine, theophylline and trigonelline; and also nonvolatile acid like chlorogenic acid.

Caffeine is known as cerebral and respiratory stimulant. It promotes formation of pepsin and hydrochloric acid in gastric juice. It is a stimulant and facilitates digestion after a heavy meal. It is a diuretic and reduces cutaneous perspiration. It is recommended in a dynamic typhoid fever, intermittent fevers, periodic nervous asthma, whooping cough and albuminuria. Coffee is a most popular beverage. It is also used in flavouring of baked good, ice-cream, confectionary and carbonated beverages. A tea is also made from coffee leaves, and flower nectar yield honey. The wood of this tree is used for making furniture.

Colchicum Linn.
Family : Liliaceae

In this genus there are about seventy five species and three are native to India. The plants under this genus are corm bearing herbs with short scape.

***Colchicum autumnale* Linn. :**

This is perennial herb up to 30 cm tall. Flowers are lavender to light pink. In autumn the flower projects out of the soil. The tube is long and the ovary remains below ground. In the spring leaves appear and the capsule is brought above ground by lengthening its stalk. The corm is conical or ovate, fleshy with numerous fibrous roots.

There are numerous varieties in it. It is distributed in the Netherlands, Denmark, England, Ireland, Poland, Spain, Italy and North Africa. It is commercially grown in Italy and Yugoslavia.

Uses : The corm constitute several alkaloids like colchicine, colchiceine, colchicerine, colchamine etc. The seeds also possess colchicine and other alkaloids.

The seeds and corms are used in medicine. The extracted colchicine is employed orally in tablet form for arthritis, familial Mediterranean fever. The corm and seeds are used to treat enlarged prostrate, gonorrhoea, dropsy, gout, rheumatism and arthritis. All parts of the plants are poisonous. Colchicine is used in plant breeding for chromosome doubling. This alkaloid is employed in animal studies to explore wound-healing and embryonic growth. The effect of colchicine on cancerous tissues has also been tested, but it is found toxic.

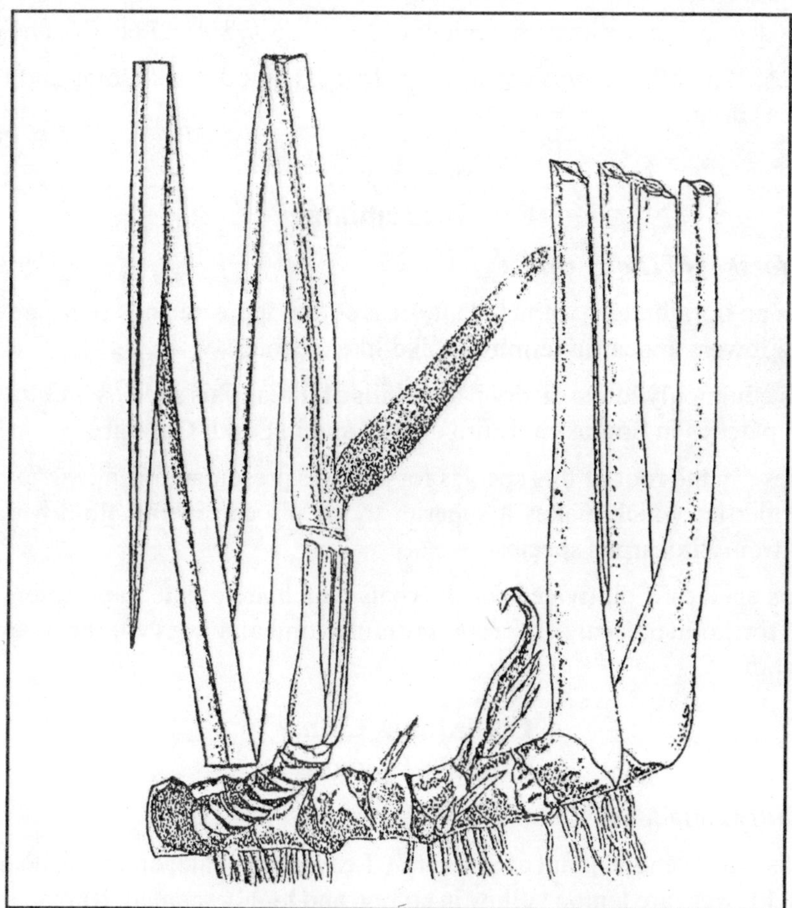

Fig. 56. Bach (Acorus calamus Linn. araceae).

Colchicum luteum Baker :

This is an annual herb. Leaves are narrow upto 40 cm long and 1.8 broad. Flowers are upto 6 cm across, yellow in colour. Fruits are 4 cm long. Corms are conical, one side flat.

The species is distributed in Afganistan, Turkistan and in Western Himalayas. In India, these plants are found in the margins of the forests- of North- Western Himalayas from Kashmir to Chambel between 800 to 3000 m altitude.

Uses : Seeds and corms contain alkaloids including cochicine. The dried corms are carminative, laxative, aphrodisiac, alternative and aperient.

Colchicum speciosum Ster. :

This species flower are rose-purple in colour, yellow spotted in the throat.

It is distributed in mountain regions of Asia Minor, Syria, Lebanon and Iran.

Uses : The dried corms constitute colchine and other alkaloids and these are sold as drug.

Coleus Lour.
Family : Labiatae

Coleus forskohlii (Poir) Briq. :

It is an upright herbaceous plant. Leaves are opposite and stems are four angled. Flowers appear in terminal spike-like racemes.

It is commonly found on dry barren hills of India. This species is cultivated in small patches in Sourastra, Kaira district and in South Gujarat.

Uses : In the root of this species forskolin is present and is now known to have properties which makes it superior to the pilocarpine alkaloid which is derived from Pilocarpus species.

This species is cultivated for the roots which are pickled and eaten. The alkaloid forskolin present in the root is useful in opthalmology for the treatment of glaucoma.

Collinsonia Linn.
Family : Labiatae

Collinsonia canadensis Linn. :

This is a perennial plant of 95 cm tall. Leaves are ovate, opposite, coarsely toothed. Flowers are lemon yellow in colour, and highly scented. Rhizomes are thick and woody.

Uses : Entire plant is useful. A tincture made from root mixed with cider is used to treat water accumulation in body. A tea made from the plant is used to cure indigestion, headaches and cramps.

Colutea Linn.
Family : Leguminasae (= Fabaceae)

***Colutea arborescens Linn.* :**

This is a hardy deciduous flowering shrub of 3.5 m tall with yellow flowers. Pods are inflated bladder like and burst on being squeezed.

The species is native to South Europe and Himalaya.

Uses : Its leaves have similar properties of senna (Cassia sp.) and are used to adulterate the latter. Leaves of these plants are purgative only if used in very strong doses. They lose their property by decoction.

Commiphora Jacq.
Family : Burseraceae

***Commiphora mukul (Hook. ex Stocks) Engl.* :**

A shrub with corked and knotty branches of 5 m height. The branches end in sharp spines. Leaves are 1-3 foliolate, alternate or fascicled, leaflets are glabrous. Plants are dimorphic. Fruits are drupe, 2 cm long, red when ripe.

This genus is widely distributed in Africa, Asia, Madagascar, Austria and Pacific Islands. In India, it occurs in the tracts of Western India. Plants are propagated through seeds or by rooted cuttings.

Uses : Oleo-gum-resin which is obtained through incisia is a mixture of resin, gum, volatile oil etc.

The oleo-gum resin of this plant is used in the treatment of arthritis and obesity.

***Commiphora absyssinica Engl. Myrrh.* :**

It is native to Southern Arabia and North-Eastern Africa. Gum resin obtained from the tree has medicinal properties.

It constitutes volatile oil containing terpenes, sesquiterpenes, esters, cuminitic aldehyde and eugenol.

Uses : It has stimulant, antiseptic and astringent properties. This is an ingredient in mouth wash.

Commiphora molmol Engl. :

This plant is also native to Africa and Arabia

The plant yield resin and gum. It has same medicinal properties as that of C. absyssinica.

Comptonia L. Herit ex Ait.
Family : Myricaceae

Comptonia peregrina (Linn.) Coult. :

A hardy deciduous sweet smelling shrub of 1.30 m tall. It bears elegant fan like foliage. The flower is a fuzzy catkin, followed by spiny brown nutlets.

This species is native to Eastern North America.

Loamy soil with plenty of organic manure and sunny situations are preferable. Plants are propagated by layering.

Uses : Leaves and new growth of these plants are used as medicine. Leaves are used to relieve toothache. A decoction of plant is used for the treatment of weakness following fever, diarrhoea, colic pain and rheumatism.

Convolvulus Linn.
Family : Convolvulaceae

Convolvulus scammonia Linn. :

This is a climbing or trailing type of plant. There are several hardy annual and perennial species under this genus. Flowers are funnel shaped, 7-15 cm across and found in various shades of colours. This species is native to Asia.

The bruised leaves of the large variety is used in infusion, in water, as a purgative. When crushed between the fingers and applied to boils, the leaves make a good maturative. From the incisions in the rhizomes a resinous purgative juice flows. Some of the species yield rosewood oil.

Coptis Salisb
Ranunculaceae

There are about fifteen species under this genus and one is native to India. These are small stemless herbs with small white flowers on slender leafless scapes.

Coptis groenlandica (Oeder) Fernald "Goldthread" :

It is a hardy evergreen perennial bog plant upto 15 cm tall. Leaves are shiny. Flowers are solitary and white. The species is native to North Temperate and Arctic region.

Moist soil rich in organic matter and shady position are ideal for these plants. Propagated from seeds and division of roots.

Uses : Entire plant when in bloom and roots are medicinal. Roots are chewed in mouth sores. A tea made from plant is used as a wash of mouth sores and eye irritations. The plant is also used to treat indigestion and restore strength after prolonged illness.

Coptis teeta Wall. "Tita", "Goldthread" or "Mouth Root" :

It is a hardy evergreen bog plant and a- small stemless herb. Leaves are compound and glabrous. Flowers are small, regular, white in colour. Fruits are many seeded folicles and seeds black. The species is native to Himalayan region and distributed in temperate regions of Arunachal Pradesh.

It requires moist soil and shady position. Propagation is by seed or division of rhizomes.

The dried rhizomes constitute the drug. It contains alkaloids of berberine group. It is a tonic and stomachic. The rhizome is administered in intermittent fevers and debility.

Corallorhiza Gagnebin
Orchidaceae

Corallorhiza maculata Raf. "Coral orchid" :

It is a low growing perennial plant without green leaves. This plant is saprophyte with much branched fleshy rhizomes. Flowers are yellow or brown or white, spotted with red or purple.

The plant is distributed in North Temperate region.

Uses : Roots and dried stalks are used for medicinal purposes. A tea made from dried stalks are given to the patients of pneumonia to relieve weakness. Roots are used to kill worms, increase perspiration and as sedative.

Coriandrum Linn.
Umbelliferae

Coriandrum sativum Linn. :

This is a hardy annual herb. Leaves are decompound. Petals are white or purplish.

The plant species is native to Mediterranean Europe and commercially cultivated in temperate and subtropical region. The species is cultivated throughout India.

Ordinary soil and open sunny situation is good. Propagated by sowing of seeds. Seeds are sown in 1 cm depth and 30 cm apart. Seeds are thrashed and dried when ripe.

The essential oil content is 0.3 to 2.2%, which contains d-and dl- -pinene, B-pinene, dipentene, P-cymene, d-linalool, aldehyde C_{10}, geraniol and acetic acid. The leaves contajn proteins, traces of fat, minerals like calcium, phosphorus, iron and carotene, fibre and carbohydrates.

Uses : This plant is important for its fruits and for essential oil. The essential oil is used for its flavour and carminative properties. The fruits are carminative, stomachic and cooling agents. It is used in curry powder, as seasoning, in confectionary and for flavouring alcoholic drinks and junkets. It cures a host of digestive disorders ranging from indigestion, nausea, dysentery, colitis to diarrhoea and chronic dysentery. Coriander chutney serves as an excellent remedy for abdominal pain caused by indigestion. Coriander leaves stimulate the appetite and act as a tonic. Coriander juice is recommended for patients who suffer from a deficiency of vitamin A, B and C. The paste of seeds ground with some water and applied to forehead can relieve a headache.

Cornus Linn.
Family : Cornaceae

Cornus florida Linn. :

This is a beautiful shrub upto 6 m tall. Bracts are white and flowers are greenish yellow. Fruits borne in dense cluster, bright red in colour, sometimes yellow.

The plant species is native to North America.

Uses : Flowers and bark of these plants are used for medicinal purpose. Root bark and inner bark preparations are used to treat aching muscles, malaria, worms, cholera, liver ailments, jaundice, sore mouth, fever and dysentery.

Corydalis Vent.
Family : Fumariaceae

Corydalis govaniana Wall. :

It is gregarious fleshy, perennial herb with coiled root-stocks. Flowers are bright yellow.

Uses : Roots are administered against muscular and gastric pains. These **are also used in syphilitic, scrorofulous and cutaneous affections.**

Coscinium Colebr.
Family : Menispermaceae

Coscinium feniestratum (Gaertn.) Colebr. :

This is a woody climber. Leaves are leathery, roundish, hairy on lower surface. Flowers are in small heads, minute, and green. Bark is yellowish.

The plant species is native to Indomalayan region. It is also found in Western Ghats.

Uses : The dried stem is medicinal, it contains berberine. Bark is antiseptic and roots have antibiotic properties. The dried stem is used to treat dyspepsia and general weakness. Stem yields a yellow dye.

Costus Linn.
Family : Zingiberaceae (= Costaceae)

Costus speciosus (Koen.) Sm. :

This is an erect perennial herb. Leaves are almost without petiole, silky pubescent beneath, upto 40 cm long. Flowers are white with red bracts and tubular in shape. Fruits are red and seeds black. The plant has rhizomatous horizontal root stock.

It is a tropical plant native to India and Malaya, found in plains throughout India and upto 1,400 m altitude.

Uses : The dried rhizome of these plants constitute the drug. It contains diosgenin. The roots are astringent, stimulant and anthelmintic. It is used to treat cough, catarrhal fever, dyspepsia, skin disease, worms and snake bite. The rhizome is rich in starch and cooked for eating.

Crataegus (Tourn.) Linn.
Family : Rosaceae

Under this genus there are about one thousand species and twe are native to India. They are thorny small trees and large shrubs. Flowers appear in terminal corymbose cymes.

These plants are mostly from North Temperate region.

Crataegus oxyacantha Linn. :

This is a hardy, spiny, deciduous shrub of 6 m tall with white flowers. There are varieties like "alba" with white flowers and Coccinea with crimson flowers in this species.

It is native to Europe. The species is also found distributed in N.E. Himalayas.

Uses : The fruits are used as a popular remedy for diarrhoea or slight phlegmasia. The liquid extract of fruit is a cardiac tonic and a remedy for organic and functional heart disease such as dyspnoea, hypertrophy, vulvular insufficiency etc.

The fruits of C. anomala, C. arnoldiana, C. calpodendron, C. occinoides, C. columbiana are edible. The flowers and fruits of C. roxburghii are eaten. The fruit of C. pinnatifida Bunge is consumed when candied, preserved, stewed, in sweet meat or as jelly.

Crataeva Linn.
Capparidaceae (= Capparaceae)

This genus are about ten species and five are native to India, and also distributed in tropical regions of the world. They are small to medium sized trees. Leaves are digitatively trifoliate.

Uses : The bark of C. nurvala Buch-Ham contains tannins and saponin. It is considered as bitter anti-periodic tonic and demulcent. For promoting appetite and as laxative, it is useful. The leaves are used as poultice. It is also useful in calculus and affections of the urinary organs.

Croton Linn.
Family : Euphorbiaceae

Croton tiglium Linn. :

This plant contains stellate hairs and scales.

It is distributed in India, China and New Guinea.

Uses : Seed oil has medicinal properties, it contains oleic acid, linoleic, palmitic, stearic and croton oil factor A.

It is a source of croton oil and is a powerful purgative drug. It is a powerful irritant and vesicant and is also a co-carcinogen.

Cupressus (Tourn.) Linn.
Cupressaceae (= Coniferae)

Cupressus arizonica Greene. :

This is a hardy evergreen coniferous tree upto 15 m tall, with small scale like leaves. The outer bark is thin and dark brownish with vertical fissures.

Uses : Fruits, leaves and stems are useful. To ease stomach gas fruits are used. Baked stems are applied to damaged skin areas and burns. Leaves are used to treat ulcers, tumors, toothache, skin inflammation and ringworm.

Curculigo Gaertn.
Family : Hypoxidaceae (= Amaryllidaceae)

Curculigo orchioides Gaertn. :

This is a stemless herb with tuberous rootstocks. Leaves are fleshy, elongate, lorate or ensiform. Root is black in colour.

This species plant is native to India and distributed in Peninsular India, Khasi Hills and subtropical Himalayas.

Uses : Flour from the root is made which is edible. The plant is used as a substitute of safed musli.

Curcuma Linn.
Family : Zingiberaceae

Curcuma longa Linn. = C. domestica Val. :

This is a fleshy rooted herbaceous perennial plant which attain a height of 80 cm. Stem short with tufled leaves. Flowers are yellow.

The species is native to India and China. It is extensively cultivated in India, China, Indo-China and Sri Lanka. In India, it is commercially grown in A.P., Maharashtra, T.N., Karnataka, Kerala and Orissa.

Uses : Turmeric contains protein, fat, carbohydrates, fibre, minerals, vitamin A, B, C and niacin. The volatile oil is orange yellow in colour and from rhizome the oil yield is 5 to 6%. About 58% of the oil is composed to turmerones and 9% tertiary alcohols.

It is a spice crop used in industries and in medicine. It is a flavourant, colourant or cosmetic. It is used in the preparation of medicinal oils, ointments and poultice. Turmeric is stomachic, blood purifier, vermicide, antiseptic, carminative and tonic. It is prescribed as an anti-periodic alternative, in case of diabetes and leprosy. Taken with warm milk or its inhalation from boiling water relieves sore throat and common cold. Turmeric powder is considered sacred and used as cosmetic by Hindu women. It is used as dye and employed as colouring material in pharmacy, confectionery, rice milling and food industries. It is used in many preparations while cooking food and also for colouring butter, cheese, pickles and food stuffs etc.

Cuscuta Linn.
Family : Cuscutaceae

Cuscuta megalocarpa Rydb. :

This is a slender annual plant, parasitic in nature, with twinning stem. Stems are white, yellow or red. It climbs over other plants. Leaves are hardly visible. Flowers appear in dense cluster, whitish in colour.

Uses : Entire plant is harvested when in flower for medicinal preparations. It is used in urinary complaints and liver diseases. Poultice made from the plant is used to treat insect sting. It is also used as contraceptive, laxative for stimulation of piles and treatment of tuberculosis.

Cyamopsis DC.
Leguminaosae (= Fabaceae)

Cyamopsis tetragonoloba (Linn.) Taub. (= C. psoralioides DC.) :

This is an erect herb with appressed, laterally attached hairs. Flowers in axillary racemes, small and purplish. [The species is native to India and Pakistan. This bean is widely cultivated as fodder (guar).

Uses : Seed is useful. It is a source of guar gum. This gum contains hydrogalactose and anhydromannose.

It is used as a binder and disintegrator in tablets. It is employed as suspensions or thickener in emulsions. The pods are used as vegetables and the plants as fodder.

Cydonia Mill.
Family : Rosaceae

Cydonia oblongata Mill. (= C. vulgaris Pers., = Pyrus cydonia Linn.) :

This is a small deciduous, non-branching tree upto 6 m tall with white or pale pink flowers. Fruits are pear or apple shaped, fragrant. These are grown for their edible fruits.

The species is native to Persia and Turkistan. It is used as a rootstock for pears. The species is cultivated in Himalayas and the Nilgiris for edible fruits. Commonly known as Quince.

Uses : The fruit is used as astringent. Jam and syrup can be made from these fruits to sweeten beverages, prescribed for chronic diarrhoea.

Cynanchum Linn.
Family : Asclepiadaceae

Cynanchum vincetoxicum (R.Br.) Linn. :

This is a temperate region plant and rarely occur at Donpur of Utterkashi above 2800 m elevation.

Uses : The paste of roots and leaves is applied with mustard oil on burns.

Cynara Linn.
Family : Compositae (= Asteraceae)

Cynara scolymus Linn. :

A hardy herbaceous perennial plant. It is a true artichoke and a valuable pot herb. The young flower heads are enclosed with bracts. The species is distributed to Mediterranean region and Kurdistan.

Uses : The root of this plant is diuretic and aperitive. It is also used as a febrifuge. The plant is rich in bitter extractive principles. It is, used for the treatment of jaundice, abdominal congestion accompanying or following intermittent fever, chronic diarrhoea.

Cynoglossum Linn.
Family : Boraginaceae

Cynoglossum glochidiatum Wall. ex Benth. :

This is a hardy perennial and a hairy bulbous herb. White hairs are there on stems and leaves. Flowers are blue. The species is native to Himalaya,.widely distributed in the barren lands, road sides and grass lands between 1500-4000 m in Himachal Pradesh.

Uses : Well drained loamy soil with adequate organic manure and sunny position are ideal. Propagation is by division of bulbs as well as from seeds. The purplish brown juice is applied as eye drops in conjunctivitis.

Cypripedium Linn.
Family : Orchidaceae

A genus of fifty two species and eight are native to India. They are terrestrial herbs with leafy stems.

Cypripedium calceolus Linn. :

It is terrestrial orchid, perennial and herbaceous in nature. The leaves originate around the stem, ovate and alternate. Flowers are dark brown, lip yellow.

The species is native to England, Europe and North Asia.

The pot compost consisting of two parts leaf mould, one part well decayed cow manure, one parts fibrous loam along with chopped sphagnum moss and sharp sand. They are cold climate plants and require mulching with leaves during winter. Semi shady and moist situation are preferable. These are propagated bv division of plants. Commonly known as yellow Lady's Stipper.

Uses : Roots and rhizomes of these plants are medicinal. A tea spoonfull of powdered roots in sugar water is said to act as sedative, prorrfote sleep and reduce pain. A tea prepared from roots is used to treat jangling nerves and headaches.

Cypripedium elegans Rehb. f.

This is a terrestrial orchid. Leaves are opposite, sessile, scape shorter than the leaves. Flowers are upto 2.8 cm across, brownish with darker brown lines.

The species is occurring in Sikkim Himalaya at an altitude of 3500 m elevation.

They grow well in a shady and airy green house. The pot compost consisting of equal parts of loamy soil, white sand and leaf mould mixed with small amount of leaf mortar or lime stone chips and little charcoal is considered very suitable. These plants are propagated by division.

Uses : Roots have medicinal properties. It is used to treat nervous disorders.

Cytisus Linn.
Family : Leguminaceae (= Fabaceae)

Cytisus scoparius Link. (= Sarothamnus scoparius (Linn.) :

This is a hardy flowering shrub upto 4 m tall with yellow flowers. Pods are brownish black, villous only at margin. There is a variety in it "Andreanus" which produce yellow and red flowers.

It is native to Europe and Western Asia. The species is naturalised in Nilgiris, Simla and elsewhere. Commonly known as common Broom.

Uses : Dried plant tops constitute the drug. It contains sparteine alkaloid and the flavone scoparin. Apart from this, presence of genisteine, sarothaminine, tannin and volatile oil have also been reported.

Sparteine has cardiac depressant, cathartic, diuretic and oxytocic properties and was used to induce and stimulate labour.

Dactylorhiza Neck. ex Navski.
Family : Orchidaceae

Dactylorhiza hatageria (D.Don) Soo. :

This is a perennial herb with palmate, tuberous roots.

It is native to North Africa, Temperate Eurasia and Alaska. Rarely occur at Chyangseel of Utterkashi between 3000-3500 m elevation and also found in sunny slopes of hills near marshy and moist places. Commonly known as Hathajari.

Uses : The decoction is given in colic pain. The root powder is used to relieve fever. The root is also used against urinary troubles. It is an astringent and expectorant.

Daphne Linn.
Family : Thymelalaeaceae

Daphne mezereum Linn. :

This plant species is native to Europe and Africa. It is bitter and corrosive. All parts of the plant including berries are highly emetic and purgative.

The bark is used to treat scurvy and syphilitic diseases, lymphatic complaints and osteocopic pains.

Daphne gnidium Linn. :

Which is known as Spurge Flax, is native to Mediterranean region and possesses very dangerous properties.

The bark applied to skin makes it red and inflammed, and if contact is prolonged, a blister forms.

Datura Linn.
Family : Solanaceae

A genus compoising of fifteen species, out of which three species namely D. innoxia Mill. D. metel L. and D. stramonium L. are important drug plants.

***D. innoxia* Mill. :**

This is a herbaceous annual, bushy in nature and attain a maximum height of 130 cm. The corolla of its flowers is 15 angled broad. Fruits have very slender prickles. Seeds are brown.

The plant has been cultivated in some European countries. In India, it is grown in Bangalore, Ahmedabad, Pune, Lucknow, Pilani and Jammu.

***D. metel* Linn. :**

It is an erect annual herb or undershrub reaches a height of about 1.8 m. Corolla of this flower is 5-angled. Flowers are whitish purple to yellowish, capsule nodding, dehiscing irregularly covered with tubercles.

It is native to Asia and Africa. Widely cultivated and naturalised in tropics. It occurs throughout India in waste place.

Seeds and leaves are source of hyoscine. The alkaloid content varies with altitude and season of year.

***D. stramonium* Linn. :**

This is a herbaceous erect, bushy annual grows upto2.5 m tall. The leaves are 18 cm long, ovate, unevenly toothed, dull green in colour. Flowers are axillary, erect, white, sweet scented. The plant produce large, prickly capsules, ovoid, deeply divided into four," filled with shiny black seeds.

It is native to Caspian region of Europe, North America and Indian subcontinent. It is cultivated in Europe and South America. In India, it has been successfully grown in Kashmir valley, hills of H.P. and Uttaranchal and also grown during winter season in northern plains. It occurs in nature in Temperate Himalayas upto 3500 m and in hilly regions of Central and Southern India.

Uses : Dried and powdered seeds and leaves of D. stramonium constitute the drug. In leaves total alkaloids range from 0.25 to 0.45% and in seeds 0.47 to 0.65%. Hyoscine content is 0.1% in leaves, 0.05% in stems and 0.1% in roots. Hyocyamine content is 0.4%, in leaves, 0.2% in stems and 0.1% in roots.

This crop is not commercially cultivated in India, since it is not an economical source of tropane alkaloids. Alkaloids of pharmaceutical importance extracted from these plants at present are hyoscyamine, hyoscine and neteloidin.

In Datura, leaves, flowers, roots and seeds are used for medicinal purposes. Formerly, it was a source of atropine. D. stramonium is antispasmodic in asthma and Parkinson's disease. The dried leaves are smoked in a pipe or home made cigarette to treat asthma. In the treatment of delirium, tremor and mania hydrocyamine hydrobromide is given. Hysociamine hydrobromide is used as a pre-anaesthetic medicine in surgery, in child birth, in ophthalmology and prevention of motion sickness.

A poultice made from the flowers is applied to wounds to reduce pain. The leaves are also applied to boils and ulcers. A decoction of flowers and roots has been used as a sedative to calm patients during setting of fractures. The plant is a dangerous hallucinogen. The whole plant of Datura metel is antiseptic, narcotic, sedative and is useful in asthma. The poultice of its leaves checks inflammation of breasts caused by excessive formation of milk. Seeds of Datura innoxia is occasionally used as a source of hyoscine.

Daucus Linn.
Family : Umbelliferae

Daucus carota Linn. :

This is a hardy biennial plant with thickened root. An erect branching, fuzzy herb, arising from a thick fleshy root. Leaves are pinnately divided. Flowers are borne in cluster. In the centre of the umbel is usually a red terminal flower. After fertilization, the peduncles all bend inwards until the fruits are ripe and then spread out again, allowing the burred mericarps to adhere to animals. The fruit is oblong with some bristles. It is a well known edible rooted vegetable. The cultivated form has much more fleshy roots than the wild. It is native to Asia and Europe.

It is a cold climate crop and require deep and well manured soil. Plants are grown from seeds. Commonly known as wild carrot.

Uses : Fruits, seeds and roots are used. To increase urine flow, seeds and roots are useful. Boiled root is applied as poultice on cuts and bruises. To stimulate menstruation, fruits are eaten and to treat intestinal worms seeds are used. It is reported to strengthen heart and brain. An appetizing drink is prepared from purple and black carrots. It has high caratene content 3500 cm 100/gm and good source of vitamin A and carbohydrats.

Delphinium Linn.
Family : Ranunculaceae

Delphimum ajacis Linn. :

This is a hardy annual plant of 140 cm height, branched with fine feathery leaves. Spikes of blue, purple lilac, salmon, salmon rose or white flowers, single or double arise in large number.

It is native to Europe and Australia.

Ordinary rich soil and open sunny situation are preferable. The plant is propagated by seeds which are sown in October-November in plains. Germination is better at lower temperature. In the hills it can be sown in March-April or during August-October. Plants take .about three months to bloom. Commonly Known as Larkspur.

Uses : The seeds of these plants have an unpleasant bitter taste and are regarded as astringent, vulnerary and emmenagogue.

Dendrobium SW.
Family : Orchidaceae

It is a large genus of ten hundred species and about eighty species are native to India. They are epiphytic herbs having bright coloured flowers.

Dendrobium alpestre Royle :

This is an epiphytic orchid. Flowers are white.

This species is occurring in the Garhwal Himalaya upto 2800 m altitude.

Uses : Pseudobulbs are medicinal. Its preparations are used to treat pimples, boils and other skin problems. The plant is also used as tonic.

Dendrobium normale Fale. :

It is an epiphytic orchid of which flowers are pinkish white in colour.

The species is occuring in Garhwal Himalaya upto 2500 m altitudes.

Uses : The entire plant is known for its aphrodisiac and tonic properties.

Dendrocalamus Nees.
Family : Gramineae (= Poaceae)

Dendrocalamus strictus (Roxb.) Nees. :

They are arborescent unarmed bamboos.

Found at Rajakho of Chhindwara District, M.P. and also occurring throughout the greater part of India.

Uses : The siliceous material from the culms is used as a tonic and astringent. The plant is also used as a raw material for paper manufacture.

Derris Lour.
Family : Papilionaceae (Leguminaceae)

Derris elliptica (Wall.) Benth. :

This is a climbing shrub. Leaves are alternate and leaflets are opposite. Flowers papilionaceous and pod indihiscent.

Distributed in West Bengal and eastern part of India.

Uses : It is cultivated for its roots, the derris or tuba root of commerce, used as an insecticide.

Dicentra Borckh. Corr. Bernh.
Family : Fumariaceae

Dicentra cucullaria (Linn.) Bernh. :

This is a hardy herbaceous tuberous rooted perennial plant upto 40 cm tall. The flowers are white or yellow.

The species is native to United States and Western Himalaya.

Uses : Tuberous roots are used for medicinal purposes, specially for veneral diseases, to increase urine flow or as a tonic. To treat skin disease, a poultice is applied.

Dictamnus Linn.
Family : Rutaceae

Dictamnus albus Linn. :

This is native to Europe and Asia.

The plant is sweet smelling, particularly the flowers. It is rich in essential oil. The plant has an acrid and bitter taste.

Uses : It is used for healing wounds, and also regarded as emmenagogue and vulnerary. Commonly known as Bastard Dittany.

Digitalis Linn.
Family : Scrophulariaceae

Under this genus there are about twenty eight species of herbaceous plants and rarely woody at base.

Digitalis lanata Ehrh. :

It is Commonly known as Woolly Foxglove or Grecian Foxglove, it is extensively cultivated because of its higher alkaloid content as compared to other species of Digitalis.

This important medicinal plant is cultivated in Kashmir and some pockets in South Indian Hills at an elevation of 1500-3000 m. Leaves of these plants contain digoxin and various other glycosides useful for treatment of cardiac ailments. It is a half hardy biennial herb. Plants attain a height upto 80 cm. Leaves are 18 to 25 cm long and 3 to 4 cm broad. Flowers are 7 to cm long, hairy, creamy, yellow or purple coloured.

Uses : To get maximum content of glycosides, age of harvesting the leaves affects to a great deal. During October-November first harvesting is done and long leaves excluding petioles are picked followed by second harvest after three and half months, and third harvest after one and half months of the second harvest. Sometimes four or five plucking is also done at intervals. Green healthy leaves are hand-plucked rejecting the yellow or brown coloured leaves and upper smaller leaves are left unharvested. After the final harvest the crop is retained in the field for seed production. Seed capsules are harvested when they turn brown.

The collected leaves are spread on the ground in layers and left for drying upto two days. Leaves should be dried in suitable temperature (40 to 50°C) within, a minimum period of time so as to prevent the loss of glycosides during drying process. Dried leaves should be packed in air tight container and stored in cool and dark place. More glycosides can be obtained from the leaves which are harvested in the spring season.

From Digitalis lanata an nearly 80 different glycosides have been reported. The principal glycosides are the tetraglycosides, lanatoside A, lanatoside B, lanatoside C and lanatosides D and E. Out of these lanatoside C and digoxin are two most important glycosides used in medicine. Other minor constitutents are diglycosides, digitalinum, verun and gitorin, the monoglycoside strospeside, saponins and the enzyme digilanidase. The seed contain viscous fatty oil and glycosides.

From D. lanata an aqueous syrup, an alcoholic tincture, a vinegar and ointments are prepared from leaves. The prepared drug of Digitalis promotes and stimulates the activity of all muscle tissues. It stimulates the digestive organs and all nervous system. It forces more blood into coronaries and improves the nutrition of the heart. It is used in case of congested heart failure and helps in the restoration and regulation of the function of the heart. This is also used to prepare ointments for application on wounds and burns. D. lanata has stronger medicinal properties than D. purpurea which is also cultivated for medicinal purposes.

Other important species with medicinal properties are D. ferruginea Linn. D. grandiflora Mill, D. lutea Linn. and D. purpurea Linn.

D. ferruginea Linn. :

This occurs in nature in the rocky hills of middle Europe and Western Asia, and commercially grown as a drug plant in Russia. A perennial plant to 2 m tall. Leaves are stipulate, and the mature leaves can grow 40 cm long and 3 cm broad, leathery, deeply veined and hairy only on the margin. Flowers are yellowish with brick red reticulation inside. The leaves are more potent than that of D. purpurea Linn. Commonly known aon Rusty Foxglove.

D. grandiflora Mill. :

This plant is native to Western Asia and Southern Europe and commercially cultivated in Russia. It is a perennial herb of 1.5 m height. Leaves are lanceolate-ovate,hairy, to 28 cm long and 6 cm wide. Flowers are yellow with brown marking. It is reported that the leaves are less potent than that of D. purpurea Linn. Commonly known yellow Foxglove.

D. lutea Linn. :

This is found in nature in the limestone hills of Middle and Southern Europe. A perennial plant upto 1.5 m tall. Leaves are oblong or lacnceolate, nonhairy. Flowers are yellow in colour appear in one-sided dense raceme. The leaves are as potent as that of D. purpurea Linn.

D. purpurea Linn. :

This plant species is native to British Isles, Temperate Western Europe and Morocco. It is commercially cultivated as a drug plant in Austria, England, Germany, Hungary, Japan. In India, it is cultivated in Kashmir, Darjeeling and Nilgiri Hills. Commonly called Foxglove or Purple Foxglove.

This is a herbaceous biennial with erect stem and attain a height of l.8 m. Leaves are ovate and hairy. The lower basal leaves are long-stalked, while upper leaves are almost without stalks. Flowers appear in erect, showy, usually one side spikes, white or purple in colour, with ciliate lobes.

Uses : Low level of active principles is reported in D. purpurea. The leaves contain at least thirty glycosides and total quantities ranged from 0.12 to 0.65%. Leaves and also seeds contain several saponins like digitonin, gitonin and tigonin. The seeds also contain digitalin.

Infusion of D. purpurea leaves are known as diuretic and sedeative. The leaf infusion is used to relieve sore throat and asthma, it is administered for the treatment of congestive heart failure. An ointment containing digitalis glycosides is used for healing wounds and burns.

Dioscorea Linn.
Family : Dioscoreaceae

It is a large genus with over more than seven hundred species. It is a perennial twining herb with tuberous roots. About sixty species are native to India. Leaves are simple or compound, alternate, thick, firm mostly the nerves are prominent triangular-ovate, shallowly or deeply cordate at the base. Diosgenin extracted from Dioscorea is used as a raw material for the industrial production of corticosteroids, including cortisone as oral contraceptives. Although there are aboutjffid species under this genus, about 700 species under this genus, about 20 species are known for their sapogenin content in their tuberous roots. A few species are important source of food. In India, Dioscorea alata L and D. esculanta (Lour.) Burk. are cultivated for their edible tubers. D. bulbifera L, D. hispida Dennst. and D. pentaphylla L. also occur almost throughout India, their tubers are edible and are consumed after processing. Some species of Dioscorea yield industrial starch. Most of the diosgenin in India, however, is produced from Dioscorea deltoidea Wall, which are found growing wild in North-Western Himalayas, specially Jammu & Kashmir, Himachal Pradesh and Uttarachal. D. prazeri Prain and Burkill occurring in Eastern Himalayas, in northern parts of Bihar, N. Bengal, Nepal, Sikkim and Assam is also rich in diosgenin. Mexican species of D. floribunda Mar. & Gel. and D. composita Hemsl. are cultivated in India apart from D. deltoidea Wall. Certain clones of Mexican Dioscorea species are having more than 5% diosgenin, while similar clones of D. deltoidea are yet to the developed.

Dioscorea floribunda Mar. & Gal. is native to Southern Mexico and Guatemala. It is a climbing, nonhairy vine with slender or stout stems and flowers are dark green in colour. It is commonly known as Alambrills.

Dioscorea floribunda is commercially cultivated in Karnataka, Kerala, Tamil Nadu, Maharashtra, Meghalaya, Assam, Andaman Island and Goa. Two high yielding variety with more diosgenin content (3 to 4%) have been developed at Indian Institute of Horticultural Research, Bangalore from Dioscorea floribunda namely "FB(C)-1 and "Arka Upkar".

Uses : Dioscorea is a source of steroidal sapogenins, like diosgenin. D. floribunda is rich in glycoside saponin which yield sapogenins on hydrolysis. The important sapogenins that are found in different species of Dioscorea are diosgenin, yamogenin, botogenin and kryptogenin. Apart from this some minor sapogenins like pannogenin and tigogenin have also been reported. However, diosgenin and its beta-isomer are most important commercially. These species also contain alkaloids, proteins and carbohydrates.

Diosgenin is used for synthesis of cortisone, which is useful in the treatment of rheumatic arthritis and are in the preparation of sex hormones. These are also used against asthma, nephritis, allergic manifestation and shock. Cortisone and hydrocortisone are also prescribed for sciatica, skin diseases and contact dermatitis. D. deltoidea tubers are used locally in Garhwal for washing clothes. D. bulbifera L. tubers are used in dysentery and piles.

D. composita Heml. :

This robust climber with thick fleshy branching tuber. Flowers are small, sessile, purplish. This species is native to southern Mexico and known as "Mexican Yam" or "Barbarco". Its properties are same as that of D. floribunda which contain 4 to 6% total spagenins.

D. deltoidea Wall. :

It is twining plant and not a vigorous grower, native to India and Nepal. Rhizomes woody, containing 8 to 10% sapogenins.

D. friedrichsthalu Kunth. :

This is a native to Mexico and Central America, the tubers contain only 3% sapogenins.

Diospyros Linn.
Family : Ebenaceae

Diospyros melenoxylon Roxb. (= D. montana Roxb.) :

This is a small tree. The flowers are green in colour.

The species is native to India, obtained at Chhindi of Chhindwara District, M.P. It is also sporadically distributed on the lower slopes and ravines of the area extending upto 1000 m in Garhwall Hill in Uttaranchal Commonly known Zeylon or kendu or Tendu.

Uses : The fruits are powdered and taken in stomach disorder. It is also applied externally for piles.

Diospyros virginiana Linn. :

This is a half evergreen tree with juicy edible fruits. The tree attain a height of 18 m. The leaves are 16 cm long and 8.5 cm wide, thick and firm, upper surface dark green and lower surface pale and fuzzy. Flowers are pale yellow and fruits are orange in colour with flower sepals attached. The species is native to North America. Locally known as Persimon or North American Ebony.

Uses : Fruits, seeds and roots of these plants are useful. Green fruits are used to treat uterine haemorrhages, sore throat, dysentery and diarrhoea. For kidney stones, powdered seeds mixed with water and stained through a cloth is used as drink. A tea prepared from roots is used to treat dysentery. The plant is cultivated both for wood and fruit.

Dodonaea (Mill.) Linn.
Family : Sapindaceae

Dodonaea viscosa (Jacq.) Linn. :

This is a popular hedge plant for'gardens. The species is distributed in tropical and subtropical^areas, commonly found at Rater of Chhindwara District. Madhya Pradesh.

Uses : The leaf decoction is taken once in a day for two days to stop excess menstrual flow.

Dolichos Linn.
Family : Leguminaceae (= Fabaceae)

In this genus there are about one hundred species and eight are native to India. These are annual or perennial herbs, prostrate or sub-erect in habit.

Dolichos biflorus Linn. :

This is a climbing annual with slender pubescent stems. Leaves are trifoliate. Flowers appear on a very short pedicels in the axils of the leaves, 1-3 nos., yellow in colour. Pods narrow, flat and curved, tipped with persistent style and contain 6 to 8 seeds.

The species is distributed in warm regions.

Uses : The seeds are nutritive and very rich in protein. Vegetative parts and seeds are rich in vitamin. Several plant enzymes eeclin, strepogenin, B-sitosterol, bulbiformin, linoleic acid, polyphenols, free sugars, and isoflavones have been isolated from this plant.

Seeds are regarded as tonic, diuretic and astringent. Decoction of seeds is used for treating bowel haemorrhage, leucorrhoea and other menstrual derangements, colic pain, bladder and kidney stones and scrofula. Infusion of whole seed is administered in hypertension. Soup prepared from seeds is suggested in subacute enlargement of spleen, liver and piles. Seeds of few species are used as human food, and leaves and stems as animal fodder. Several other species are nutritive and used in traditional medicine in Ayurveda.

Dolichos catiang and D. fabaeformis are diuretic. The roots of D. falacatus are used in piles, constipation, ophthalmia and skin disease. Roots of D. lablab L. and seeds of D. minimus are poisonous. D. lablab seeds are used to stop nose bleeding and in phlegmatic disorder. Leaves of D. trilobatus are laxative. Seeds of D. pruriens Jacq. are regarded as astringent, anthelmintic, nerve tonic and aphrodisiac. Seeds and bark of Dolichos sesban are astringent and used in diarrhoea, skin eruptions and itching, to check excess menstruation and for reducing enlarged spleen. Leaves of D. trilobus are considered cooling, sedative, antibilious and tonic.

Duboisia R. Br.
Family : Solanaceae

Duboisia myoporoides R. Br. :

This is a small tree. Leaves are alternate, ovate-oblong to broad lanceolate, blunt tipped, tapering at base. Flowers borne in terminal, much branched panicles, bell shaped, corolla is rounded, white or pale lavender in colour. Bark of the tree is grey, corky, fissured and the wood is light. Fruit black and juicy, globular, containing 6-12 seeds.

The tree is native to Australia and cultivated commonly in Queensland. These plans are tried under Dehradun, Lucknow and Karnataka climatic conditions. In Karnataka, better yield of leaves with higher total alkaloid content is reported. The leaves of these plants are rich source of tropane alkaloids.

Uses : These plants are damaged by a number of insects like trunk-borer, small flea beetle and brown olive scale. Application of Furadan effectively controls the insects. Some of the virus diseases like wilt and dieback are transmitted by thrips to these plants. Control of thrips by spray of Rogor at the rate of 750 ml in 500 litres water/ha can be done, which ultimately will reduce the spread of virus.

Leaves are plucked six months after planting, three times in a year at an interval of three months leaving 25-30% of the leaves in the plants itself for subsequent flush of growth. Leaves are immediately dried in the sun and are then packed in bags. About 1000 to 1500 kg of dried leaves per hectare per year can be obtained.

Leaves are the rich source of tropane alkaloids like hyoscine, hyoscyamine, norhyoscyamine, valeroidine, tigloidine, poroidine, isoporoidine, isopelletierine and anabasine. The plant is a superior source of tropane alkaloids, that what is extracted from Datura hyoscyamus species. The leaves of these plants contain 2-4% total alkaloids, of which more than 60% in hyoscine and about 30% hyoscyamine are present.

There is another species Duboisia leichardtii F. Moeller which contain 1.5-3% total alkaloids with more than 60% hyoscyamine and 30% hyoscine.

The alkaloids of these plants are used in the modern medicine for their antispasmodic, anticholinergic, sedative, antiemetic and mydriatic properties. It is administered as a premedication before anaesthesia and employed for sympatholytic paralysis. In combination with other drugs, it is prescribed for the treatment of duodenal and gastric ulcers. To dilate pupil of eye it is useful in ophthalmology. It is also used for treatment of respiratory and intestinal disorders. The leaf and fruit extracts of these plants were used in early ages for dilation of pupils. The extract was administered to relieve night sweats of consumptives and painful tenesmus caused by inflammation of urethra and neck of the bladder. It was also used for controlling motion sickness, sea sickness as well as shock cases in military personnel. The soft wood of this tree is used for carving and engraving woodcuts.

D. leichardtii F. Moeller :

This is an another species of small tree where the bell shaped flowers are borne on terminal panicles, and the corolla is long and pointed. Fruit is over with 6 to 12 seeds.

The leaves contain 2.7 to 3% total alkaloids, and most important are hyoscine and hyoscyamine.

Dryopteris Adans.
Polypodiaceae (Aspidiaceae)

Dryopteris filix-mas Schott. (= Aspidium filix-mas SW) :

This is a temperate zone terrestrial fern and there are many varieties in this. Plant is 0.5-1.8 m long.

These are native to Europe, Central and North America and Asia. Commonly known as "Male Fern".

Uses : The plant contain oleoresin, filicic acid, flavaspidic acid, filmaron and albaspidin.

Rhizomes and stipes are used in medicine and said to be anthelmintic. The decoction of the rhizomes of these plants are effective in treating taenia and worms lodging in the alimentary canal.

Dryopteris cristata (Linn.) Gray :

This fern is native to England and North America. Fronds are erect, stiff, to 80 cm in height. Grows better in wet woods, thickets, bags and swamps.

Well drained sandy loam soil rich in organic matter, humid and semi-shady location are preferable. /For medicinal purposes, the dried roots are used. To induce perspiration, to break a fever, for treatment of intestinal worms and to clear phlegm from the chest these roots are reported to be useful. Commonly known as created shield fern.

Dyssodia Cav.
Family : Compositae (= Asteraceae)

Dyssodia papposa (Vent.) Hitche :

It is an aromatic annual of 80 cm tall. Flowers are yellow in colour.

The species is native to Mexico and South-West America.

Uses : Mature leaves and seeds are useful. A tea prepared from leaves is used to treat diarrhoea. To treat stomachache fresh leaves are chewed. Seeds soaked in warm water and stained water is given to babies against severe stomach colic.

Echinacea Moench.
Family : Compositae (= Asteraceae)

Echinacea purpurea (Linn.) Moench. :

This is a hardy herbaceous perennial plant upto 1.8 m tall. Leaves are lance-shaped, alternate and leaf margins are toothed. Flowers are purplish-red.

The species is native to North America.Commonly known as Purple corn Hower.

Uses : Roots of these plants constitute the drug. It is used- to treat blood poisoning, snake bite, skin diseases, syphilis and rabis.

Echinocereus Engelm
Family : Cactaceae

Echinocereus enneacanthus Engelm. :

This is a cactus plant with fleshy cylindrical and spiny stems. Flowers are purple in colour. Fruits are edible, about 2.6 cm long.

The species is native to Mexico and Texas.

Uses : Flowers of these plants are used to treat excessive water retention in the body and intestinal worms.

Elettaria Maton.
Family : Zingiberaceae

Elettaria cardamomum (Linn.) Maton. :

This is a perennial herbaceous plant with erect stems upto 4 m tall. Leaves are large, narrow, with strong median nerves. It flowers on leafless shoots from the rhizome. Flower stocks appear from near the base of the stem and lies on the ground. Flowers are white or pale green in colour. Fruits are pale green to yellow in colour, ovoid, 3 celled with many triangular brownish-black seeds.

The species is native to India, Ceylon, Guatemala and El Salvador. It occurs wild in moist forest and hilly regions of Karnataka and Kerala. It is commercially cultivated in Kerala, Karnataka and Chennai. Commonly known as Chhote Elaichi.

Uses : Seeds contain essential oil which constitute eucalyptol, sabinene; d, - terpineol and acetate, borneol, limonene, terpinene etc. It also contains crude protein, starch, minerals, vitamins A, B and C.

Seeds are used as drug and flavour. It promotes digestion and is administered with purgatives, and as a flavouring agent. The tincture of cardamom is administered to relieve winds from stomach. It has stimulant, carminative, stomachic, diuretic and digestive properties. It is reported to be a cardiac stimulant, and checks nausea and vomiting. Seeds are chewed to prevent bad smell in the mouth, indigestion and nausea etc.

Emblica Gaertn.
Family : Euphorbiaceae

A small genus of five species and three are native to India.

Emblica officinalis Gaertn. :

This is a deciduous tree with small leaves which are very closely set in pinnate fashion. Flowers are pale green with small dense clusters. Male and female flowers borne on the same tree. Fruits are fleshy, roundish, rather indistinctly marked with six lobes, pale green or yellowish in colour, 1.5 to 2.5 cm in diam.

At is distributed in Madagascar and East Asia. Commonly occurring at Chímtipur of Chhindwara district, Madhya Pradesh and also found thoughout tropical and subtropical India.

Uses : The fresh or dried fruit is used as drug. The fruit poultice is used to stop bleeding from cuts. The fruit powder (about 30 gms) is used as a coolant and laxative. Fruits are good liver tonic. The fruit juice is useful in indigestion, jaundice, anaemia, heart complaints. It is a rich source of vitamin-C. Dried fruits are useful in diarrhoea and dysentery. Seeds are administered against asthma, stomach disorder and bronchitis. It is one of the ingredients of triphala and chyavanprash.

Embelia
Family : Myrsinaceae

Embelia tsj eriam-cottem A.DC. :

This is a small tree. Leaves are narrow at both ends upto 12 cm long, lower surface reddish and hairy on nerves. Flowers are pale green in colour, minute in size. Fruits are small, globose, pointed and red when ripe.

It is found in Eastern India and Deccan Peninsula.

Uses : The dried fruit has me.dicinal value. It has antibacterial and antitubercular activity. It is laxative and effective in the treatment of tapeworms.

Ephedra Linn.
Family : Ephedraceae

A genus of forty five species and nine species are native to India. Several species of Ephedra possess medicinal properties like E. equisetina Bunge, E. gerardiana Wall (£. vulgaris Rich,), E. intermedia Schrenk and Meyer, E, major Host. (E. nebrodensis Tineo) and E. sinica Stapf.

They are evergreen thick stemmed shrubs. Twigs are with fine longitudinal ridges or ribs. Leaves are minute. Flowers appear in round or elongated catkins. Fruits are red in colour, sometimes yellow and globular in shape. E. equisetina is hollow stemmed upto 2-Sm tall native to Inner Mongolia between 1500 to 1800 m altitude. E. gerardiana is a dwarf shrub, distributed in the Northwest Himalayas in Northern India, West Pakistan, Tibet and China. E. intermedia is a shrub 70 cm tall, occurs in inner Mongolia to Pakistan. E. major is a tall shrub, native to Spain, Sicily, Afganistan and Pakistan. E. sinica is a shrub of 50 cm in height and occurs in Northern China and outer Mongolia.

Uses : Green branches and twigs "are high in ephedrin, while basal stem contains very low alkaloid. Roots and fruits also do not contain alkaloid. Stems which are less than 1.25 cm in diameter are harvested and sun dried. Dried stems are then processed, packed and stored in dry atmosphere.

The total alkaloid content varies from 0.8 to 2.5 depending on species, place of cultivation, time of the year and stage of harvest. The plants contain ephedrine and pseudoephedrine in varying quantity. For pharmaceutical trade ephedrine content must not be less than 1.25%. The plant has astringent taste and is strong in pine odour. Ephedrine is administered in cases of rhinitis, asthma, hay fever and emphysema. Ephedrine is given orally for treating epilepsy, nocturnal enuresis, myasthenia gravis and urticaria accompanying angioneurotic edema. To relieve congestion arid swelling ephedrine salts are prescribed as nasal spray. This is also employed subcutaneously to prevent hypotension during anesthesia. Pseudoephedrine is also taken orally which is very effective as nasal decongestant. However, many preparations from ephidra are used to treat malarial and other headache, fever, wild & arugh etc.

Epilobium (Dill.) Linn.
Family : Onagraceae

Epilobium latifolium (Dill.) Linn.:

A herbs with irregular purple or white flowers.

The plant is used as a vulnerary and detergent. The dried and powdered plant parts are used against haemorrhages.

Epipactis SW.
Family : Orchidaceae

Epipactis helleborine (Linn.) Crantz. :

A terrestrial orchid with stout stem. Leaves are variable. Inflorescence are dense or loose, few to many flowered with large floral bracts. Flowers are 2.6 cm, greenish flushed with purple, bell shaped.

It is native to North America, Asia and Europe.

Uses : The roots of these plants are medicinal which cure insanity.

Epipactis latif olia Wall. :

Plants stem are stout, not tufled, to 95 cm long. Leaves are broadly elliptic to 14 cm long and 8 cm broad. Inflorescence are rather dense, sometimes twisted. Flowers are greenish, variously marked with yellow, white or purple, upto 1.6 cm across. Lip is broad and column is very short.

It is occurring in Kashmir and Sikkim between 2200 to 3800 m altitude.

Uses : Rhizomes are useful. These are used to treat nervous disorders.

Equisetum Linn.
Equisetaceae

Equisetum debiole Roxb. :

This is an aquatic plant. Leaves are green narrow, rush like.

Commonly found in Garhwal Himalayan region upto 2500 m elevation.

Commonly known as Horsetai / or Foxtailed As paragus.

Uses : It posesses cooling and diuretic properties and is used to treat gonorrhoea.

Equisetum ramoissimus (Roxb.) Hauke :

These plant is found at Rajakho of Chhindwara District, Madhya Pradesh.

Uses : The plant paste (about 20 gms) is applied twice a day for seven days in bone fracture.

Eria Linn.
Family : Orchidaceae

Eria muscicola Linn. :

This is a small epiphytic plant. Pseudobulbs are globose and crowded, to 6 mm broad. Leaves are only 3 to 4, petioled, oblanceolate upto 4 cm long and 5 mm broad. Inflorescence erect, 4 to 7 flowered raceme. Flowers are very smalh to 6 mm across, yellowish green in colour.

This species occurs in Jharkhand and Khasia Hills between 1200-1500 m altitude.

Plants are grown in small pots or bushels in a compost consisting of equal parts of osmunda fibre, fibrous loam and sphagnum moss.

Uses : Pseudobulbs are used in medicine. It is used to treat chest, heart, lung, eye, ear and mental problems.

Erigeron Linn.
Family : Compositae (= Asteraceae)

Erigeron canadensis Linn. :

This is a hardy herbaceous perennial plant and attain a height of 22 m. Leaves have bristles. Flowers are of different colours.

It is native to North America. Commonly known as Canadian Fleabane.

Uses : Mature leaves and flowering tops are useful. The oil extracted from the plant is used to treat pulmonary problems, haemorrhoids, and diarrhoea. A decoction of the tops of plants are used to cure sore throat and dysentery.

Eriodictyon Benth.
Hydrophyllaceae

Eriodictyon californicum Greene. :

This plant is occurring in Southern Oregon to Central California. Leaves constitute the drug. The extract of leaves contains flavone derivatives, eriodictyol, eriodictyonone, tannins, resin and volatile oil.

Uses : It is used as a flavour to mask bitterness of the medications containing quinine.Commonly known as yerba santa.

Erodium L'Herit.
Family : Geraniaceae

Erodium cicutarium (Linn.) L'Herit. :

This is a hardy biennial hairy herb, with compound leaves. Leaflets are without any stem. The flowers are purple.

It is native to Mediterranean region. Commonly known as common stork's Bill or Heron's Bill.

Uses : Mature leaves and blooming plants are useful. A tea prepared from leaves are used to increase perspiration and urine flow, to treat uterine haemorrhage and water retention. This plant is also effective to treat pains of rheumatism.

Eryngium (Tourn.) Linn.
Family : Umbelliferae

Eryngium yuccifolium Michx. :

This is a hardy perennial erect herb of 1.40 m tall with spiny-toothed leaves. Leaves are long and narrow, 70 cm long and 2.8 cm broad. Flower heads surrounded by spiny, coloured bracts. Flowers are small, whitish to pale blue in colour.

It is native to temperate region of America. Commonly known as Button snake root.

Uses : Roots of these plants constitute the drug. Roots are used to treat kidney trouble, bladder problem, respiration ailments, rheumatism, liver ailments and rattle snake bite.

Erythrina Linn.
Leguminosae (Papilionaceae)

Under this genus there are about one hundred species and eight are native to India. Trees are with prickly branches. Flowers are large coralred in colour.

Native to tropical or subtropical region.

Erythrina herbacea Linn. :

A half hardy herbaceous perennial upto 1.60 m tall, sometimes it grows tall like a tree. Leaves have prickly stems and three leaflets. Flowers are scarlet upto 6 cm long, pods are 13.6 cm long, and seeds are bright red with black spots.

The species is native to West Indies. Commonly known as coral Bead.

Uses : Roots are eaten to increase perspiration. A decoction of mature plant with full bloom increases the urine flow and perspiration. The same decoction is considered laxative.

The seeds of E. arborescens contains alkaloids. The bark of E. stricta in powdered from is used to treat biliousness, fever, rheumatism, itch, asthma, leprosy and epilepsy. The flowers of this species is given as antidote to poison.

Erythronium Linn.
Family : Liliaceae

Erythronium grandiflorum Pursh. :

This is a hardy bulbous perennial with two leaves arising from the base. Flowers are yellow in colour.

The species is native to North-West America.

Compost mixture consisting of equal parts of leaf mould, loamy soil, well decayed cow manure and sand is suitable. Semi-shady location is preferable. Bulbs are planted 8 cm deep. Annual top dressing with decayed manure is beneficial. Commonly known as Yellow Fawn Lihy.

Uses : Bulbs are crushed to make poultice for treating boils.

Erythroxylum R. Br.
Family : Erythroxylaceae

Erythroxylum coca Lam :

This is small evergreen tree or a tall shrub. Branches are slender with red-brown bark. Leaves are oval to elliptic, alternate upto 12 cm long and 6 cm broad. Flowers are fragrant, yellowish or white. Fruits are with thin flesh, scarlet when ripe, single seeded.

It occurs in nature in the uplands of Colombia, Ecuador, Peru, Bolivia and Western Brazil. It is cultivated commercially in Peru and Bolivia.

In this species there are several varieties like var. coca, novogranatense and spruceanum. Commonly known as cocazine tree.

Uses : Coca leaves contain 0.7 to 2.5% total alkaloids of which cocaine, cinnamylescaine and and fi-truxilline are important. Other alkaloids are tropacocaine, valerine, hygroline and cuscohygrine.

Cocaine is administered under strict governmental control. It is a stimulant for the central nervous system. The drug prepared from this plant reduce the sense of pain and is used in nasal and oral operations and in cases of inoperable cancer. Cocaine solution is used in earache. It is an ingredient of ointments and suppositories for relief of haemorrhoids, neuralgia and urticaria. An infusion of leaves is used as sedative for the nerves, pectoral, stomachic and asthma. Leaves are smoked or chewed to relieve colds, catarrh and asthma. In food and beverage decocainized extracts of coca leaves are used for flavouring ice-cream, candy, soft drinks and liqueurs.

Eucalyptus L'Herit.
Family : Myrtaceae

A genus of five hundred species, mostly Australian. Many species are cultivated in India. These are evergreen aromatic trees.

Eucalyptus dives Schau. :

It is native to New South Wales. Essential oil is extracted from leaves which is around 4%. The oil contains piperitone, and B - phellandrene, p-cymene, and-terpinene and terpinolene. Commonly known as Broad - Leaved Peppermint.

E. fruticetorum F. Muell. (- E. polybractea R.T. Baker) :

This occurs in Western New South Wales and Victoria. Leaves contain 1.5 to 2.9% essential oil, which constitute cineole, cuminal, phellandral, cryptone and macropone. Commonly known as Blue Mallee.

E. macrorryncha F.V.M. :

This occurs in Victoria and New South Wales, Australia.

Mature leaves are useful which yield 15 to 25% rutin. For pulmonary haemorrhage, capillary haemorrhage and for hypertension, rutin is used. Commonly known as Red Stringy bark.

E. radiata Sieb, ex DC. :

It is distributed in Southern New South Wales. Leaves contain essential oil 4 to 8.5%. The oil constitute cineole, geraniol, cineole, geraniol, terpineol, citral and methyl cinnamate. Commonly known as Narrow leaved beppermint.

Uses : medicinal use, eucalyptus oils, E. dives, E. radiata and E. fruticetorum were preferred since oils of these species are free from objectionable low boiling aldehydes which are pronounced in E. globulus Lab. and E. smithii. These oils are also used as flavour.

Eugenia Linn.
Family : Myrtaceae

Eugenia caryophyllus (Spregal) Bullock et Harrison. :

An evergreen tree with straight trunk attaining a height upto 15 m. The flower buds have a slightly cylindrical base and is surrounded by plump ball like unopened corolla which is surrounded by the four toothed calyx. The flowers are fleshy purplish. Fruit is oval, one seeded.

It is native to India, Sri Lanka, Madagascar, East Africa, Indonesia and Haiti. The plant grows in Kerala and Chennai.

The unopened, green fully grown buds are picked. These are airdried which ultimately become dark brown. This is used as a spice and is the clove of Commerce. Commonly known as clove or laung.

Uses : The leaves, unopened buds and unripe fruits are aromatic and contain essential oil. The clove buds contain 15% volatile oil and also protein, carbohydrate, minerals, vitamins A, B and C. The essential oil of the dried flower buds contain 90% eugenol and also eugenol acetate, caryophyllene, acetyleugdnol, vanillin, furfural and methyl-n-amyl ketone.

Clove is highly aromatic and impart warming qualities. It is used as a culinary spice. It has expectorant and antiemetic properties. It was also used as anodyne and antiseptic in dentistry. It is carminative and stimulant and used in flatulence and dyspepia. Clove is an ingredient of toothpastes and mouth washes. The clove oil possess antiseptic and anti-bacterial properties and hence used in many pharmaceutical preparation.

Eulophia R. Br.
Family : Orchidaceae

A genus of two hundred species and about twenty six species are native to India. Terrestrial herbs with fleshy tubers.

Eulophia campestris Wall. :

This is a terrestrial orchid with deformed pseudobulbs. Leaves are 2 in number, linear, upto 42 cm long and 2 cm broad. Inflorescence many flowered to 45 cm long. Flowers 4 cm in diameter, are pale pink with darker lines, petals are narrower than sepals, and the lip is as long as sepal.

The species is distributed in Andaman Islands, North Bengal, Tarai and Duars and also plains of Deccan.

Uses : Tubers of these plants are medicinal. It is aphrodisiac. Tubers are used to treat pulmonary and cardiac problems.

Eulophia herbacea Linn.

This is a terrestrial orchid with tuberous roots. Stem is upto 15 cm long, slender and sheathed. Leaves are linear-lanceolate, many nerved upto 40 cm long and 10 cm broad. Scape is 95 cm long, inflorescence 9-12 flowered. Flowers are 2.5 cm long, white with purple nerves.

It is found in Dehradun, Mussoorie, Garhwal, West Bengal, Orissa and Karnataka.

It is used in the preparation of Ayurvedic medicines and considered as general tonic.

Euonymus Linn.
Family : Celastraceae

Euonymus atropurpureus Jacq. :

This is a deciduous dwarf tree upto 8 m tall. Leaves are mostly oval shaped and variegated. Branches are green and flowers are purplish. Fruit is four lobed, bright red in colour.

The species is native to Britain and Siberia. Commonly known as American Spindle Tree.

Uses : Seeds, bark root and stem are useful. A decoction of inner stem bark is drunk for uterine discomfort and also as an eye lotion. For facial sores, a poultice of pounded root and stem bark is prescribed. Bark is used to treat syphilis, indigestion, constipation and to stimulate liver. To destroy head lice, oil extracted from seeds is used.

Eupatorium Linn.
Compositae (= Asteraceae)

A genus of one thousand species and about seven are native to India. These are Herbs or undershrubs.

***Eupatorium cannabinum* Linn. :**

This plant is native to America, Europe and Asia. The plant is also found in temperate Himalayas and Khasia Hills.

The leaves of these plants are aperitive, emetic and vulnerary and are recommended for cachexy, dropsy, liver complaints. Externally used for skin diseases, inveterate ulcers, swelling of the legs and scabies. The roots are pungent bitter in taste and with no odour, it has purgative effect.

***Eupatorium perfoliatum* Linn. :**

This is a perennial shrub upto 2.5 tall. Stems ae slightly hairy. Leaves are opposite. Flowers borne in flat heads, purplish to white in colour. The species is native to America. Commonly known as Boneset or Creswort.

Uses : Leaves and flowering tops of these plants are useful. It is used to treat malarial fever and worms, aches and pains. A tea prepared from leaves is used to cure coughs, cold and chest ailments.

***Eupatorium purpureum* Linn. :**

A perennial plant of 4 m tall. Stem hollow. Leaves are arranged in whorls and 5 to 7 leaves are there in one whorl. The flowers are purple in colour.

The species is native to America. Commonly known as Sweet Joe-Pye weed or Gravel root.

Uses : All plant parts and roots are useful. This plant is used for curing fever, bladder ailments and to induce vomiting. The roots are used to cure urinary infection, as tonic, stimulant and also to increase urine flow.

Euphorbia Linn.
Euphorbiaceae

A large genus of about two thousand species and sixty species are native to India. Herbs or shrubs or small trees with milky juice.

Euphorbia hirta Linn. :

This is a herb of erect habit upto 60 cm tall with greenish yellow or white flower. Stems covered with yellowish hairs. Leaves are small in opposite pairs.

It is found throughout India and common in warmer region.

Uses : The plants are used to treat colic troubles, dysentery, cough, asthma, worms and vomiting. It has vermifuge properties. It is also eaten as vegetables.

Euphorbia parviflora Linn. :

This is an erect annual herb. The leaves are variable in shape. Flowers are white or pale purple in colour.

It is native to India.

Uses : The dried leaf infusion is used to treat diarrhoea, dysentery and leucorrhoea.

Euphorbia pilosa Linn. :

A herbaceous plant upto 95 fern tall, perennial in habit with yellow flowers. There is one variety "major" which is 50 cm tall and produce yellow flowers.

It is native to Europe and North Asia. The species is found in moist temperature zones of Himalaya.

Uses : The roots are purgative and vermifuge. It is also used to treat fistular sores.

Euphorbia nivulia Linn. :

It is a small tree, prickly and fleshy with yellow flowers. Fruits are reddish or yellowish.

It is native of Himalaya.

Uses : The latex of this plant possess vesicant wormicidal and purgative properties.

Euphorbia royleana Boiss. :

It is a shrub with 5-8 angled branches, spinous and flowers are yellowish.

Abundantly occur at Massu of Utter Kashi between 500-1500 m in the rocky and exposed valleys. The latex is filled in the hollow cavities of decayed teeth to check the infection. The latex is cathartic and anthelmintic.

Uses : The leaf extract of E. an tiquorum L. is used in dropsy and as nerve tonic. Powdered leaves of E. branharti is used as poultice on boils. The juice of E. neriifolia L. is used in earache and asthma. The juice of E. nivulia is diuretic. The white juice of E. tirucalli is applied in rheumatic pain and toothache.

Euphrasia Linn.
Scrophulariaceae

Euphrasia officinalis Linn. :

A herbaceous plant with white or lilac flowers and purple veins. It is native to temperate Himalayas.

Uses : This plant is believed to be effective for ophthalmia ailment.

Fagopyrum Moench.
Polygonaceae

A genus consists of fifteen species and three are native to India.

Fagopyrum aibotrys (D. Don) Hara :

It is an erect hardy annual herb.

The species is native to Asia, commonly found at Donpur between 2200 to 2800 m of Utter Kashi, Garhwal Himalaya.

Uses : The paste of crushed leaves is applied on the insect bite. To inhale the smoke of seeds is believed to be good for headache. It is also grown for seed for pheasant and poultry feeding and for ploughing and digging as a green manure.

Fagopyrum esculentum Moench. :

This is grown as a minor grain crop in the hilly regions of North India and Nilgiris and also found wild in different parts of India. The plant is an erect herb. The species is native to Asia. It is largely cultivated in North America for its seed in which there is floury endosperm.

Uses : The leaves contain 3% rutin. It is said that these leaves are vasoconstrictor which stops capillary bleeding and also reduces capillary fragility. This plant is also used as green fodder, and a good honey plant. The flour from the seed is used mostly for making bread, pancakes and porridge.

Fagus Linn.
Family : Fagaceae

Under this genus there are about ten species all are deciduous trees.

Fagus grandiflora Ehrh. :

This is a tall hardy deciduous tree, grows upto a height of 38 m. Leaves are silky when immature, bluish green above upto 15 cm long. Fruits are 4 cm long with prickles.

It is a good seaside tree. The species grow well in organic rich moist sandy, chalky and gravely loam soil. Plants are multiplied by seeds.

Mature leaves and tree bark are useful. A concoction of leaves is applied to burns, scalds and frost bite. Fresh bark preparation is used to cure rashes. Timber is used for making tools etc.

Fagus sylvatica Linn. :

It is a perennial deciduous tree upto 28 cm tall, native to Europe. It forms a homogeneous forests. It is grown in Kulu Valley, Nilgiris and elsewhere for ornaments.

Uses : Beech creosote is used in medicine for treating tooth decay. The wood of this plant is hard and used in making artistic material. Oil is extracted from the nuts. The nuts are edible.

Ferula Linn.
Family : Umbelliferae

A genus of one hundred and thirty species and four are native to India. They are perennial herbs.

Ferula asafoetida Linn. :

A herb with massive tap roots or carrot shaped roots. The species is native to Afghanistan, Iran and Kashmir.

Hing is extracted from tap root of 4 to 5 years old plant. It is a dried latex exuded from the living taproot. Hing is acrid and bitter in taste and emits a pungent smell.

Uses : Hing contains 40-60% resin, 25% gum and 12-18% volatile oil. From the gum resin essential oil is obtained. The oil contains secondary butylpropenyl disulphide, disulphides, trisulphide, pinene and terpene. The pungent and disagreeable odour of the oil is due to presence of sulphur compound.

It has antibiotic properties and inhibits microbial growth. Hing is used medicinally and said to stimulate intestinal and respiratory tracts and nervous system. It is also used to treat chronic bronchitis, asthma, whooping cough and is given as enema for intestinal flatulence. It is also employed in hysterical and epileptic affections, in cholera and as sedative in heart diseases. Hing is used also for flavouring curries, sauce and pickles.

Ferula narthex Boiss.:

This is a perennial herb. Lower leaves simple upto 80 cm long, upper leaves divided into segments and young leaves are densely hairy. Flowers appear in large terminal clusters, small and yellow in colour. Fruit is 1 cm long, roots are robust carrot-shaped.

This species is native to Kashmir in India, Tibet and Afghanistan.

Oleo-gum resin from rhizomes and roots constitute the drug. It contains 25% gum and also volatile oil and resin.

The resinous and scented substance obtained by the incision in the living rhizomes or roots of the plant.

Uses : It is a stimulant, antispasmodic, carminative, expectorant and laxative. It is used in indigestion, whooping cough, flatulent colic, infertile convulsions etc. It is a stimulant for nervous and respiratory system and prescribed in bronchitis and pneumonia in children. It is used as condiment in Persia.

Other species like Ferula foetida Regal and Ferula rubricaulis Boiss are distributed in Iran and Western Afghanistan and also contain gum; 4 to 20% volatile oil, mainly isobutyl-propanyl disulfide; 40-65% resin containing asaresinol ferulate and free ferulic acid. Rhizomes and roots of these species constitute the drug, and these species possess same medicinal properties as that of Ferula nor the x Boiss. It is a source of the drug asafoetida.

Foeniculum Mill.
Family : Umbelliferae

Foeniculum vulgare Mill. :

This is a hardy herb of 80 cm tall. Flowers are yellow. The seeds are small, oblong, ellipsoidal upto 6 mm long, straight or curved, greenish yellow in colour.

The species is originated from Mediterranean Europe, and Western Asia. The plant is commercially cultivated in different parts of the world. It is cultivated as a cold season crop in North India, Maharashtra, Gujarat, Karnataka, Punjab, Haryana and Rajasthan. It requires a fairly mild climate and propagated by seeds.

On steam distillation of seeds volatile oil is obtained. The essential oil content varies from 2.5 to 6.5% which contains d-pinene, camphene, d-L-phellandrene, dipentine, 50 to 60% anethol, fenchone, methyl chavicol, aldehyde and anisic acid. The seed constitute protein, fat, crude fibre, minerals, vitamins A, B & C. Funnel seeds also contain 9 to 12% fixed oil.

Uses : The fruits and essential oil is used as a flavour and said to be carminative. The leaves are used in culinary and leaf stalks as vegetables, which has diuretic properties. Funnel is regarded as a tonic and stomachic. The juice made from roots is febrifuge, purgative, sudorific and carminative. The decoction of leaves strengthen the sight. Seeds of these plants facilitate digestion and the essential oil made from seeds are aromatic. It cures asthma and windy colic. It is useful in diseases of chest, spleen and kidney.

Fragaria Linn.
Family : Rosaceae

A genus of twenty species and four are native to India. These are perennial scapigerous herbs with creeping stolons.

***Fragaria vesca* Linn. :**

It is a low perennial herb. Flowers are white in colour. Fruits are small, scarlet and edible. The species is native to Europe and North America. It is vegetatively propagated by runners. Commonly known as Alpine Strawberry.

Uses : The infusion of leaves of these plants is diuretic and diaphoretic. The fruits are refrigerant and diuretic and sometimes laxative. It is recommended for gout. The root is known as astringent and useful in genito-urinary diseases.

***Fragaria virginiana* Duchesne. :**

A low growing perennial herb with edible fruits. Leaflets are three. Flowers are with fine white petals. Fruit is a small, light scarlet in colour. The species is native to America. Commonly known as Verginia Strawberry.

Uses : Leaves, fruits and roots are useful. Tea prepared from leaves is used for diarrhoea. Fruits are used to treat kidney stones and gout. Roots are used to increase urine flow and to treat gonorrhoea.

Fraxinus Linn.
Family : Oleaceae

Under this genus there are about seventy species and six are native to India. These are deciduous trees, rarely shrubs.

***Fraxinus americana* Linn. :**

This is a hardy deciduous tree, grows to 60 cm tall. The tree bark is ash grey in colour and furrowed. Leaves are upto 25 cm long with seven leaflets.Commonly known as white Ash or American Ash.

Seeds are winged, occur in clusters upto 8 cm long, narrow and flat. Average life of the tree is 300 years. It is native to America.

Uses : Organic rich moist soil, chalky or gravely situation and sunny location are ideal for this species. They are suitable for seaside planting. This species is propagated by stratified seeds and also by grafting on common species.

Leaves, seeds and inner bark of trunk, roots, stems and branches are useful. A strong tea made from leaves is given to women after child birth. Seeds are aphrodisiac and also increase the urine flow and appetite. The bark is also taken internally to increase Urine flow and perspiration. The bark is used as a poultice on sores.

***Fraxinus excelsior* Linn. :**

It is a deciduous hardy tree, grows very tall and attains a height of 46 m. It has large pinnate leaves with accessory buds in axils. The petioles are grooved, and water is said to enter its groove which is absorbed by the leaves.

It is native to Europe, Asia and Himalayan region. Commonly known as European Ash.

Uses : The leaves and bark of this plant are aperitive, diuretic and febrifuge, it has been regarded as quinine substitute. The fruits are acrid, bitter and recommended for dropsy.

Fumaria Linn.
Fumariaceae

***Fumaria officinalis* Linn. :**

It is a hardy annual climber which attain a height of 1 m. Flowers are purple tipped crimson.

The species is native to Europe.

Rich ordinary soil and sunny situation are good and plants require support to climb. The plants are propagated by seeds.

Uses : This plant is slightly stimulating. It is administered as tonic and depurative, regarding it as hyposthenic. It is effective in treating skin diseases. This is applied in poultice form, to scurvy complaints.

Furcellaria Linn.
Algae

***Furcellaria fastigiata* Lam. :**

It is a red alga.

This is occurring in Arctic Sea and North Atlantic Ocean, Mediterranean and Black Seas. Commonly known as Danish Agar.

Uses : This is used as a substitute for agar. .Furcellaran, a gum is useful. It is a sulphated lectan. It is used now for foams and disintegrating tablets, suspensions and emulsions.

Galega Linn.
Leguminasae (= Fabaceae)

***Galega officinalis* Linn. :**

A hardy perennial herb upto 25 m tall. It is sometimes cultivated as fodder plant.

The species is native to Southern Europe.

It can be grown in ordinary rich soil in sunny situation. Plants are propagated by seeds or division of roots. After every two to three years, these are to be replanted. Commonly known as Goat's Rue.

Uses : This plant is sudorific for pestilential poison, epilepsy, snake bites and intestinal parasites.

Galium Linn.
Rubiaceae

***Galium aparine* Linn. :**

It is a hardy herbaceous climbing plant stems four angled with hairy joints with small reflexed hooks on the stem. The schizocarp is also provided with hooks. Leaves are arranged in whorls of eight. Flowers are white in colour. Fruits are very bristly.

The species is native to temperate region.

Ordinary rich moist soil and sunny position are suitable. Plants are propagated by seeds or division. Commonly known as Badstraw.

Uses : During flowering, entire plant is useful. It is used to treat vitamin C deficiency, fever and also to stimulate appetite aiuTincrease urine flow.

Garcinia Linn.
Guttiferae

Garcinia hanburyi Hook. :

It is evergreen woody perennials.

The species is native to South West Asia. Commonly known as Gambogi.

Uses : From the trunk bark, gum resin is obtained. It contains 75 to 90% resin containing chiefly gambogic acid and 18 to 28% gum. It is a powerful hydragogue cathartic.

Gardenia Ellis.
Rubiaceae

It has about two hundred sixty Palaeotropical species and eight are native to India. These are trees or shrubs, armed or unarmed.

The compost consisting of one part each of loamy soil, leaf mould, cow manure and charcoal is ideal. Well drained situation and sunny location are preferred. Application of liquid manure occassionally is beneficial.

Uses : The leaves a/nd buds of G. gummifera L. are used to treat nervous disorders of diarrhoea due to dentation and rubbed on gums to allay irritation. It occurs in W. Peninsula.

The leaves and fruits of G. campanuiata Roxb. are cooked and eaten. The fruit is cathartic. The resin of this species is given in corpulence and enlarged spleen.

The fruits of G. jasminoides Bills "Cape jasmine" are emetic diuretic and stimulant. It is used to treat jaundice and kidney disorder. The fresh flowers yield 0.8% essential oil.

The root preparation of G. turgida is used as a remedy for indigestion in children. Crushed roots are applied to the head against headache.

Gaultheria Kalm. ex Linn.
Family : Ericaceae

A genus of erect or scandant shrubs consisting of two hundred fifty species and fifteen species are native to India.

Gaultheria fragrantissima Wallich. :

A much branched evergeen shrub of 4 m tall. Leaves are leathery, broad, toothed margins, dotted with glands. Flowers are greenish white, small and appear in axillary branches. Fruit are enclosed in blue sepals and round in shape.

The species is distributed in the hilly regions of India between 1800 to 3000 m altitude.

Uses : The oil obtained from fresh plants is used for the treatment of rheumatism. This oil is added to certain ointment, are liniments to counteract their irritating effects. It has stimulant and carminative properties, and used to treat hookworms.

Gaultheria procumbens Linn. :

It is a woody creeping perennial plant. Leaves are aromatic, shiny and waxy. Flowers are drooping, arises singly from leaf axils and white in colour.

The plant is native to Eastern-North America. Commonly known as canada tea & creaping intergreen.

Uses : Leaves of these plants contain volatile oil which constitute 99% methyl salicylate. It also contains paraffin, aldehyde, ketone, ester and secondarv alcohol.

Entire plant in full bloom and the leaves are useful. The plant is a source of wintergreen oil which is used to reduce swelling of tissues, counterirritant to relieve gas and colic and it is also used as a flavouring agent. The plant is used to treat tooth ache, rheumatism, dysentery; and also to delay menstruation and promote milk flow in nursing mothers. Methyl salicylate is employed in liniment and ointments for pain relief in rheumatism, sciatica and lumbago.

Gaura Linn.
Onagraceae

Gaura parviflora Dougl. :

This is a hardy and hairy biennial plant, attain a maximum height of about 2 m. Leaves are 12 cm long, hairy and soft. Flowers are small, reddish in colour and appear on spikes.

Sunny well drained position is good for their growth. Plants are grown from seeds. Commonly known as Velvetweed.

Uses : Entire plant when in bloom is crushed to make poultice to treat arthritis and muscular pains.

Gelidium Lam.
Gelidiaceae
(Algae)

A plant body or thallus In Gelidium is composed of narrow, compressed or tubular axis with similar branches arising from it. There are several species of Gelidium which are useful like G. amansii Lam., G. cartilaginium Gaill, G. corneum, G. liatulurn, G. lingulatum, G. pacificum Okam, G. pristoides and G. sesquipedale. The agar of this algae is used in pharmaceutical industry. The dry matter of Gelidium contains ash 6 to 8%, nitrogen 3 to 4% fibre 18-20% galactans 25% and petosans 3 to 5 %. It contains polysaccharides which is a tasteless, colourless, odourless, transparent mucilaginous substance. This agar serves as a bulk laxative. In pharmaceutical preparations it serves as carrier for medication, ingredient in capsules, a base and disintegrator in tablets, suppositories, emulsions and lubricants.

G. amansii Lam. is native to Japan, G. cartilaginium Gaill is distributed in off coasts of South Africa, Mexico and America. C. corneum is found in South Africa, Morocco, Spain and Portugal, while G. liatulum, G. lingulatum and G. pacificum occur in Japan. G. pristoides is distributed in South Africa. G. tesquipedale is found in Morocco and Portugal. These grow in bunches hanging down from the rocks near low-tide mark.

Gelsemium Juss.
Loganiaceae

Gelsemium sempervirens (Linn.) Ait. :

This is a much branched, perennial, evergreen woody climber. Leaves are aromatic, lance shaped and short stalked. Flowers are tubular, attractive yellow and fragrant. Rhizomes are cylindrical with peculiar odour, brown or yellowish orange in colour.

It is native to America and Mexico. Commonly known as caro line yellow Jasmine or False Jasmine.

Uses : Flowers, roots and rhizomes are useful. Root and rhizome preparations are used to treat fever, pain and also used as a depressent of central nervous system. A tea prepared from flowers is said to cure coughs, stomach pains, shortness of breath, pleurising and help in child birth. It is used to treat migraine. It has also been reported that it is dangerous and toxic.

Genista Linn.
Leguminasae (= Fabaceae)

Genista tinctoria Linn. :

It is a hardy deciduous flowering shrub, native to Britain. It grows well in dry bank and stony ground. Open situation and ordinary rich soil is preferred. Commonly known as Dyer's Greenwood.

Uses : Seeds and flowers of these plants are purgative and are effective in sciatica. Flowers cure dropsy and scrofula. Poultice prepared from branches, flowers and pods have been used in cold abscesses, oedema and scrofulous tumours. The wash from the ashes has been recommended for dropsy.

Gentiana Linn.
Family : Gentianaceae

A large genus of about five hundred species and sixty five species are native to India, mostly Himalayan. Which are herbs or undershrubs.

Gentiana kurroo Royle :

This is a small perennial herb upto 40 cm height with tufted decumbent stems. Flowers are blue with white spots, to 5.2 cm long and 3.2 cm in diameter. Fruits are 3, cm long.

The species is native to North-Western Himalaya between 1800-4000 m altitude in Garhwal, Kashmir and adjoining hills.Commonly known as Indian Gentian.

Uses : The whole plant is a bitter tonic. The dried rhizome is useful for medicinal purposes. It has a pleasant flavour and is agreeable to taste. It is used for stimulating gastric secretion and improving appetite. It is an ingredient of many tonic for stomatic preparations. The roots are also used to treat fever, abdominal pains and is known as blood purifier.

Gentiana lutea Linn. :

This is a hardy perennial plant of 1.5 m tall. Flowers are yellow in colour. he species is native to Sourthern and Central Europe. It can be suitably cultivated in the North- Western Himalaya region of India at an altitude of 2800-3800 m. Commonly known as Yellow Gention.

Uses : The whole plant is a bitter tonic. The dried rhizome is useful for medicinal purposes. It has a pleasant flavour and is agreeable to taste. It is used for stimulating gastric secretion and improving appetite. It is an ingredient of many tonic for stomatic preparations. The roots are also used to treat fever, abdominal pains and is known as blood purifier.

***Gentiana lutea* Linn. :**

This is a hardy perennial plant of 1.5 m tall. Flowers are yellow in colour. The species is native to Sourthern and Central Europe. It can be suitably cultivated in the North-Western Himalaya region of India at an altitude of 2800-3800 m. Commonly known as Yellow Gentian.

Uses : Dried rhizome and roots constitute the drug. It contains glucosides likes gentiopicrine; also gentiomarine and anarogentine. The tannin like principle gentiamarine; gentisin, gentisic acid and the trisaccharide gentianose.

As antimalarial, gentiopicrine is used. Bitter tonic is prepared from roots which is used in dyspepsia and anorexia.

***Gentiana calesbaci* Linn. :**

Uses : Decoction of flowers of this plant is recommended for treatment of pneumonia and as a tonic and sudorific.

***Gentiana verticillata* Linn. :**

Uses : The flowers are used to treat intermittent fever.

***Gentiana villosa* Linn. :**

It is a perennial plant with ascending stems and lance-shaped leaves. Tubular shaped, greenish white to purplish-green flowers appear in the top of the plant in cluster.

The plant is native to America. Commonly known as Sampson's Snake root.

Uses : Roots and rhizomes of these plants have medicinal properties. It is used to treat indigestion, rheumatism and also as tonic for nervous distress.

Geranium Linn.
Family : Geraniaceae

A genus of about four hundred twenty species and twenty five are native to India, Mostly are herbs or undershrubs. Several perennial species are cultivated.

Geranium acellatum Cam. :

This is a prostrate annual herb with pink flowers.

The species is occurring in the moist places in the Himalayan ranges upto 2500 m altitude.

Uses : The plant has astringent and diuretic properties.

Geranium maculatum Linn. :

This is a perennial herb with single stem and thick rhizome upto 65 cm tall. Leaves are deeply cleft upto 18 cm across. Flowers are rosy purple to white in colour upto 3.5 cm acrossjjhe species is native to America. Commonly known as Spotted geranium or Alum Blowm.

Uses : Leaves and rhizomes of these plants before flowering are useful. Leaves are applied as poultice for healing of wounds. A tea preapred from leaves is used for sore throat and dysentery. The roots and rhizomes are antiseptic in nature and are used to reduce swellings, treat diarrhoea, diuretion and to haemostat from cuts, scratches and wounds.

Geranium nepalense Sweet :

This is a herbaceous plant and pubescent. Roots arises from the nodes. Flowers are pale purple in colour.

It occurs in the subalpine zones at Kantal, Ghuttu and Bhilagona valley upto 2700 m altitude.

Uses : The entire plant has medicinal value and is said to be diuretic and astringent. It is used to treat renal diseases.

Geranium robertianum Linn. :

This is an annual or biennial plant upto 50 cm tall.

The species is native to North America. Commonly known as Herb robert or Fixgeranium.

Uses : The plant is regarded as an excellent astringent to be applied to the wound. The crushed plants are used externally in some erysepalous affections. The plant is also effective in haemmorrhages and sore throats.

Geranium wallichianum D. Don. ex SW.:

This is a perennial herb, hairy, may attain a height of 1.5 m. Flowers are blue in colour.

The species is occurring at Belcha of Utter Kashi between 2500-3500 m. It is also found in temperate zones at Bhagirathi Valley, Kedarnath, Ponwali and Harsil of Garhwal Himalaya.

Uses : The root paste is given in the stomach disorders of infants, and this is also applied on the cut bruises.

Gerbera Gr.
Family : Compositae (= Asteraceae)

Gerbera languinosa Benth. :

It is a small herbaceous perennial plant with spreading leaves from base. Flowers are white and red.

Normally, found in the Garhwal Himalaya upto 2000 m elevation.

Uses : The white cotton like coating on the under surface of the leaf is used as tincture and for stounching wounds.

Geum Linn.
Family : Rosaceae

A genus consists of forty five species and only one species is native to India, Which are erect herbs.

Geum elatum Wall. :

This is a hardy perennial tufted herb to 50 cm tall; fibrous and woody perennial root stock; leaves are radical spreading. Flowers bright yellow on corymbose cymes.

Abundantly occur in alpine and sub-alpine pastures and open slopes of H.P. between 3200-4500 m elevation.

Uses : Roots of these plants possess medicinal properties.

Geum rivale Linn. :

This is a hardy perennial herb of 1.5 m tall. Leaves are three loabed. Flowers are purple. It has a thick rhizome.

This plant is native to Britain. Commonly known as water Avens.

Uses : Above ground parts of these plants before flowering, and roots are useful. Roots and tops of the plants are chopped and mixed with milk and sugar and taken internally for indigestion, run down conditions and dysentery. For treatment of haemorrhages, stomach ulcers, intestional worms and menstrual disorders a root decoction is used.

***Geum urbanum* Linn. :**

This is a hardy perennial flowering, soft herb. Flowers are pale yellow in colour.

The species is native to Australia and also commonly found in the valleys of Garhwal Himalayas between 2000 to 3500 m elevation. Commonly known as Avens or clove wort.

Uses : Roots are having aromatic clove like odour and possess medicinal properties. The plant is astringent, sudorific, cordial and febrifuge.

Gillenia Moe.
Family : Rosaceae

***Gillenia trifoliata* (Linn.) :**

It is a hardy perennial plant upto 1.2 m tall. Stems are reddish and hairy. Flowers are 2-5 cm across, red or white in colour.

The species is native to North America. Commonly known as Moench Indian Psysic.

Uses : Roots of these plant constitute the drug. It is regarded as tonic in small doses. Root bark is used to treat intestinal worm, rheumatism, dyspepsia, chronic diarrhoea and fever. Dried roots are laxative.

Ginkgo Linn.
Family : Ginkgoaceae

***Ginkgo biloba* Linn. :**

This is a tall tree of 40 m. Leaves are fan shaped to 15 cm across. The fruit is ill smelling, yellow, with an edible sweet kernel or nut within, upto 2.8 cm long. The species is native to Eastern China. Commonly known as Maiden hair tree or "Ginkgo'.

Uses : Fruit and seeds are medicinal. After removing nauseous flesh, oil is extracted from seeds. The seeds of this tree is edible. Fruits are used to soothe stomach trouble and to treat bronchial complaints, gonorrhoea and worms. The poultice is applied to cut and wounds. Nuts are used to release phlegm from the throat and to calm upset people. The timber is useful.

Glechoma Linn.
Family : Labiate

Glechoma hederacea Linn. :

This is a hardy perennial creeper. Leaves are silvery. Flowers are light blue in colour.

The plant is native to Europe. Commonly known as Grounding.

Uses : Ivy leaves are a popular remedy for binding cautery ulcers and blisters, they give a pleasant sensation of freshness. Ground Ivy is astringent, diuretic and tonic, useful in kidney diseases and as an anti-scorbutic. Fruit has slightly bitter taste with purgative properties.

Gliricidia Kunth.
Leguminasae (= Papilionaceae)

Under this genus there are about ten species distributed in tropical America and West Indies. In this genus there are small trees and shrubs.

Gliricidia sepium (Jacq) Walp. :

This plant species is widely cultivated in India as ornamental shade tree. For green manuring it is very useful. In some places flowers are used as vegetable.

Uses : Leaves, seeds and bark of these plants possess insecticidal properties and also considered poisonous to rats and rodents. The seeds yield fatty oil.

Gloriosa Linn.
Liliaceae (= Smilacaceae)

Gloriosa superba Linn. :

This is a tuberous rooted, perennial herbaceous climber and plant may climb upto 4 m. Flowers are orange, yellow or red.

The species is native to tropics. It occurs in the humid hilly districts of India. It is native to Tropical Asia and Africa. Commonly known as Glory Flower.

Uses : The tubers possess abortifacient, stimulant and anthelmintic properties. It is used to treat leprosy.

Glycyrrhiza Linn.
Leguminosae (= Fabaceae)

***Glycyrrhiza glabra* Linn. :**

This is perennial herb or subshrub for subtropical and temperate zone. The plant attain a maximum height upto 1.8 m. The underground stem grow horizontally upto 2.2 m length, highly branched consisting of short tap root with large number of rhizomes. The diameter of root varies from 0.85 to 2.4 cm, grey-brown exterior and yellow interior. Leaves are alternate, pinnate, with 10 to 18 ovate, yellow green leaflets covered with soft hairs on underside. Flowers appear in axil of terminal and axillary leaves in raceme, pea-like, lavender to purple in colour. Seed pod is 2 to 2.6 cm long containing 4 to 6 seeds. Different varieties of this species are in cultivation namely typica, a Spanish liquorice; glandulifera, Russian liquorice and violacea a Persian liquorice.

It is distributed in Southern Europe, Syria, Iran, Afghanistan, Russia, China, Pakistan and Northern India. This plant is cultivated in Russia, UK, USA, Italy, France, Germany, Spain, China and Northern India, and also collected from wild source for its utilization. The semi-arid regions of Gujarat, Haryana and Rajasthan in India are suitable for cultivation of this crop. Commonly known as Liquorice.

Uses : The liquorice roots are used both in medicine as well as for flavour. It is known as demulcent, expectorant, laxative and sweetener. Various preparations of liquorice is administered in cough, gastric and duodenal ulcers, cardiac asthma, hypertension, rheumatoid arthritis and dermatitis. Liquorice is sweet in taste and free from bitterness. It is also used in flavouring as sweetener and as a preservative. The extract is used for flavouring cigarettes, chewing and snuff tobacco.

Gossypium Linn.
Family : Malvaceae

***Gossypium hirsutum* Linn. :**

This is a shrub with palmately lobed leaves.

It is native to South America. This species is cultivated and naturalized throughout tropics and subtropics. Commonly known as short stable cotton.

Uses : Seeds are used for extraction of oil. This is a semidrying oil which constitute fatty acid, glycerides, phospholipids and phytosterol.

he oil is used in pharmaceutical industry as emolient, lubricant and as a vehi e for injections. The fibre obtained from the seeds is utilised in the manu acture of textiles and rubber-type fabrics. Oil from the seeds is edible.

Gracilaria Grev.
Family Gracilariaceae
(Algae)

Gracilaria conferioides Grev. :

This is mostly occurs in the coasts of southern hemisphere and harvested commercially in South Africa. It grows below lowtide mark on the Atlantic side of the Cape. The best plants are found where there is sheltered water and bottom is sandy or sand with rocks.

Agar is used in drugs which is a polysaccharide, is colourless, tasteless, odourless and transparent mucilaginous substance. The dry matter of Gracilaria contains ash 4-7%, nitrogen 0.7 to 1.5%, fibre 4.5%, galactans 22-23% and pentosans 2.0 to 2.2%. Gracilaria agar is a complex of several gelling material. The total agar content lies between 55 to 65% on a dry weight basis.

Uses : This is generally in pharmaceutical industry as suspending agent, suppositories, lubricants, emulsions and ingredient of capsule. It has laxative properties. The agar produced from this species serves extremely well in the manufacture of sweets.

Gratiola Linn.
Family : Scrophulariaceae

Gratiola Officinalis Linn. :

It is a herb with opposite, entire or toothed leaves.

The species is native to temperate zones. It grows well in moist places. Commonly known as Hedge Hyssob.

Uses : The leaves are emitive and also purgative. It is effective in skin diseases, worm complaints and intermittent fever. Strong dose may cause serious accidents and this plant should be administered with much care and caution.

Grewia Linn.
Family : Tiliaceae

A genus of about one hundred and sixty species and forty four species are native to India, which are shrubs or trees and many yield edible fruits. Few species have valued for timber.

***Grewia asiatica* Linn. (= *G. elastica* Royle) :**

This is a tree with densely crowded yellow flowers. Bark is grey coloured. Leavs are ovate, rounded and cordate.

It is occurring in the Sal forests of India and commonly found in Garhwal of Uttaranchal plains.

Uses : The bark and fruits have medicinal value. Fruits are edible. The bark yield strong white fibre, wood is used for making sports goods. Commonly called Phalsa.

***Grewia optiva* J.R. Dramm ex. Prurret. (= *G. oppositifolia* Roxb.) :**

This is generally occurring at Arakot of Utter Kashi between 600-1800 m elevation.

Uses : The extraction of bark is given to pregnant women for smooth delivery. The extraction is also given in constipation. The ripe fruits are eaten. The bark is used as a detergent for cleaning hair of ladies.

***Grewia sclerophylla* Roxb. :**

This is a small shrub which may attain a height of 2.4 m. Leaves are ovate or acuminate. Flowers are pale-yellow.

It is found in the sub Himalayan tract of Tehri Garhwal. Commonly known as Gurhheli.

Uses : The fruit and bark are used and medicine.

Grindelia Willd.
Family : Compositae (= Asteraceae)

***Grindelia squarrosa* (Pursh.) Dunal :**

This is a coarse hardy perennial plant of 95 cm in height. Leaves are 8 cm longand toothed. Flowers are 5Tcm in diameter, sticky, and yellow in colour.

The species is native to America. Commonly known as Gum Plant.

Uses : Leaves, flowering tops and roots are useful. A tea prepared from flowering tops is used to treat skin rashes, due to toxic and poisonous plants, urinary ailments and stomachache. Tea prepared from root is given for liver trouble. For rheumatic joint pains a poultice of plant is applied. For lung diseases, small pox, pneumonia and gonorrhoea, a decoction of the flowering top is administered.

Guaiacum Plum.
Family : Zygophyllaceae

A genus consists of twelve species native to warm America and West Indies, which are evergreen shrubs and trees in this genus.

Guaiacum officinale Linn. :

This is a very attractive tree. Flowers appear in clusters at the end of branches, which are blue in colour. The flowers fading to pale-silvery with age.

Uses : The resins from this plant is used as fat stabilizer. This is used to prevent loss of flavour and ransidity in preserved and dehydrated food. The resin in innocuous. The fruits are edible and the bark-fibre made into ropes.

Gutierrezia Lag.
Compositae (= Asteraceae)

Gutierrezia sarothrae (Pursh.) Britt. & Rusby :

This is a perennial shrub of 80 cm tall. Leaves are slender with smooth margins. Flowers appear in dense cluster, sticky and yellow.

The plant is native to America. Germany known as Broom Weed.

Uses : Plant in full bloom stage is used to treat snake bite, stomachache, malaria and rheumatism.

Habenaria Willd.
Family : Orchidaceae

A large genus of about six hundred species of which one hundred are native to India, which are terrestial plants, leafy herbs, flowers appear in spikes or racemes.

Habenaria acuminata Linn. :

This is a terrestrial orchid of 80 cm tall. Leaves are erect, linear-lanceolate, to 12 cm long and 2 cm broad. Inflorescence elongate, lax flowered, yellowish green in colour, lip purple, trilobed.

Native to Annamalai Hills, Kerala and Garhwal Himalaya.

Uses : Tubers are used as a tonic because of its nutritive properties.

Habenaria commelinifolia Wall. ex. Lindl. :

This is a robust plant of 95 cm length. Leaves are oblong, finely acuminate to 18 cm long and 4 cm broad. Inflorescence 25 cm long, many flowered, white in colour.

It is occurring at Rajakho of Chhindwara District of M.P., Kumaon, Central India, Orissa, Chota Nagpur and Karnataka.

Uses : The root decoction (about 20 ml) is taken twice a day for one month in urinary trouble.

Habenaria edge worth :

This tuber is useful which is considered as a blood purifier and rejuvenator.

Habenaria goodyeroides Don. :

This plant is 80 cm long, leaves are elliptic upto 18 cm long and 8 cm broad. Inflorescence to 25 cm long, many flowered. Flowers are sweet scented, olive green in colour. Tubers are ellipsoid or oblong and hairy.

It occurs in Bihar, Orrisa, Sikkim, Manipur, Garhwal, Mussorie and Khasia Hills at 1200 m altitude.

Uses : The tubers are used as tonic.

Habenaria griffithi Hook :

This is a tuberous rooted terrestrial orchid. Flowers are white, pinkish in colour.

The species is distributed in Garhwal between 1800 to 2500 m altitude.

Uses : The tubers of these plants possess nutritive properties and are used as a tonic.

Habenaria intermedia D. Don. :

This is a semi-terete orchid of 28 cm long. Leaves are rounded at the base, long acuminate to 12 cm long and 6 to 8 nerved. Inflorescence 4 to 6 flowered. Flowers are 5 cm across, white or greenish.

It is native to Kashmir to Kumaon, Sikkim and Khasia Hills between 800-2500 m altitude.

Uses : It is one of the eight drugs of the 'Ashtravargra group well known m Ayurveda. Tubers are used as general tonic. It is an ingredient of chyavanprash.

Haematoxylon Linn.
Family : Leguminosae (= Fabaceae)

Haematoxylon campechianum Linn. :

This tree is native to Tropical America and cultivated in India. It is an ornamental thorny tree.

Heart wood contains haematoxlin used in dyeing. It is sudorific.It is used in medicine as a tonic and astringent, in fluxes and towards the end of dysentery.

Hamamelis Linn.
Family : Hamamelidaceae

Hamamelis virginiana Linn. :

This is a hardy deciduous shrub of 6 m tall with forked branches and smooth brown bark. Leaves are almost round, with short stalk. Flowers are yellow thread like. Fruits appear in cluster. Seeds are shiny black.

The species is native to Canada and America.

Uses : Leaves, twigs and bark constitute the drug. Leaves contain gallotannins and some condensed catechins and proanthocyanidins. Bark contains gallotannins, condensed tannins and also proanthocyanidins of cyanidin and delphinidin.

Leaves, twigs and bark of these plants are useful. It has astringent and homeostat properties. The extracts from the twigs, leaves and bark are utilised in several medicinal lotions for sprains, bruises and shaving lotions. Distilled extract of plant is used to treat haemorrhoides. The bark is also applied to skin tumors and skin inflammations. It is also used in eye lotions. The bark is chewed to freshen the mouth.

Hedeoma Pers.
Family : Labiatae

Hedeoma pulegioides (Linn.) Pers. :

This is a much branched annual of 60 cm tall. Leaves are small, hairy, erect and ovate. Flowers are 0.8 cm across, scented and blue in colour.

The species is native to America. Commonly known as Pennyroyal.

Uses : Plant in full bloom is useful. This plant is said to be a stimulant and useful to reduce spasms and convulsions. A tea prepared from above ground parts of plants is used for delaying menstruation, induce perspiration and also to treat cold, pneumonia, headache and itching of eyes.

Helenium Linn.
Family : Compositae (= Asteraceae)

Helenium amarum (Raf.) H. Rock. :

This is a hardy herbaceous perennial of 95 cm long. Leaves are narrow and smooth. Flowers are yellow, appear in profusion.

The species is native to West America. Commonly known as sneeze weed.

Helenium autumnale Linn. :

This is a perennial plant up to 1.8 m tall. Flowers are yellow in colour.

There are varieties like pumilum which is 90 cm tall and striaturn with yellow and brown flowers of 1.40 m tall.

The species is native to North America.

Uses : Leaves and flowers of these plants are useful. It is used to treat intestinal worms and colds.

Helenium hoopesn Gray.

This is a herbaceous perennial of 80 cm tall. Leaves are upto 30 cm long. Flowers are 8.5 cm across and yellow in colour.

The species is native to North America.

Uses : Roots of this plant constitute the drug. A tea prepared from roots is used to cure diarrhoea, intestinal worms and stomachache. By rubbing the dried and ground roots, pains due to pulmonary diseases and rheumatism can be relieved.

Helianthemum (Tourn.) Mill.
Family : Cistaceae

Helianthemum canadense (Linn.) Michx. :

This is a hardy evergreen perennial hairy shrub growing upto 50 cm in height. Leaves are small upto 2.8 cm long. Inflorescence is cincinnus. Flowers are yellow in colour. Flowers contain no honey and is homogamous, with sensitive stamens which move upwards when touched. Commonly known as Rock-rose.

Uses : Entire plant is useful when in bloom. The decoction of plant said to cure diarrhoea, sore throat, skin diseases and syphilis.

Helianthemum vulgare Gaertn. :

This is a trailing shrub with yellow flowers. The species is native to Europe.

Uses : Plant grow in a sunny bank or rockeries in light sandy soil. Propagated by seeds or by vegetative means. It is reported to be an astringent and vulnerary.

Helianthus Linn.
Family : Compositae (= Asteraceae)

Helianthus strumosus Linn. :

This is an annual plant of 4 m tall. Leaves are rough and pale green in colour.

The species is native to America. Commonly known as Dahlia Sun Flower or Many However peremial sun flower.

Uses : Seeds and roots are useful. Crushed root is applied to bruises. Seeds are used to clear phlegm and increase urine flow.

Helianthus tuberosus Linn. :

A tuberous roots of this plants are used as vegetable.

The plant is native to North America. Commonly known as Jerusalem Artichoke.

Uses : The tuber contain the carbohydrate, insulin which on hydrolysis yield levulose. Levulose sugar is useful in the diet of diabetic patients.

Helicteres Linn.
Family : Sterculiaceae

Helicteres esora Linn. :

This is a shrub with spreading habit. The bark is grey coloured. Fruit is cylindrical and pubescent.

The species is native to Asia and Australia. In India, it is found at Chimtipur of Chhindwara District of M.P. and also common in the hill slopes upto 1500 m altitude at Garhwal Himalaya.

Uses : About 20 gms seed powder is used thrice in a day for a week in the treatment of dysentery and stomach pain. Bark is also used to treat diarrhoea and dysentery. The juice of the root is given in diabetes and prescribed for intestinal injection.

Helleborus Linn.
Family : Ranunculaceae

Helleborus foetidus Linn. :

This is a hardy perennial plant with fibrous roots. Each shoot from the stock take several years to reach maturitv and flower.

The plant is native to Europe. Commonly known as Helleborne.

Uses : Long ago, this plant was used for brain diseases, epilepsy, nervous diseaes, worms and scurvy. Now-a-days, it is rarely used except in desperate cases where a strong repubion is needed.

Hemidesmus (Linn) Schult.
Periplocacea (Asclepiadaceae)

Hemidesmus indicus (Linn.) Schult. :

This is a prostrate or twining undershrub. The rootstock is aromatic. The species is native to India commonly occurs from upper Gangetic plains eastwards to West Bengal to M.P. and Southern India. Commonly known as country sarsaporilla.

Uses : The root decoction (about 10 ml) is used thrice a day for one month as blood purifier and in skin diseases.

Heptatica Mill.
Family : Ranunculaceae

Hepatica acutiloba DC. :

This is a hardy perennial herb of 30 cm tall. Leaves are liver shaped, with 4 to 6 lobes. Flowers are white to purplish in colour. The plant is native to temperate Eurasia.

Uses : Leaves and roots of these plants are useful. The plant is acrid and caustic. It is used as a gargle for sore throats, in poultice for hernia, skin diseases, diabetes, urinary and lung diseaes. Tea prepared from roots and leaves is used to treat dizziness and coughs. A distilled water from it has been used as a cosmetic to freshen and tone up the skin and used to combat the effects of wind and sunburn and to remove freckles. The roots are chewed to relieve persistent cough.

Heracleum Linn.
Family : Umbelliferae

Heracleum maximum Bartr. :

This is a coarse hardy perennial herb with large feathery shaped green leaves. Plant attain a maximum height of 4 m. It has woody appearance. Flowers are white or purple.

The species is native to tropical mountains. Commonly known as Cow Parsnib.

Uses : Roots of these plants constitute the drug. The roots are prescribed to take internally for asthma, indigestion, diarrhoea, colic and gas, Powdered roots and its preparations are rubbed on the body parts to treat rheumatic pains, gum trouble and heart palpitations.

Herniaria Linn.
Family : Illecebraceae (Caryophyllaceae)

Herniaria glabra Linn. :

This is a hardy trailing perennial herb. Foliage is moss-like, forming mats which turn deep bronze red in winter. Leaves are dark green in colour and 2.6 cm long. Flowers are apetalous.

This species is native to Europe. Introduced into gardens of India for its attractive foliage. Commonly known as "Rupture - Wort" or "Burst Wort".

Uses : It is used as a stomachic and also recommended fur urinary diseases.

Heuchera Linn.
Family : Saxifragaceae

Heuchera americana Linn. :

It is a hardy perennial herb of 60 cm tall. Leaves are toothed. Flowers appear in long spike, red in colour.

The plant is native to North America. Commonly known as Alam root.

Uses : Roots and leaves of these plants are useful. To treat diarrhoea an Tnh^ion of root is used. For sore-throat a decoction of root is prescribed. Leaf and root poultice is applied on wounds, skin abrasions and skin lesions.

Hibiscus Linn.
Family : Malvaceae

A genus generally consisting of two hundred five species and about forty species are native to India.

The seeds of H. abelmoschus Medic. "Muskus Okra" are used as tonic, stimulant, antispasmodic, stomachic and carminative. This is also administered as tincture. The leaves and roots are applied as poultice, and the paste made from seeds with milk is applied for itches. The oil obtained from seeds is used in perfumery. Green pods, tender leaves and shoots are eaten.

Uses : The seeds of H. cannabinus L "Ambari Fibre Plant" are considered appetising, aphrodisiac and stomachic. Seeds are used externally as poultice for pains and burns. The juice extracted from flower is mixed with sugar and black pepper and given for treating biliousness. The plant also yields quality fibre. This is mainly cultivated as fibre crop in A.P., Karnataka, Maharashtra, M.P. and Bihar.

Holarrhena R. Br.
Family : Apocyanaceae

Holarrhena antidysentrica (Linn.) Wall. :

This is a dwarf tree of 12 m tall, latex milky white. Leaves are thin, ovate, upto 35 cm long, nerves are conspicuous and leafstalk small. Flowers appear in large terminal branches, white and fragrant. Fruits are linear, slender, cylindric upto 60 cm long and 2 cm thick, dark grey with white specks. Seeds are 1 cm long. This species is native to India and occurring throughout the country upto 1500 m altitude.

It is very commonly found in Madhya Pradesh and Maharashtra and sporadically in the tropical zones of Garhwal. Commonly known as "Kurchi".

Uses : It contains primarily conessine alkaloid, and other alkaloids are concuressine, conessimine, isoconessimine, holonamine, kurchessine, kurcholessine, kurchicine and a photoplasmic poison like emetine.

Bark and seeds of these plants are useful for medicinal purpose. Conessine has been used in amoebic dysentery and vaginitis. The bark of the plant has digestive, tonic and febrifuge properties. The seeds are effective in diarrhoea fevers, jaundice and stone in. bladder. The alkaloid conessine retard the growth of tubercular bacillii.

Humulus Linn.
Family : Cannabidaceae (= Moraceae)

Humulus lupulus Linn. :

This is a twining and hardy perennial climbing herb, growing to 8 m in length. The leaves are 13.6 cm long with stems about 8.6 cm long, three lobed, smooth. Inflorescence is cymose, dioecious. Fruit is membranous and cone like.

This species is native to Europe. It is cultivated in Western Himalaya. Commonly known as Hop.

Uses : Fruits of this plant are useful. This plant has tonic properties and also regarded as diuretic, diaphoretic and anthelmintic. It stimulates the appetite and helps digestion. It is said to increase vital energy and vigour of the organ. It is used as a sedative and depurative in homoeopathic medicine. It is also used for treating fever, intestinal worms and rheumatism. The fruit catkin is used for brewing.

Hydnocarpus Gaertn.
Family : Flacourtiaceae

A genus of about forty five species and eight species are native to India. Which are trees with alternate leaves. The berries are globose, many seeded with hard rind.

Hydnocarpus kurzu (King) Warb. :

This is a tree of 16 m tall or more with drooping branches. Leaves are leathery, pointed at tip and upto 25 cm long. Flowers appear in small axillary clusters, small, yellow in colour. Fruits are round, 8 cm across, brown in colour, with many seeds.

It is native to evergreen forests of Assam and Tripura. Commonly known as Spongy berry tree or chaulmoogra.

Uses : The seed yield oil, certain preparations of oil are injected for the treatment of leprosy. The bark contain tannins and is used to cure fever.

Hydnocarpus laurifolia Sle. (= H. weightiana Blume) :

It is native to China, Southeast Asia and Western India. It is found in Western Ghats of India.

Uses : Seed oil is used in medicine. The oil is rich in cyclopentenyl monocarboxylic acids, chaulmoogric acid, hydnocarpic and garlic acid. It also consists of lower homologues of chaulmoogric acid, olic and palmitic acid.

The oil is used to treat leprosy, chronic skin diseases and rheumatic pains, strains, tubercular laryngitis and inflammations.

Hydrangea Linn.
Family : Hydrangeaceae

Hydrangea arborescens Linn. :

This is a hardy deciduous flowering shrub of 4 m tall. Leaves are large and opposite. Flowers are white or greenish in colour and appear in clusters at the top of the branches.

The species is native to East America. Commonly known as Hydrangeaceae.

Uses : Roots of these plants constitute the drug. It is regarded as tonic and laxative. Roots are used to treat kidney stones and increase urine flow.

Hydrastis Ellis. ex Linn.
Family Hydrastidaceae

Hydrastis canadensis Linn. :

This is a low growing perennial plant upto 40cm height. Plant has one stem and two, five or seven-lobed leaves on tne top of the stem. Flowers are greenish white and without petals. -Fruits are not edible and look like raspberry.

The species is native to America. Commonly known as Golden Seal or Yellow Puccoon.

Uses : Dried rhizomes and roots have medicinal properties. It contains hydrastine, berberastine, berberine, canadine and reticuline alkaloids.

Leaves, tops of plants and roots are useful. It is diuretic and homostat in uterine haemorrhage. Roots are chewed to heal mouth sore. Leaf infusion is used to treat liver ailments and gastric disorders and stomach trouble. A tea prepared from root is a tonic and appetizer. Powdered root is applied on wounds and cuts. It is also used in eve lotions.

Hygrophila R. Br.
Family : Acanthaceae

Hygrophila auriculata (Schum.) Heine. :

This is an unbranched stout erect herb which may attain a height of cm. Stems are four angled and hairy. Leaves are whorled, 6 at a node, each possess sharp yellow coloured spines on its axil. Flowers are purple blue, about 4 cm long, 2 lipped, lower lip folded, 8 in a whorl.

It grows in marshy and moist situation and found throughout India. Commonly known as Talmakhana.

Uses : Entire plant including roots constitute the drugs. Seeds and roots are diuretic. Leaves are used to treat cough and urethral discharges. Seeds are also employed in the treatment of veneral diseases. The drug from plants is used to cure diseases of urinogenital system, jaundice, dropsy and rheumatism.

Hyocyamus Linn.
Family : Solanaceae

A genus of as twenty five species and four are native to India.

Hyocyamus niger Linn. :

Plant species is occurring in the Western Himalayas from Kashmir to Garhwal between 1800-3500 m elevation. Its leaves and seeds contain tropane alkaloids. This is an erect, biennial herb, densely covered with glandular hairs and attain a height of 1.8 m. Leaves are 18-25 cm long, coarsely dentate and pinnately lobed. Upper leaves are smaller and divided into many segments. Flowers are 4-5 cm in diameter, yellowish pale, some borne solitary at the place of branching of stems, others borne on terminal cymes. Fruits are globose, 1.4 cm in diameter, containing numerous black coloured seeds of 1.6 mm long, marked with fine conspicuous reticulation. It is distributed from Scandinavia and Southern England to Mediterranean and Siberia, North Africa, Western Asia and India. It is commercially cultivated in America.'Tn India, it is cultivated in Nilgiri Hills, Kashmir, U.P., M.P., Maharashtra and Punjab.

Although these plants are inhabiting in temperate regions, areas with equitable climate, throughout the year are also suited for their cultivation.

Uses : All parts of the plants contain alkaloids. The leaves and flowering tops of this plant contain 0.1% to 0.15% total alkaloids. From this crop, important alkaloids like hyosyamine and hyoscine are extracted. The drug is useful for relieving certain painful spasmodic conditions of muscles, and in nervous irritations as in hvsteria and cough. It dialates the pupil of the eye. This has sedative effects on brain. The drug is also administered against asthma and whooping cough. It soothes the cough and induces sleep. The seeds of these plants have medicinal properties, they are pasted and applied locally on pains. This plant, however, is most violent, bitter narcotic poison. Externally it may be used as a poultice, it is maturative and narcotic.

Hyocyamus muticus Linn. :

A plant is native to Egypt, Sudan, Syria and Afganistan. It is cultivated in America and India. A perennial herb, attain a maximum height of K'm. Flowers are funnel shaped, yellow or greenish white. Seeds are kidney shaped. They prefer tropical climate and light medium organic rich soil for cultivation and propagated from seeds.

The alkaloid content of this plant is more and used as a source of hyoscyamine and trophane alkaloids. The total alkaloids content ranges from

C0.6 to 1.8%. This is used as poison, as narcotic, sedative and antiseptic and mydriatic. The leaves are also smoked for intoxication. This species is also useful for the production of atropine. Commonly known as Egyptian henbane.

Hypodematium Kunze.
Family : Aspidiaceae (= Hypodematiaceae)

Hypodematium crenatum (Forsk.) Kuh. :

This is a, fern, commonly occuring at Mt. Abu and Gwaparnath of Rajasthan.

Plant is used in gynoecological disorders and for getting relief from insect bite.

Hyssopus Linn.
Family : Labiatae

Hyssopus officinalis Linn. :

This is a hardy evergreen shrub upto 80 cm tall. Leaves are narrow, sessile and aromatic. Branches are erect. Flowers are bluish-purple. Nutlets are dark brown, narrow and smooth.

This is native to Southern Europe and temperate zones of Asia, it occurs in Himalaya from Kashmir to Kumaon. It is cultivated in Southern France for its essential oil. Commonly known as tyssop.

Uses : The herb contain volatile oil, codine, tannins, carotene, xanthophyll, fat, sugar and ursolic acid. It also yield a greyish yellow dye. The yield of oil is 0.15 to 0.30% from fresh herb. Hissop oil is colourless, aromatic, camphoraceous odour, warm and slightly bitterish in taste. The oil consists of ketone-1-pinocamphone.

Leaves and flower tops constitute the spices which has aromatic odour, and used for flavouring food. Shoots and flowers, infused in water, are largely used as an expectorant. Hissops oil in one or two drops promotes expectoration in bronchial catarrh and asthma. Preparation from flowers are used for asthma, as a stomach stimulant and pectoral to loosen the incise glairy matter. Externally used as a resolvent particularly for ecchymosis of the eyelids and eye. Flowers and leaves on distillation yield oil for perfumery and flavouring liquors. It is also used to treat cold, cough and lung complaints. A tea prepared from plant is used to treat urinary, pulmonary, digestive trouble, nervous disorder and toothache.

Ilex Linn.
Family Aquifoliaceae

A large genus of about four hundred species of shrubs and trees, and about twnety two species are native to India.

Ilex aquifolium Linn. :

This is a hardy dwarf shruby tree, attain an ultimate height of 10 m. Leaves are dark green. Wood is white in colour. Flowers are dioecious. Berries are red. There are several varieties in it like "albo-pictum", "belearica", "ferox", "inesme" etc. The plant is native to Europe and Asia.

This is a good seaside plant. Well drained sunny banks, exposed slopes and ordinary rich soil is good for this species. They are propagated from stratified seeds. Generally known as common holly or Aunt Mary's tree.

Uses : Holly leaves are febrifuge, sudorific and antiarthritic. Timber is used for cabinet-making.

Ilex opaca Ait. :

This is an evergreen tree upto 18 m in height. Leaves are with spiny teeth upto 15 cm long, green above and yellowish green below. Male and female flowers borne in separate trees. Fruits are bright red berries.

This species is native to America. Known as American Holly.

Uses : Ripe fruits, stem and root bark are useful. Decoction of bark and fruit is~laxative and used to treat worms, fever, coughs, pleurisy, constipation, gout, rheumatism and tumors. To treat colds, coughs and tuberculosis root bark decoction is useful.

Ilex vomitoria Ait. :

This is an evergreen shrubby tree, may attain a height of 10 m. Leaves are shiny upto 6 cm long. Flowers appear in clusters, white in colour. Fruits are red or orange upto 2 cm in diameter. Generally called Youpon.

Uses : Leaves are useful. Infusion of leaves is laxative and narcotic. It is also used as tonic.

Illicium Linn.
Family : Illiciaceae

Illicium verum Hook. :

This is an evergreen tree upto 14 m tall. Leaves are upto 15 cm long and 2.5 cm broad, eliptic to oblanceolate. Flowers are solitary, white to red in colour.

Fruits are reddish brown, star shaped. Seeds are brown, compressed, ovoid, smooth and shiny. The carples have an anise-like aroma. It is native to Southern China and Southeast Asia.. This plant is cultivated in China and Indo-China, Known as Star Anise or Ansaphal.

Uses : Fruits contain 2.5 to 5% essential oil which constitute 80 to 90% anethole, and also chavicol methyl ether and safrole. The oil is colourless or pale yellow with characteristic aromatic odour and taste.

It is used as a flavouring agent and as carminative. The fruit is chewed to sweeten the breath and to help digestion. It possesses carminative and stomachic properties and is used to treat flatulence, dysentery and spasmodic affection of the intestinal canal. Oil is used for flavouring candy, confectionery, tobacco, liquors, pharmaceutical preparations and chewing gum. The oil is stimulant, stomachic, expectorant, diuretic and carminative.

Impatiens Linn.
Family : Balsaminaceae (= Geraniaceae)

Impatiens balsamina Linn. :

This is a hardy annual plant upto 80 cm tall, usually branched and bushy and bear large number of flowers. There are different colours in flowers.

The species is native to tropical Asia. Commonly known as Balsam or Balsam Apple.

Uses : The leaves are used as vulnery and detergent.

Inula Linn.
Family : Compositae (= Asteraceae)

Inula helenium Linn. :

This a hardy herbaceous perennial and native to Asia. Ordinary rich soil, moist sunny location are preferable. Commonly known as Alacompane. Propagated by seeds or division of roots. Plants require annual top dressing with well decayed manure.

Uses : It is used in pharmacy, they are bitter and aromatic.

Ipomoea Linn.
Family : Convolvulaceae

This is a genus of over five hundred species and sixty species are native to India. These are herbs or shrubs usually twining, sometimes prostrate, floating or erect.

Ipomoea hederacea Auct. non Jacq. :

It is a spreading herbaceous twining plant. Stems are retorsely covered with hairs. Leaves ovate-cordate, 3 lobed, 8-14 cm in diameter, flowers in axillary stout peduncled, upto 5 cm long, blue coloured tinged with pink. After plucking, flowers become purplish in colour. Corolla funnel shaped, stamens 5, filaments hairy at base, stigma lanceolate. Fruit is a capsule, 3 celled, dull brown, glabrous. Seeds are small, glabrous, black in colour with a brownish spot at the lulum. Plant is widely distributed in tropical and warm temperate regions of the world.

It is found throughout India upto 2000 m elevation.

Uses : Seeds are used in medicine. Seeds contain lf£% crude resinous matter, a fixed oil, small amounts of saponin, mucilage and tannin. It has nauseating acrid taste and disagreeable odour.

Dried seeds are used as purgative. In over doses the drug cause symptoms of irritant poisoning. It is also used in medicine as extract, a tincture and a compound powder. Large number of species are found in India and many of them have laxative or purgative properties.

Ipomoea orizabensis Ledenois :

The Plant species is native to Mexico. The roots and resin are used in medicine. The root contain 6 to 18% resin of which main constituents are jalapin, ipuranol, ipurganol and also sitosterol. It is hydragogue cathartic. Commonly known as mexican scammony.

Ipomoea purga Hayne. :

This is distributed in nature of Central America, Mexico. Dried roots and resin have medicinal properties. The roots contain 10 to 20% resin, containing convolvulin, also isolated scammonin. It is a drastic hydragogue cathartic. Commonly known as Jalop Plant.

Iris Linn.
Family : Iridaceae

A genus of about three hundred species and twelve species are found in India. Plants are rhizomatous herbs.

Iris germanica Linn. :

This is a tall bearded iris which are common rooted perenpial plant upto 80 cm tall. Flowers are purple and lilac, fragrant.

The species is native to Southern Europe. Commonly known as Flag tres or Blue Iris.

Uses : It possesses acrid principle and used to treat irritation in sores. The rhizome of this plant is diuretic and purgative. The dried root stock is used for various purposes.

Iris kumaonensis Well. ex. Don. :

It is a perennial herb which flowers between May-July. It has creeping rhizome. Flowers are blue in colour. Freqently occur in the temperate zones, on open grassy slopes of Northern Himalaya between 3500-4500 m.

Uses : Roots are used against kidney trouble and fever.

Jacaranda Juss.
Family : Bigoniaceae

A genus consists of fifty five species and native to warm South America and West Indies, which are shrubs and trees in it.

Jacaranda acutifolia Humb. & Bonpl. :

It is an elegant tree with finely divided foliage. Flowers are blue in colour, appear in loose pyramidal panicles. Commonly known as Jacaranda.

Uses : The leaves and bark of this plant are used to treat syphilis and blennorrhagia. Powdered leaves are used for vulnerary. The bark infusion is used to treat ulcers.

Jasminum (Tourn.) Linn.
Family : Oleaceae

It is a perennial flowering plant widely cultivated and esteemed for their attractive, fragrant flowers. Jasmine is a climbing or spreading shrub with opposite or alternate, 3 foliolate or pinnate leaves, sometimes with only one leaflet and often green angled stem. Flowers are generally about 3.4 cm in diameter in many flowered clusters and sometimes solitary on the ends of branchlets. This genus comprised of three hundred five species dispersed throughout the warm parts of Europe, Asia, Africa and Pacific region. Large number of species, however, are centred around the area comprising of Himalayas, China and Malaysia.

Uses : Some of the species like J. auriculatum Vahl., J. grandiflorum L., J. officinale L., J. pubescens Willd, J. sambac (Soland) Ait are very rich in the essential oil content and producing oil by solvent extraction method ranges from 0.40 to 0.45 per cent. The jasmine oils are composed of terpenes, alcohols, aldehydes, ketones, esters, ethars, phenols, acids, nitrogenous and sulphur compounds which are glycerides of higher fatty acids. The oils are extensively used in the manufacture of cosmetics, soaps, confectionary, perfumes, perfumed tobacco, syrups, aerated water, ointments, disinfectants and detergents. The different plant parts like leaf, stem, bark, root, seed and fruit are very useful and most important in the pharmaceutical beverages. These parts contain manitol. Green leaves and stems contain glucocides which are hydrolysed by emulsion, but do not yield any odorous compounds. The roots of J. angustifolium, J. humile L., J. officinale L and J. scandans are said to be used in the treatment of ringworms. Leaves of J. multiflorum and J. sambac are applied as a poultice for indolent ulcers. The leaves of J. officinale are chewed for relief in case of ulceration of the mouth. The juice of the stem off, malabaricum is used for cataract in the eye. Lotion made from flowers of J. officinale and J. sambac are used against headache and eye troubles. The milky juice exuded from the incisions in the J. humile bark is effective in destroying unhealthy lining walls of chronic sinuses and fistulas. A decoction of J. sambac leaves are used in fevers. Leaves of J. lemonii and J. rottlerianum are said to be blood purifier. In controlling whooping cough a leaf preparation of J. lemonii is very effective, while for pneumonia J. angustifolium leaf preparations are very effective for controlling spasm, bronchitis, asthma, night sweats etc. The fragrant oil of J. flexile can be used as irritant, sedative, neuralgia and mflammative. The leaves of J. ritchiei are reported to be useful in toothache. The seeds of J. arhorescens are eaten in time of scarcity. A yellow dye is extracted from roots of J. huinile. The flowers of J. sambac contain a yellow pigment, used as a substitute for saffron.

Jeffersonia Bart.
Berberidaceae (= Podophyllaceae)

Jeffersonia diphylla (Linn.) :

This is a hardy perennial herb of 60 cm tall. Two identical, opposite leaf segments appear in the top of the stalk. Flowers are 2.8 cm. in diameter and white in colour. Fruit is a capsule and large.

The species occurs in North America. Commonly known as Twin leaf or Jeffersonie.

Uses : Roots of this plant constitute the drug. It is used to treat some throat, rheumatism, muscular spasms, syphilis, cramps, ulcers, mild cases of scarlet fever and also used as expectorant and to increase urine flow.

Juglans Linn.
Family : Juglandaceae

***Juglans regia* Linn. :**

It is a hardy deciduous nut-bearing trees upto 25 m tall. Leaves are alternate, addpinnate. leaflets are oblong and ovate. Flowers are green in colour. Fruits are nut, green, ellipsoid.

It is distributed in Asia and Himalaya, commonly found at Chewan, Utterkashi between 1800-2500 m. It is wild throughout the valley upto 3500 m of Garhwal.

Uses : The decoction of leaves are used as astringent, in injections for leucorrhoea and in baths for lymphatic affections. The leaves are good remedy for scrofula. The outer covering of nut is applied in the laceration of toes. The bark of root and stem is used as tooth brush, for the treatment of toothache. The bark is also anthelrnintic. The oil from nut is also anthelmintic and used to treat skin diseases. The inner bark of Juglans cinera is used as purgative in extract form. Timber of this tree is used for making gun stocks, furniture and veneering. Average life of walnut tree is 325 years.

Juniperus Linn.
Pinaceae (= Cupressaceae)

This is a genus of about sixty five species and six are native to India. Which are evergreen aromatic tress or shrubs.

***Juniperus communis* Linn.** :

It is a dwarf evergreen plant, low and spreading in habit. The needles are straight, sharp pointed, ridged and nearly at right angles to the branchlets. Tree trunk with reddish-brown bark and tends to shred. Fruits are fleshy, round, dark purple in coloiir, berry like upto 14 cm in diameter, covered with waxy block. Seeds have £ gin like aroma.

The species is native to Europe. It occurs in Himalavas from Kumaon, Westwards between 1500 to 4500 m. altitude. Commonly known as Common Juniper.

Uses : The fruit contain volatile oil, sugar, resin, juniperin, fixed oil, protein, wax, gum, pectins, organic acids and salts. The yield of oil is 0.8 to 1 kg.

Fruits, foliage and wood are useful. Fruits are stimulant, tonic, diuretic, stomachic and diaphoretic. They help in digestion, dissipate flatulence, increase the appetite, induce sweating and increase mucous secretion, relieve gas and colic. It has been used in catarrhal affections and diseases of the bladder and lungs.

The berries are also used in cases of aphony, laryngitis and pharyngitis, pulmonary catarrh and asthma. The wood yields cedar oil and is used to cure chronic disorders of the genitourinary tract. Poultice made from needles and twigs are employed to treat wounds. It is also used in making gin.

Juniperus virginiana Linn. :

This is an evergreen tree upto 20 m in height. Leafy twigs are rounded, four angled and slender, younger leaves are needle shaped, occur opposite to each other, while the adult needles are scale like. Cone dark blue, highly aromatic. Berries are fleshy.

The species is native to North America. The plant is grown in rocky slopes and barren areas. Commonly known as Red cedar.

Uses : Leaves, twigs and fruits are useful. Boiled leaves, twigs and fruits together are used to treat bronchitis by inhalation of steam. Bark preparation is used to treat skin rash. For coughs, decoction of boiled fruits and leaves is used. A tea of boiled leaves is prescribed for convalescence and weakness. For menstrual delay and to induce perspiration the decoction of plant is used. It is also used for making pencil, the timber is useful.

The dried fruits of J. macropoda Boiss on steam distillation yield volatile oil. Wood is used for making pencils and pen holders. The species is native to West Himalaya. Leaves and twigs of J. wallichiana are sold as incense and known as insect repellent.

Kigelia DC.
Family : Bignoniaceae

A monotypic genus distributed in warm Africa and Madagascar.

Kigelia pinnata DC. :

This is a spreading tree. Inflorescence are borne on old wood, hanging down on very long stalk upto 3 m long. Flowers are deep chocolate-red in colour. Commonly known as Etuatree.

The roasted seeds are eaten. Fruits are used to treat syphilis and rheumatism. Fruits are also used as purgative. The bark of this plant is useful for treatment of veneral diseases and rheumatism.

Kleinhovia Linn.
Family : Sterculiaceae

A monotypic genus and the species is Kleinhovia hospita WMahot Fleur", an evergreen tree. Rose-pink flowers appear in large terminal, much-branched showy clusters. (The species is native to Tropical Asia.

Uses : The leaf juice is used for eye wash. A decoction of leaves is used to treat scabies and skin eruption. The bark and leaves are used as hair wash to destroy lice and killing eels. Young leaves and flowers are eaten.

Krameria L. ex. Loefl.
Family : Krameriaceae

Krameria triandra Ruiz. & Pavan :

The plant species is native to Bolivia and Peru.

Roots of this plant constitute the drug. It contains 8 to 20% catechin tannin, leucoanthocyanadins, calcium oxalate and kramic acid.

It is a primary sensitizer dermatologically. It is used for cleaning wounds and as poultice for haemorrhoids. Tincture of the roots is used as oral and mucosal astringent and for treating dysentery and diarrhoea.

Kydia Roxb.
Family : Malvaceae

A genus consists of three Indian species.

Kydia calycina Roxb. :

This is a quick growing tree. Flowers are white or pink and distributed in tropical Himalayas and southwards to Peninsular India.

Uses : The leaves of these plants are applied as paste in body pains. Leaves are chewed for formation of saliva. The leaves of these plants also used as poultices for skin diseases. The bark-fibre is used for cordage and ropes.

Lactuca Linn.
Family Compositae (= Asteraceae)

In this genus there are about two hundred species and about twenty species are found in India, which are glabrous or hispid, milky herbs. Flower heads are sessile or peduncled, yellow, purple or blue in colour.

Laetuca Linn.
Family : Lactucaceae

Lactuca sativa Linn. :

The leaves of these plants are eaten as vegetables.

It is distributed in South Europe and West Indies.

Uses : The plant contains several medicinal substances. The juice prepared from stem covering is known in medicine as thridace.

Lactuca scariola Linn. :

This is a herbaceous plant of 80 cm height. Leaves are cleft, with deep lobed. Flowers are yellow in colour.

This species is distributed in Europe.

Uses : Whole plant and milky juice of the stem are used for medicinal pTirposes. The juice of this plant is sedativc and said to be nerve tonic. Tea made from leaves is given to new mothers to hasten the milk flow. The whole plant is administered to increase urine flow and to soothe sore and chapped skin.

Lactuca virosa Linn. :

This is a wild lettuce, slightly narcotic.

Its juice was recommended in angina pectoris.

This species is distributed in tropical Asia and Europe. It is commonly known as 'Acrid Lettuce'.

Lagerstroemia Linn.
Family Lythraceae

This is a genus of about sixty species of showy flowering trees and shrubs, of Southern and Eastern Asia, Africa, New Guinea and Australia. The leaves are usually set in opposite pairs in two rows and have smooth entire margins. The calyx-lobes and petals usually six each in number and petals are wrinkled or crisped.

These plants grow well in warm humid climate and are easy to cultivate. Propagation is done by means of seeds, layering and soft wood cuttings.

L. speciosa (Linn.) Pers roots are astringent and is considered a stimulant and febrifuge. The bark and leaves are purgative and seeds are narcotic.

Thc fruit is used as an application for aphthous ulcers in the mouth. The gum of L. parviflora is said to be sweet and edible.

Laggera Sch. Bip. ex Oliv.
Family Compositae (= Asteraceae)

Laggera alata (D. Don.) Schulz :

This is a herbaceous plant with purplish or pink flowers. It is native to India and occurring at Kareem of Chhindwara District, M.P. and also found in tropical Himalaya.

Uses : The leaf and root powder (about 30 gms) is taken thrice a day for one week in snake bite.

Laminaria Lam.
Family : Laminariaceae

Laminaria hyperborea :

It is a brown algae. The coarse leathery leaf-like algae attached to stone by conspicuous holdfasts. In Laminaria the blade develops numerous large patches with many sporangia in each. The zoo-spores form the sporangia, grow into the oogonial or antheridal filaments. The fertilized egg or oospore resulting from the union of the sperm and egg grows into a bog Laminaria. It occurs in England.

Uses : It contains 12-14% water, 6 to 6.5% raw protein, 0.6 to 6.7% fat, IB-16% ash and 55% carbohydrate and vitmain B.

Algin is used in pharmaceutical purposes. It constitutes carbohydrate material. Algin and alginales is utilised as suspending agents, protective colloids, emollients. They accelerate disintegration of tablets, and gene body and spreadability to ointments and stability to emulsions.

Lannea A. Rich.
Family : Anacardiaceae

A genus of about seventy five species and three species are found in India.

Lannea coromandelica (Houff.) Merrill. :

It is a deciduous tree and commonly found in the lower valleys of Garhwal Himalaya and also at Sidhauli of Chhindwara District, M.P.

Uses : The bark is astringent and is used as a lotion for leprous ulcer. The powdered fruits are used for healing wounds. The bark decoction is used to treat toothache and gum trouble.

Lannea nudicaulis Less. :

The species is found at Kaream of Chhindwara District M.P. The leaf juice is applied four times in a day for one week in conjunctivitis.

The root decoction (about 40 ml) mixed with sugar is taken thrice a day for one month in urinary troubles.

Larix Mill.
Family : Pinaceae

Larix laricina (Du Roi) K. Koch. (= L. americana) :

This is a hardy tree of 30 m in height. Branches are horizontal. Bark is reddish in colour. Bluish green needles 4 cm long. Cone is 3 cm across.

The species is native to America. Commonly known as "American Larch".

Uses : Tree bark and gum are medicinal. The bark is regarded as laxative and tonic. A poultice of bark is used for wounds and bruises. The bark is also used to treat rheumatism. The gum exude from the tree bark is chewed against indigestion.

Laurus Linn.
Family : Lauraceae

Laurus nobilis Linn. :

This is a medium sized evergreen hardy tree, attain a height upto 35 m. Leaves are dark green, aromatic, elliptical, tapering to a point at base and tip of leaf. Male and female flowers are borne on separate plants.

Flowers are insignificant, yellow in colour. Berries are dark purple, ovoid upto 1.8 cm long, black, coarsely wrinkled and contains only one seed.

This species is native to Southern Europe. Laurel is also grown in Indian Tea gardens.

Uses : The leaves on stem distillation yield 1-3% essential oil. The leaves contain protein, fat, carbohydrate, minerals, vitamins A, B and C. The principal constituent of volatile oil possess cineol which is 50%. It is colourless, strong and aromatic. It has camphoraceous odour and a cooling taste.

Every part of the plant is stimulant and can be employed in some cases of general atony and debility. Leaves and fruits are prescribed for lack of appetite. The oil extracted from this plant is used to prepare ointment for external application in rheumatic pains and sores. The berries are used to treat dropsy, buer and diarrhoea. The leaves are used as a condiment and flavouring agent in foods and confectionary. It is used for culinary purposes.

Lavandula Linn.
Family : Labiatae

Lavandula officinalis Chaix. :

This is a hardy evergreen fragrant flowering shrub, upto 120 cm tall with grey foliage. Flowers are highly esteemed for their fragrance in dried state.

The species is native to Mediterranean region. Commonly known as Lavender.

Uses : This is a sweet smelling plant and its essence contains a hydrocarbon. The plant is a component of various preparations, which is a stimulant, sternutatorv and tonic.

Lawsonia Linn.
Family : Lythraceae

Lawsonia inermis Linn. :

It is a much branched shrub. Branches are usually 4-angled which are ending in a sharp point. Leaves are 3 cm long, opposite, acute and sharp pointed. Flowers are small, fragrant, appear in terminal large branches, white or pinkish in colour. Fruits are pea sized, round, many seeded. It is grown as hedges in the gardens and farms.

This plant is found throughout India, chiefly in drier parts of peninsula. It is commercially cultivated in Rajasthan, Punjab, Madhya Pradesh and Gujarat. Commonly knonwn as Henna.

Uses : Leaves, bark and seeds of these plants are used for medicinal purpose. The leaves are astringent. The decoction of leaves is used as gargle in sore throat. Paste of leaves are applied for relieving headache and burning sensation in feet. They are also applied on boils, burns and skin diseases. Leaves are used for treatment of typhoid and haemorrhagia. In Unani and Ayurvedic medicines, barks and seeds of these plants are also employed. The orange dye derived from this plant is used for colouring palms, nails, hair, beard and feet. The dye is also used for textile and in hair dye. The oil extracted from flowers are valued in perfumery.

Leonurus Linn.
Family Labiatae

Leonurus cardiaca Linn. :

This is a perennial plant of 2.5 m tall. The leaves are lobed and dentate. The flowers are very fuzzy, white to pink in colour.

The species is native to temperate Eurasia.

Uses : Above ground parts of the plant at flowering time is harvested for medicinal purpose. This plant is considered stimulant and tonic and increases urine flow. It is taken as infusion for treating asthma and heart palpitations. It is also effective to treat rabies.

Leucas (Burn.) R. Br.
Family : Labiatae

A genus of about one hundred and ten species and forty are native to India. These are woolly or villous herbs or undershrubs. Flowers are usually white in colour.

Leucas cephalotes Spreng. :

This is an erect annual herb upto 60 cm tall with white flowers. Occur commonly in the tropical zone on the lower foot hills of Garhwal.

Uses : The entire plant has stimulant and insecticidal properties. Flowers are used to treat cold and cough.

Leucas lanata Benth. :

A tuffled, woolly, tomentose, perennial herb with white flowers in axillary loose whorls.

Comrnonly distributed in grassy slopes, rocky boulders and dry field borders upto 3000 m of H.P.

Uses : The decoction of the leaves and apical shoots with gur is used as an abortifacient upto 3 months of pregnancy. Infusion of flowers are used in skin diseases. It is also used to treat piles.

Linaria (Tourn.) Mill.
Family : Scrophulariaceae

Linaria vulgaris Mill. :

This is a hardy, perennial herb of 80 cm tall, bushy and having erect branches. Each years growth arising from an adventitious bud upon the summit of the root. The flower is closed at the mouth, honey is secreted by the nectary at the base of the ovary and collects in the spur. Flowers are yellow.

It is native to Northern hemisphere specially Mediterranean region and Europe.

Uses : Infusion of flowers are used in skin diseases. It is also used to treat piles.

Lindera Thu.
Family. Lauraceae

Lindera benzoin (Linn.) Mirssn. :

A tall deciduous shrub may attain a height upto 6 m. The leaves are aromatic and alternate upto 14.5 cm long. Flowers appear in dense cluster, greenish yellow in colour. Fruits are bright red berries. The bark is aromatic. It is native to Malaysia and Himalaya.

Uses : Twigs, ripe fruits and barks of these plants are useful. A decoction of these parts are used to treat intestinal worms, dysentery, coughs and colds, and also to stimulate blood circulation and to increase perspiration.

Linum Linn.
Family Linaceae

Linum usitatissimum Linn. :

It is herbaceous annual with fibrous stem. Plants attain a height of 60 cm. flowers are delicate blue, mauve and white.

It is native to Europe, Russia, Eastern and Northern Africa, Western Australia, North-Central United States, Argentina, India and Iraq. It is commercially cultivated in Madhya Pradesh, Bengal and Bihar in India. It is also grown in Argentina, Russia, Ireland and Belgium.

Uses : The oil contains unsaturated fatty acids chiefly linolenic acid and also linoleic. The seeds hold a mucous containing acetic acid and salts, extractive starch, resin wax, a colouring matter, gum, albumin and principally a fine oil.

Seed and oil is used in medicine and serve as a demulcent and emollient. It is used in the form of powder or flour. It is widely used as a sedative drink, for treatment in genito-urinary diseases, and also as poultices. It is also famous for linen fibre. Flax is the fibre of this species obtained by retting of the softer tissues in water. The stem fibre is used for making cloth, twines and wrapping paper.

Liparis L.C. Rich.
Family Orchidaceae

A genus of about two hundred and sixty species and forty five are found in India, which are epiphytic and terrestrial plants under these perennial herbs.

Liparis odorata Linn. :

This is an orchid with very stout pseudobulbs. Leaves are 4.5 with sheaths below. Inflorescence erect, with a stout peduncle upto 40 cm tall, many flowered. Flowers are pale yellow green with purple lip, about 3 cm across.

The species is distributed in Malaya peninsula, India, Sumatra and Java.

Uses : Pseudobulbs of these plants are used to treat cancerous ulcers, gangrene, fever and dropsy.

Liparis paradoxa Rchb. :

The plant species is with tufted stem, base sometimes pseudobulbous. Leaves are l4 cm long and 3 cm broad. Scape stout, angular, to 25 cm long. Flowers are coriaceous upto 2 cm across, yellow brown in colour.

It is distributed in Kumaon, Naga and Khasia Hills, Sikkim, Chota Nagpur and Karnataka.

Uses : It is one of the eight drugs of the "Ashtravargra group" and used as a general tonic.

Liparis paradoxa Rchb. :

It is an orchid with greenish flowers.

It occurs in Mussorie and Chakrata between 2500 to 3000 m elevation. The bulbs of these plants are used as tonic.

Liquidambar Linn.
Family : Hemamelidaceae (= Altingiaceae)

Liquidambar styraciflua Linn. :

This is a hardy deciduous flowering tree of 50 m tall. Leaves are downy, very fragrant, palmate. Flowers are inconspicuous, greenish yellow. The fruit is contained in woody, spiky globular capsules. The seeds are winged.

It is native to North America.

Uses : Leaves and twigs, bark and gums from the tree trunk have medicinal properties. Twigs are chewed to clean the teeth. Resin is used to treat wounds and fevers. To treat herpes and skin inflammations and toothache gum is used. Bark and leaf preparations are prescribed to treat diarrhoea and dysentery. Boiled leaves are used to treat sore feet and cuts. This tree is also useful for treating scabies. The wood is useful which is known as satin walnut.

Liridendron Linn.
Family : Magnoliaceae

Liridendron tulipifera Linn. :

This is a hardy deciduous tree may grow to over 50 m in height. Flowers are tulip-shaped which are greenish yellow, orange with 6 cm long. Leaves are bright green turning gold in autumn. The fruit is cone like of samaroid capsules, upto 8.5 cm long, brown in colour.

The species is native to North America, China and Indo-China.

Uses : Leaves, bark from trunk, roots, branches and buds are useful. To treat scalds, burns and inflammation, an ointment is made of the buds crushedin grease. A poultice is made from crushed leaves to treat headache. The inner bark is a stimulant. A decoction of root bark is applied warm to infected tooth for relief of pain. It is also used to treat hysteria, malaria and rheumatism. It yields useful timber which is known as canarv white wood.

Lobelia Linn.
Campanulaceae (= Lobeliaceae)

It is a genus of over two hundred species and twenty two are found in India. The plants are herbs or shrubs often laticiferous.

Lobelia inflata Linn. :

It is a half hardy, branching, herbaceous annual, attain an ultimate height of 1 m. Small flowers violet to pinkish white in colour borne in the axil of the leaves. The base of the flowers inflate to form a seed capsule. Seeds are many in a capsule, small, shiny and brown.

It is native to Eastern U.S. and Canada and cultivated in Europe.

Uses : The dried leaves and flowering tops are used in medicinal preparation. Plants are harvested at full bloom stage. It is used as an emetic, in bronchial troubles and pulmonary complaints. It also serves as antiasthmatic and as an expectorant in cases of laryngitis. It has toxic effects and should be used in caution.

Lobelia nicotianaefolia Hyne. ex Roth. :

It is a herbaceous plant with stout and hollow stem, grows to a height upto 6 m. Leaves are very large upto 60 cm long in lower part of the stem, main nerve of the leaf is whitish. Flowers appear in very large terminal branches, white in colour. Fruits are upto 1.5 cm in diameter. Seeds are many and small, yellowish brown in colour.

The species is distributed in Deccan peninsula.

Aerial parts of the plant is dried in shade and used as medicine. It is used to treat asthma and bronchitis. It causes irritation of mucous membrane, toxic, sometimes may be fatal.

Lobelia siphilitica Linn. :

This is an unbranched hardy perennial herb upto 2 m tall. The leaves are 18 cm long. Flowers appear in long spike, blue in colour.Qhe plant is widelv distributed in America.

Uses : Plants and roots of this species combined with roots of Podophyllum JjeTfoTum L. (Ground Lemon) are used to treat syphilis and gonorrhoea. Above ground portion of the plant has been used to increase urine flow, to treat dysentery, to induce vomiting and also to clear the chest of phlegm.

Lonicera Linn.
Family Caprifoliaceae

Lonicera caprifolium Linn. :

This is a deciduous twining plant. The flower is visited chiefly by hawk-moths at night. The flowers open in the evening. Flowers are mucilaginous and highly fragrant, creamy yellow tinged red.

The species is native to Southern Europe.

Uses : Leaves are astringent, used in a decoction to prepare detersive gargles. Infusion of flowers is prescribed in the treatment of slight pulmonary catarrh.

Lycopodium Linn.
Family : Lycopodiaceae

Lycopodium clavatum Linn. :

This is a low growing evergreen perennial plant. Stems clothed with scale like, dark green leaves. The leaves are often with bristle at the apex. Foliage is ornamental. The plant produce yellow spores and no flowers. Spores appear in short cylindrical spikes. Capsules kidney shaped enclosing a powder known as lyxopodium or commonly called vegetable sulphur or vegetable brimstone. The spores are very light, fine, odourless and highly inflammable. Water takes a long time to wet it. It contains sugar, a waxy substance insoluble in water and soluble in ether. It is said to be a particular nitrogenous principle, yellow combustible and liable to putrification.

The species is occurring in colder and temperate regions of old and new world.

Uses : It contains fatty oil and minor amount of palmitic and linoleic.

This plant spores which are known as. 'vegetable sulphur', are used in medicine for many purposes. It is used as a local dessicative to prevent chapping the children's skin and to dust the bed of patients who develop sores from the prolonged rubbing of sheets. It has a mobile powder which has the property of greatly diminishing friction between bodies, and hence, in pharmacy it is used to dust pills to prevent them from sticking together. It is administered as a decoction, as an antirheumatic, diuretic, anti-epileptic and anti-diarrhetic. It is orally taken for urinary and digestive complaints.

Lyonia Nutt.
Family : Ericaceae

Lyonia ovalifolia (Wall.) Drudo :

This is a hardy decidous flowering shrub. Flowers are white and fragrant. It is native to Himalayan region.

Common]y found at Gamari, Utterkashi between 600-3000 m. This species is also found in Khasi Hills.

Uses : The extract of leaves is rubbed on itching. The infusion of young leaves and buds are used in skin diseases.

Maclura Nutt.
Family : Moraceae

Maclura pomifera (Raf.) Schneid. :

This is a hardy deciduous spiny tree upto 15 m tall. The leaves are 14.5 cm long, egg shaped, bright green. Spines are there in the leaf axils. Flowers are yellowish green, inconspicuous. Fruits are round, golden yellow in colour, upto 14.5 cm in diameter.

It is native to North America.

Uses : Roots of this plant constitute the drug. A decoction of boiled root is used to bathe irritated eyes.

Macrocystis C.A. Agardh.
Family : Laminariaceae
(Algae)

Macrocystis pyrifera C. Agardh. :

This is a brown algae and is an extremely valuable plant.

It occurs in Southern California and Australia. This algae grows for the most part in densely compact beds in water from 8 to 28 m. The rate of growth depends largely on temperature. The reproductive bodies or spores^ are produced throughout the year. In a sea temperature between 19° to 25° C it is subjected to a black rot bacterial disease.

Uses : Algin is used for pharmaceutical purposes. This species is the main source of algin. It is used as suspending agents and emollients and also used in ointments and emulsions.

Madhuca J.F. Gmel.
Family : Sapotaceae

A genus consists of about eighty species and four are native to India.

Madhuca indica Gmel. (= Bassia latifolia Roxb.) :

It is a large deciduous tree upto 25 m tall, with a spreading habit. The leaves are clustered at the end of the branches. The fleshy, tubuJ-a1\ creamish white flowers with intoxicating smell appear at the end of branches during March-April. The flowers are stalked and drooping. Fruits are ovoid berry, turning yellow when ripe with 1-4 brown ovoid shining seeds.

The species is native to India. It is found in Kaream of Chhindwara District, M.P. and also occur in all the plains and lower hills of India and common in sub-mountainous regions of the Himalayas. Commonly Known Mahua.

Uses : The leaves, flowers, bark and seeds have medicinal value. Leaves are astringent. The ash of leaves mixed with ghee is applied on burns and scalds. The flo\vers have cooling and nutritive and antibacterial propcrtics and are used to treat bronchitis and cough. The spirit made from flowers is considered to be nutritive and tonic. The bark preparation is used for treating itch, ulcers and bleeding gums. Bark is also administered to cure diabetic problem. Seeds help in formation of milk and promoting of milk flow. Oils extracted from seeds are said to be laxative and used to treat skin diseases. Seeds contain #0% yellow oil. Mahua Fat or Mahua Butter is usually used for edible and cooking purposes, but mainly in the manufacture of soap.

Magnolia Linn.
Family : Magnoliaceae

Magnolia virginiana Linn. (= Magnolia glauca Linn.) :

This is a hardy shrub may attain a height of 6 m. The leaves are upto 18 cm long, green above and white or grey below. The flowers are highly fragrant, white upto 8.5 cm in diameter. Fruits are red and 6 cm in diameter.

The species is native to Eastern America.

Mature leaves, cones, ripe seeds and root bark are useful. A warm infusion of cones, bark and seeds are used for rheumatism. Bark preparations are also used to treat lung and chest diseases, dysentery, fever and malaria. A tincture of leaves is used to treat rheumatism and gout.

Malaxis Soland. ex SW.
Family : Orchidaceae

Under this genus there are about three hundred species and seventeen are distributed in India. These are terrestrial herbs, rarely epiphyikior lithophytic.

Malaxis cylindrostachya (Linn.) O. Kfze. :

This is an orchid with stem like pseudobulb, clothed with leaf sheaths. Leaves are solitary, oblong or rounded to 12 cm long, blunt and plicate. Inflorescence densely many flowered upto 48 cm long. Flowers are pale yellowish green, short leaved.

It is distributed in Kashmir and Sikkim.

Its pseudobulb preparation is considered as tonic.

Malaxis muscifera Ridl. :

This is a terrestrial orchid, pseudobulb stem like, rather small, clustered. Leaves are paired, sessile or short stalked upto 10 cm long. Inflorescence upto 45 cm long, many flowered raceme. Flowers are very small, yellowish green in colour.

It. occurs in Kashmir and Sikkim.

Uses : Pseudobulbs are used in medicine. Its preparations are considered as tonic and rejuvenating drug.

Malaxis wallichii Linn. :

Uses : Pseudobulbs of these plants are medicinal. It is said to cure tuberculosis and enhances sperm formation. It is an ingredient of Chyavanprash, an Ayurvedic medicine.

Mallotus Lour.
Family : Euphorbiaceae

Mallotus philippenis (Lamk.) :

It is a much branched dwarf tree. Young parts of the plants are densely covered with minute red hairs. Leaves are borne on a long stalk, prominently veined. Leaves are alternate uptc 16 cm long and 8.5 cm broad, ovate, acuminate, entire or slightly toothed. Flowers are small, dioecious. The males are clustered, sessile or very shortly pediculate, appear in erect terminal spikes, yellow, 9 to 16 cm long. Sepals 4-5, lanceolate, acute and stamens numerous, buds globosely ovoid. Female flowers are sessile in short spikes. Fruits 9-14 mm in diameter, 3-lobed, roundish, covered with bright red coloured powderv indumentum, when ripe. Fruits are dehiscent 3 valved capsule, globose. Seeds are 6 mm in diameter, subglobose and black.

This is distributed throughout the tropical part of India in Assam, W. Bengal, Uttar Pradesh, Punjab and Andaman Islands. It is found at Karrapan of Chhindwara District of M.P. and common throughout the lower valleys of Garhwal in moist slopes upto 1200 m elevation. It is also reported growing in Burma, Pakistan, Singapore, Malaya Islands, China and Australia.

Uses : It is sold in the market as Kamela, the red glandular and hairy substances reported from the fruits for preparation of drugs. The crude powder from the exterior of the fruits is found to be anthelmintic and this activity is due to rottlerin and isoallorottlerin. This is used to destroy thread worm, hook worms and round worm. The powdery coat of fruit is mixed with gur, is used once in the morning as purgative. The powdery coat of fruit mixed with water and is taken for one week to relieve colic pain and expel bile. The glands on fruits are bitter pungent, purgative, cathartic, vulnerary, detergent and carminative. It is also used to treat bronchitis, enlargement of spleen, jaundice and piles. Leaves and roots are useful in treatment of skin diseases. The decoction of bark is used in abdominal pain. The grounded roots are rubbed on the painful parts in articular rheumatism. Paste prepared from the grounded seeds are applied to wounds and cuts.

Kamela is used for dying silk and wool, and also-for colouring soaps, oils, ice-cream and drinks. The oil extracted from seeds is used in paints, varnishes and painting work. The rottlerin and its pentapotassium derivatives are employed for colouring food stuffs, lemonades, lime juice and other beverages.

Malva Linn.
Family Malvaceae

Malva parviflora Linn. :

This is an annual plant of 1.4 m tall. Pink flowers of small sizes appear in clusters in leaf axils.

The species is native to America.

Uses : Mature leaves and flowering tops are useful. It is used to treat swollen glands of throat, fever, pneumonia, headache and also to induce perspiration and menstrual flow.

Mangifera Linn.
Family : Anacardiaceae

Mangifera indica Linn. :

This is a large tree with deep tap root system. The tree is spreading, evergreen with a dense rounded or globular crown. Leaves are simple, petiolate, alternate, irregularly placed along the branchlets. The fruit is a large drupe derived from one carpel of the flower.

It is native to South-East Asia, Indomalayan region. Distributed in different parts of India, and is in commercially cultivated in different states of the country. India is the only country where truly wild forests of mangoes are found in N.E. India.

Uses : The fruit is a rich source of vitamin A and C apart from minerals and other vitamins. The ripe fruit contain nearly 25% total soluble sugar, 0.4 to 0.8% acid and 2% protein. The kernel contain 10.5% protein. Mango seed oil is extracted from mango kernels (8 to 15).

The bark is astringent, it is used in diptheria and rheumatism, ft possesses a tonic action on the mucous membrane. The fruit possesses outstanding nutritional value. The kernel juice is snuffed once a day for three days to stop nasal bleeding. The unripe mango is useful in opthalmia and eruptions.

Manilkara A dans.
Family : Sapotaceae

A genus of about seventy five species and five are native to India.

Manilkara archras Fosberg :

It is a slow growing evergreen tree upto 9 m tall. Leaves are elliptic to ovate. Flowers are solitary in leaf axils, sepals 6, petals 6. The fruit is a drup, egg shaped or round. The skin of the fruit is rough. Fruit is edible and delicious. Each fruit contains 3 to 5 black shining seeds,

The tree is native to Mexico. It is cultivated in different states of India. There are several varieties in it.

Uses : The plant bark has several medicinal uses. The ripe fruits are eaten and tender leaf tips are also edible. The latex is used for making chewing gum. The fruits are used for making mixed jam, jellies, industrial glucose and pectin.

Manilkara hexandra (Roxb.) Dub. :

This is a tree with edible fruits.

The species is native to Central and Peninsular India occurring at Junapani Forest near Rater of Chhindwara District, M.P.

Uses : The stem bark is used as tonic and astringent. This species is used as a root stock for sapota.

Marrubium Linn.
Family : Labiatae

Marrubium vulgare Linn. :

This is a hardy perennial herb of 80 cm tall. Leaves are ovate, whitish above and grey below, fuzzy, aromatic.

The species is native to England.

Uses : Leaves and stems are harvested before blooming, and bark of the plant constitute the drug. The plant is laxative and it is used to treat kidney ailments, menstrual irregularity, asthma and diarrhoea. A tea prepared from dry bark is used to treat colds and debility. Leaves are also mixed with honey and given for coughs.

Marsilea Linn.
Family Marsileaceae

In deep water, the leaves float, while in shallow water they stand erect. Leaves of these plants are petiolate with four lobes, resembling those of four-leaved clover.

Marsilea minuta Linn. :

This is a fern found at Alwar, Ajmer, Mt. Abu and Sirohi of Rajasthan.

Uses : The decoction of leaves along with ginger is used in cough and bronchitis.

Matricaria Linn.
Family : Compositae (= Asteraceae)

Matricaria chamomilla Linn. :

This is a much branched aromatic herb with white or vellowish flowers.

The species is distributed in Europe and also found in Punjab and appear in Gangetic Plain.

Uses : Extracts of chamolile is made from flower of these plants. Distilled water, syrup or tincture is made of this. It is a remedy for windy colic using between 4 to 8 gms to little of water. It is used to treat indigestion, dyspepsia, diarrhoea and fever. In case of benin neurosis this is administered. It produces essential oil which is used in perfumery.

Megacarpaea DC.
Family : Brassicaceae (= Cruciferae)

Megacarpaea polyandra Benth. :

A robust perennial herb upto 1.5 m tall, with radish like odour. Roots are thick and annulate. Flowers are yellow and fragrant.

It is occasionally found on open sunny and loose slopes of Northern Himalayas between 3600-4500 m.

Uses : Roots are used as medicine against fever. Root is a febrifuge and tonic. The leaves are edible and used as vegetable.

Melaleuca Linn.
Family : Myrtaceae

Melaleuca cajuputi Powell. :

This is an evergreen tree of 15 m or more in height with white bark. Leaves are aromatic, deep green and upto 25 cm long. Flowers are creamy-white in colour.

It occurs in Malaysia, Burma and Australia.

Uses : Leaves are useful for medicinal purpose. The oil is extracted from the young twigs and leaves which primarily contains cineole, apart from this it also constitute pinene, benzaldehyde, terpineol and its esters.

It is used as anthelmintic and used to expel worms from intestine. Leaves are externally applied on neuralgia and rheumatism and internally administered for colds and coughs. Oil is used externally to treat scabies and other parasitic skin diseases. Oil is put into the tooth cavity to relieve toothache. It also cures headache and earache.

Melastoma Linn.
Family : Melastomaceae

Melastoma malabathricum Linn. :

It is a shrub in which mauve-purple flowers are produced in terminal, corymbose panicles.

The plant is distributed in Indomalayan region, occurring at Junapani forest near Rater of Chhindwara District, M.P. and also in other parts of India in humid places. Commonly known as Indian Rhododendron or Malabar Goosaberry.

Uses : The decoction of flower (about 30 ml) is taken in the morning for one month in spermatorrhoea. The leaf juice is taken in diarrhoea and dysentry. The leaves and roots yield a pink dye. Fruits are also edible.

Melilotus Linn.
Family : Leguminosae (= Fabaceae)

Melilotus officinalis Lam. :

A herbaceous plant commonly known as "Melilot" or "Cock's Head", native to Europe and Asia. This is emollient, resolutive, anodyne and carminative and mostly used externally. In a fomentation on the abdomen, it is good for windy colic; in baths, for gout and rheumatism; in a rectal injection, it is vermifuge and carminative.

Uses : The infusion and distilled water have been recommended for eye diseases.

Melissa Linn.
Family Labiatae

Melissa officinalis Linn. :

A hardy herbaceous perennial of 1.5 m tall. Leaves are lemon scented. Flowers are white or yellow.

The species is native to Mediterranean region. Commonly known as Balm.

Uses : The plant is stimulating and anti-spasmodic. The tea prepared from Lhese plants is used for flatulence, lack of appetite and indigestion. It is also useful in nervous complaints, hysteria, cardialgia, spasms, vertigo, nigra and in all cases of general atony.

Meniospermum Linn.
Family : Menispermaceae

Menispermum canadense Linn. :

A hardy deciduous flowering climber, perennial and woody in nature which attain a height of 5 m. Leaves are lobed. Flowers are greenish yellow which appear in clusters. Fruits are small black berries.

It is native to North America. Commonly known as Maple Vine.

Uses : Rhizomes and roots of these plants are useful. Roots have tonic properties and increase urine flow, given after prolonged illness to overcome weakness. A tincture of fresh roots is used to treat skin infections, gout, rheumatism, syphilis, and it is also considered as laxative.

Mentha Linn.
Family : Labiatae

A genus of about twenty six species, and seven are distributed in India. Which are aromatic perennial herbs.

Mentha arvensis Linn. :

An erect branched perennial herb with running rootstocks and rigid branching stems upto 85 cm tall. Leaves are 6 cm long, leaf stalk small or none, margins toothed. Flowers are lilac in colour.

I is cultivated in Tarai and Haldwani area in U.P., Jammu & Srinagar and some parts of Punjab. Commonly known as Japanese Mint or Podina.

Uses : The oil yield is 5% by distillation of leaves, which contain 40-50% menthol and 50-60% dementholised oil.

The oil is antisepctic, carminative, refrigerant, stimulant and diuretic. The plant is used to treat liver and spleen diseases, asthma and jaundice.

Menthol is used in medicine for stomach disorders and in ointments for headache. The infusion of leaves is used in indigestion and rheumatic pains.

Mentha cardiaca Gerard. ex Baker. :

It occurs in America. Flowering tops contain essential oil, which is used as a flavour and carminative. Commonly known as Scotch Mint.

Uses : The thin pastc of leaves is applied to rheumatic and painful swollen joints. It is antiseptic, stimulant and administered in digestive complaints and fever. It is also used to treat diarrhoea, bronchitis and dysentery. It has carminative properties.

Mentha piperita Linn. :

A hardy aromatic herb. This species is a cross between M. spicata and M. aqualica. Plants grow upto 95 cm tall and glabrous. This plant has strong odour and a more aromatic taste. It is rich in essential oil. Leaves are long petioled, opposite, lanceolate.

This species originated in Europe. It is commercially cultivated in America, France, South Africa, Yugoslovakia, Hungary, Hngland, Thailand, Vietnam and Bulgaria. In India, it is commercially cultivated in Tarai region. Commonly known as Peppermint.

Uses : Flowering tops contain essential oil of 0.4 to 0.8% which constitute 66% free methanol. It is used as carminative and flavouring agents.

The peppermint oil has a fine odour and possess carminative, antiseptic, preservative and gastrostimulant properties. There are several pharmaceutical preparations of mint. It is used in lotions and fomentations, externally as a resolvent for bruises and scabies. The plant distilled in water or syrup made from it are recommended for treatment of vomiting in children, and for gout. The essence is also applied for toothache and swollen gum. It is administered in scrious nervous disorders, in headaches, colics, nervous vomiting and tympanitis, hiccups, flatulence and periodic discharges with nervous symptoms.

Mentha spicata Linn. :

A perennial herb of 1.5 m tall. Flowers are borne in long spikes, white or pink in colour.

The species is native to North America. Commonly known as spearmint.

Uses : Above ground part when in bloom is useful. Essential oil is extracted from flowering tops which varies from 0.30 to 0.50%. Essential oil contains 56.8% carvone and also -pinen, -phellandrene, 1-limonene, octyl alcohol and dihydrocarveol.

It is used as carminative and employed as flavour. The dried tops of this plant is used as a stimulant and to prevent vomiting. Tea made from this plant is used to treat neuralgia, indigestion and diarrhoea, and also to wash sores.

Menyanthes Linn.
Family Gentianaceae

Menyanthes trifoliata Linn. :

It is a hardy aquatic perennial with a thick rootstock. Leaves are trifoliate. Flowers are dimorphic, appear in a cluster on a long stalk, pink coloured, fragrant. Rhizomes are creep ing.

This plant is native to North Temperate region in Europe. Commonly known as Buckbean.

Uses : The plant is harvested at full bloom stages. The leaves and rhizomes of these plants contain bitter principle, they are used as a tonic and digestive. Preparation from it is also used as sedative for the nervous system. It is also used to treat skin eruptions, worms, scurvy, jaundice and constipation.

Mercurialis Linn.
Family Euphorbiaceae

Mercurialis annua :

A herbaceous plant which has unpleasant odour and almost fetid.

The species is native to Mediterranean region, Europe and Africa.

Uses : The plant has viscous, slightly acrid and salty taste. From the seeds, purgative pills are prepared. The plant is said to have hypnotic or diuretic properties.

Mesua Linn.
Guttiferae

About three species distributed in Tropical Asia and India, out of which One species is native to India.

M. ferrea Linn. :

The plant species is occurring in Eastern Himalayas, Assam, West Bengal, Western Ghats and Andaman Islands. Flowers are white. Commonly known as Nagkesur tree.

Uses : The seed oil is used to treat rheumatism and skin diseases. Dried flowers are used for preparation of perfumed ointments. Seed oil is also used for making soap.

Michelia Linn.
Family Magnoliaceae

Under this genus there are about fifty species and distributed in Tropical Asia and China. These are trees or shrubs.

Michelia champaca Linn. :

It is cultivated throughout India and native to E. Himalayas, Assam and Western Ghats. These trees possess handsome foliage and the flowers are highly fragrant. Commonly known as Champaka.

Uses : The bark of M. champaca Linn. diuretic, febrifuge and stimulant. The leaf juice is used to treat colic. The oil extracted from flowers are employed in perfumery industry. This is also used in cephalalgia, ophthalmia, gout and rheumatism. Flowers and fruits are considered stimulant, diuretic, stomachic and antispasmodic. Seeds and fruits are used for healing of cracks in feet.

Microstylis Nutt.
Family : Orchidaceae

Microstylis muscifera (Linn.) Ridl. :

Plants are variable in size ranging from 14 to 50 cm in length. Stem is tuberous and sheathed. Leaves are 15 cm long and rounded. Inflorescence upto 30 cm long. Flowers are 4 mm in diameter, yellowish green in colour. The speciesoecurs in Temperate Himalaya from Kashmir to Sikkim between 3000 to 3500 m altitude & It is a terrestrial orchid. Commonly known as Adder's Mouth ordird.

Uses : The roots of these plants are regarded as tonic and lactagouage.

Mimusops Linn.
Family : Sapotaceae

Under this genus there are about fifty species and three are native to India.

Mimusops elengi Linn. :

A Native to Peninsular India, the fruits of which are edible when ripe, but of poor quality. The tree is attractive, elegantly shaped with highly fragrant flowers. Commonly known as "Indian Medlartree" or "Bakul tree".

Uses : The bark of this tree is used to treat diarrhoea and dysentery. A decoction of bark is used as gargle. The fruit and bark possess tonic and astringent properties. The powdered seeds are pasted with oil and used as suppositories in case of obstinate constipation especially for children. The snuff is prepared from flowers.

M. elata Allem. :

This is a Brazilian milk tree, flowers are edible, but strangest of all is the vegetable milk which exudes in abundance when the bark is cut, it has about the consistence pf thick cream.

Uses : It is used as milk or for glue. Commonly known as Cow Tree of Para.

Mirabilis Linn.
Family : Nyctaginaceae

Mirabilis multiflora (Torr.) Gray. :

This is a half hardy tuberous rooted perennial, much branched, grow upto l^m tall. The flowers are purple, around 6 cm long.

The plant grows between 1500 to 3500 m elevation in the foothills.

Uses : Roots of this plant constitute the drug, Powdered root is used to treat stomachache.

Mitchella Linn.
Family : Rubiaceae

Mitchella repens Linn. :

This is a hardy evergreen trailing herb. Leaves are roundish and shiny green. Flowers are fragrant, white tinged with purple. Berries are scarlet. The species is native to North America. Commonly known as Chequer Berry.

Uses : Plants with fruit is useful. A tea prepared from this plant is used to treat insomnia, diarrhoea and also to ease child birth.

Mitragyne Korth.
Family : Naucleaceae (Rubiaceae)

Mitragyne parviflora Korth. :

A native to Tropical Africa and Asia. Found at Kaream of Chhindwara District, M.P. and also in different parts of India.

Uses : The stem bark and root decoction is taken in fever. Wood is used for furniture and implements.

Momordica Linn.
Family : Cucurbitaceae

Momordica charantia Linn. :

An annual plant with slender climbing or trailing habit. Leaves pubescent and suborbicular. Flowers monoecious, solitary and yellow. Fruits differ in sizes, colour, surface characters, pendulous, fusiform, ribbed with numerous tubercles, generally yellow in colour when ripe. Seeds are compressed in the pulp of fruits, brownish in colour.

Uses : Fruits contain protein, fat, carbohydrate, mineral matter and moisture. It is rich in vitamins like riboflavin, thiamine, ascorbic acid and as corbegenin. Green fruit contain luteolin. Seeds yield essential oil. Fruit contain bitter glycosides, cucurbitacins. The plant is of specific importance because of its hypoglycemic and antidiabetic activity.

The entire plant has medicinal properties and used for preparation of indigenous medicine. The fruit is regarded as stomachic, carminative. The leaves have purgative, emetic and bitter properties. The fruit is used for the treatment of rheumatism, gout, diseases of liver and spleen and useful in diabetes. It is also used in leprosy, piles and jaundice.

Monarda Linn.
Family : Labiatae

Monarda fistulosa Linn. :

A hardy aromatic herbaceous perennial, upto 1.25 m tall, soft and fuzzy. The leaves are narrow, grey-green in colour. Flowers appear in dense clusters, purple, upto 2.6 cm long.

This species occurs in North America.

Uses : Entire plant in full bloom stage, mature leaves and roots are useful. A decoction of root is given to drink for stomach disorders. A tea made from leaves and flowers is used for bronchial trouble. Lotion of boiled leaves are applied on pimples or skin eruptions.

Monotropa Linn.
Family : Monotropaceae

Monotropa uniflora Linn. :

A small succulent plant, waxy, upto 35 cm in height which is parasitic in nature. Leaves are tapered, smooth-edged. The flowers are bell shaped, single and appear at the top of the stem.

The species is native to North Temperate region. Commonly known as Pipe Plant.

Uses : Entire plant and roots are useful. Fresh juice of the plant is used for eye ailments. Powdered roots is used to treat convulsions and epilepsy. Dried plants are also used to induce sleep and relieve pain.

Morina Linn.
Family : Dipsaceae (= Morinaceae)

Morina longilfolia Wall. ex DC. :

A hardy perennial herb with white to crimson flower upto 95 cm tall. Flowers pale purplish and fragrant.

It is native to Himalay commonly occurring at Chyangeel of Uttarkashi, above 3500 m in the grass slopes. Commonly known as "Whorl Hower".

Uses : The roots are also aromatic and applied locally as poultice on boil The root paste is applied on wounds to kill the worms. The infusion of roc is dropped in pused ear. At flowering time when persons come in contac with the plants it causes dizziness and giddiness.

Moringa (Burm.) Adans.
Family : Moringaceae

About ten species distributed in Mediterranean region and India and two are native.to India.

Moringa oleifera Lamk. :

A quick growing tree, widely cultivated all over the plains of India. The leaves and green pods are used as a vegetable.

Uses : The leaves of these plants are used to treat scurvy and catarrha affection. Paste of the leaves is applied externally on wounds. Flowers an tonic, diuretic and cholagogue. The seeds are antipyretic. The oil extractec from seed is used to treat gout and rheumatism. The roots of the young tree and root bark are vesicant and rubefacient. The roots are also used as condiments. Flowers, tender leaves and pods are eaten.

Morus Linn.
Family : Moraceae

Under this genus there are about ten species and two species are native to India. In this genus there are shrubs and trees.

Morus alba Linn. :

It is a tree of 16 m tall with a trunk up to 65 cm in diameter. Leaves are irregularly toothed, pale green and 15 cm long. Fruits are berries of 6 cm long, edible, sweet, white to purple. It is native to China, Europe and North America. There are several varieties in it like "globosa" which is compact and "pendula" which is weeping type. Commonly known as white Mulberry.

Uses : Ripe fruits and roots are useful. A tea made from root is used to treat diarrhoea. Fruit preparation is given for high fever. Fruit is also mild laxative.

Morus nigra Linn. :

This is a tree of l5 m tall. Leaves are with toothed margin, dull green in colour, may attain a length of 25 cm. Fruits are black or purple, upto 2.6 cm long, edible.

It is native to West Asia. Commonly known as Black common Mulberry.

Uses : Ripe fruits and bark are useful. From fruits, cough syrup is made and is also used to treat fever. To expel intestinal worms, bark is used.

Morus rubra Linn. :

It is a tall tree of 25 m in height. Top is broad and rounded. Leaves are rough above and fuzzy beneath, upto 15 cm long. Fruits are dark purple in colour.

This species occurs in Central America. Commonly known as American Mulberry.

Uses : Fruits, bark and sap of the tree are useful. Fruit is laxative and used to treat fever. To expel intestinal worms outer bark is used. A tea made from inner bark is also used as laxative for children. Sap is useful for treating ringworm.

Mucuna Adans.
Family : Leguminasae (= Fabaceae)

Mucuna pruriens (L.) DC.

This is a twining annual plant with slender and long branches covered with short hairs. Leaves are trifoliate and elliptical in shape. Flowers are short stalked, large, papilionaceous, purple in colour and appear in axillary and pendulous racemes. Pods are upto 14 cm long, slightly curved and covered with hair. Seeds in pod varies between 5 to 7.

It is distributed in tropical and subtropical region.

Uses : Seeds and roots are medicinal. The seeds contain L-DOPA (L 3:4 dehydroxy phenylalamine acid), Glutathione, Lecithin, gallic acid, glycosides, micotine, prurienine, prurienidine, dark brown viscous oil, mineral matter. Seeds and roots are effective nerve tonic. These are considered adphrodisiac and diuretic. These are also used to treat kidney trouble and dropsy.

Murraya Koen. ex Linn.
Family : Rutaceae

A genus consists sixteen species of trees and shrubs and three are native to India.

Murraya koenigi (Linn.) Spreng. :

This is a small tree upto 6 m tall with attractive habit and graceful leaves.

The species is native to India. It commonly occurs in the foothills of Himalaya, Assam, Sikkim, Kerala, Tamil Nadu, A.P. and Maharashtra. Commonly known as Curry Leaf Tree or Limbleea Tree.

Uses : The leaves are aromatic and contain protein, carbohydrate, fibre, minerals, carotene, nicotinic acid and vitamin C. It is rich in vitamin A and calcium. The leaves contain very high amount of oxalic acid. The leaves also contain crystalline glucoside, koenigin and a resin. Fresh leaves on steam distillation yield 2.5% volatile oil. The oil is light yellow in colour, aromatic with spicy note.

It is mainly used for seasoning curry and other food preparation. From the seeds, fat oil is produced. The leaves, bark and the roots possess carminative, tonic and stomachic properties. It is also used in Ayurvedic system of medicine. The juice of root is taken to relieve pain associated with kidney. The leaves yield essential oil 0.04%, bright yellow in colour. The paste of leaves is applied on utricaria. The bark is used externally to treat eruptions and bites of poisonous insects and reptiles.

Murraya paniculata (Linn.) Jack :

This is an evergreen small tree with white fragrant flowers. It is commonly found in the shady ravine slopes of upto 1600 m altitude in Garhwal Himalaya.

Uses : The leaves contain essential oil of 0.01%. The leaves of these plants possess stimulant, tonic, stomachic, astringent and refrigerant properties. The leaves are used to treat cough, rheumatism and hysteria.

Marrubium Linn.
Family : Labiatae

Marrubium vulgare Linn. :

This is shrub of 1.5 m tall. Leaves are fuzzy, ovate-round, upper surface white and lower surface grey. Foliage is aromatic when fresh. It has a strong, slightly musky odour, and bitter, astringent, slightly naucous taste.

It is a native to temperate Eurasia. Commonly known as White Horehound.

Uses : Leaves and small stems are cut before blooming and dried in shade, and the bark of the plant can be collected any time for medicinal purposes. The plant contains a volatile oil, gallic acid and a bitter principle which is soluble in water and in alcohol. The plant juice is rich in tannin.

It is recommended for chest complaints. It is a strong depurative. The infusion has been recommended for irritation in the bronchia, chronic catarrh, mercurial salivation, chlorosis and anemia. Leaves are combined with honey and given in cough. Tea made from dried bark is used to treat cold and debility. The plant is also useful to treat asthma, diarrhoea, kidney ailments and menstrual irregularities.

Musa Linn.
Family : Musaceae

Musa paradisiaca Linn. :

This is a monocotyledonous perennial and attain a height upto 15 m, with rhizome and false aerial stems. The inflorescence arises from rhizome and appear at the top of aerial stem.

This species is native to Tropical Asia and occurring at Chimtipur in District, M.P. Commonly known as Plantain.

Uses : The root and stem decoction (about 20 ml) are taken in the morning for one month as tonic in blood and veneral diseases.

Myrica Linn.
Family Myricaceae

This is a hardy perennial shrub of 15 m tall with waxy branchlets and small catkin flowers of no beauty. Leaves are lance shaped, green and highly fragrant. Fruits are berries, greyish in colour.

This species is native to America. Commonly known as "Candle Berry".

Uses : Leaves, fruits and root bark are useful. A decoction of boiled leaves is given to expel intestinal worms. Leaves are rich in vitamin C and used to treat scurvy and fever. The fruit provide wax and used in candle making. Root bark decoction is used to cure uterine haemorrhage, spongy gums, cankers of mouth and throat, jaundice and dysentery.

Myrospermum Jacq.
Family Leguminasae (= Fabaceae)

Myrospermum perniferum Linn. :

Leaves are dotted. Fruits are winged, indehiscent.

The plant is native to Tropical America and West Indies.

It is diuretic and recommended for urinary diseases. The plant is also a stimulant used in chronic catarrh.

Myroxylon Linn. f.
Family : Leguminasae (= Fabaceae)

Myroxylon balsamum (Linn.) :

It is an attractive evergreen tree upto 35 m tall with spreading crown. Bark is rough and when injured a yellow-brown resin is excreted which is pungent and fragrant. Flowers are yellow. Fruits are 13.6 cm long with a wing of 2.6 cm wide. Seeds are kidney-shaped. It is grown as ornamental tree in parks and gardens.

It is native to Argentina, Brazil, Paraguay, Venezuela, Bolivia, Colombia and Peru. This species is commercially cultivated in Columbia. The plant is also successfully grown in India. Commonly known as to be Balsam.

Plate 1.

Uses : It contains 75 to 80% resinous material like cinnamic and benzoic acid, esters of a complex alcohol, tolu- resinotannol. It also contains free cinnamic acid 12 to 15% and free benzoic acid 89r. It constitute benzyl benzoate, benzyl cinnamate and traces of vassillin and yields 1.5 to 7% volatile oil by steam distillation.

Alstonica scholaris

Artemisia

Plate 2.

Bauhinia purpurea

Bombax ceiba

It is an ingredient of tincture benzoin. Resin is used in lozenges for the relief of throat irritation, and coughs. For the treatment of respiratory ailments it is heated and the vapour inhaled. The extract is used in chewing gum, soft drinks and candy ice-cream. Oil is used in the formulation of perfumes and flavouring soaps. Wood is hard, heavy, fragrant and strong and is used in cabinet works, furniture etc.

Nardostachys DC.
Family : Velerianaceae

Nardostachys grandiflora DC. N. Jatamansi D. Don :

This is a perennial herb grows to a height of 85 cm. Lower leaves are upto 25 cm long, and upper leaves are smaller. Flowers appear in bunches, small and pinkish white in colour. Rhizomes are woody and long, covered with aromatic fibres, hence rhizomes are very fragrant.

It is native to India and Bhutan. This species is distributed in Himalayan hills between 3000 to 5000 m elevation. Commonly known as "Spokenard".

Uses : The dried rhizomes and roots have medicinal properties. It has tonic, antispasrnic and stimulant properties. It is used to treat convulsion, certain types of fits, palpitation of heart and constipation. It improves urination, menstruation and digestion. It is also used to colour the oil for hairs.

Nasturtium R. Br.
Cruciferae

Nasturtium officinalis R. Br. (Syn. Roripa nasturtium var. aquaticum) :

It is a hardy succulent aquatic perennial. Leaves are alternate, divided, light green in colour. Stems are hollow and thick. The species is native to Britain.

The plant grow in flowing water of ditches, slow streams and ponds. Water must flow through beds, and not be stagnated. They are propagated from seeds or division of plants. Seeds are sown in the nursery during July to November which is kept wet and later on seedlings are transplanted. The beds are flooded when plants have grown to about 5 cm. The plants are spaced at a distance of 15 cm x 15 cm.

It has a pleasant and pungent flavour. This plant is largely used for salads and is a rich source of vitamin A and C. This plant is given to be chewed for a simple or ulcerous stomatitis. The plant is used in antiscorbutic juices, syrup and tincture. For the treatment of kidney, heart trouble and tuberculosis the plant is eaten.

Nauclea Linn.
Family : Rubiaceae

About forty five species distributed in Tropical Asia and out of which six are native to India. Which are trees or shrubs mostly.

Uses : The decoction of bark, leaves and wood of Nauclea orientalis L. is used as vulnerary. The fruit is edible. The leaf paste is applied on boils. It produces yellow fragrant flowers.

The bark of Nauclea sessilifolia is considered as tonic and astringent and used to treat fever, bowel complaints. The wood of this plant also possess tonic and depurant properties.

Nelumbium Jass.
(Nelumbo Adams)
Family : Nymphaeaceae (Nelumbonaceae)

Nelumbium nucifera Gaertn. (= N. speciosum Willd) :

This is a lovely aquatic with ornamental flowers and foliage. Flowers are large upto 40 cm across. Rhizomes are thick, wide spreading, horizontal. It is distributed all over India and extending as far as north of Kashmir. This species is also distributed in China and Japan.

Uses : The underground rhizomes of N. nucifera are eaten as vegetable, usually cooked, fried or preserved in vinegar or sugar. They are also used as starch. The nut like seeds when ripe, are eaten either raw, roasted or boiled. In China, the rhizomes are served up in summer with ice. The filaments of N. nucifera, known under the name of "Kinjaka" are astringent and cooling, and prescribed in the burning of the body, bleeding from piles and menorrhagia. The rhizomes are applied as a paste in ringworm and other cutaneous affections. The seeds are used to check vomiting and given to children as diuretic and hypothermic. The leaves of these plants form cool bed sheets, useful in fever, accompanied by much heat and burning of skin. A Sherbet of this plant is used as hypothermic in small-pox and is said to stop eruptions.

Nepeta Linn.
Family : Labiatae

About two hundred fifty species of erect or prostrate herbs and thirty five are found in India of this plant species.

Nepeta cataria Linn. :

It is a hardy, erect, herbaceous perennial upto 1.5 m tall. The stem is downy and whitish. Leaves are more or less heart shaped, opposite, coarsely dentate upto 85 cm long. Flowers are tubular, small, appear in clusters, whitish to pale purple in colour. Plants when rubbed give a disagreeable odour.

It is native to Europe and Temperate Western Himalaya. Commonly known as catmint.

Entire plant is harvested when in flower. The plant is slightly acrid and bitter in taste, and regarded as aperative and sudorific. Dried leaves and stalks are smoked to relieve respiratory ailments. A tea made from the plant is used to treat stomach ailments, nervous conditions and treating cold.

The plant poultice is used to reduce swellings. The plant has also been used to delay menstruation. Leaves are chewed to relieve toothache.

Nepeta elliptica Royle ex. Benth. :

This is an annual erect herb with purple flower. Flowers are gynodioceous.

The plant is distributed in the alpine zone between 2800 to 3500 m altitude at Bhagrathi Valley, Herkeedun and Bhilagana valley of U Garhwal. It is native to Temperate Eurasia, North Africa.

Uses : The infusion of seeds are used as carminative and useful to treat dysentery and uterine disorders.

Nepeta graciliflora Benth. :

This is a hardy herbaceous perennial. Leaves are toothed, heart shaped. The species is native to Himalaya, occurring at Gamari of Utterkashi above 2800 m.

Uses : The leaf paste is applied on forehead in headache. The paste is also applied in swelling due to injury.

Nepeta hindustana (Roth.) Haines. :

It is an erect herbaceous plant upto 80 cm tall with purplish flowers. Sporadically occur on the moist slopes of foot hills upto 1800 m in Garhwal.

Uses : The decoction of the plant is used in sore throats. Leaves and flowering tops have carminative, refrigerant and stimulant properties. The plant is administered as cardiac tonic and also in fever.

Nerium Linn.
Family : Apocynaceae

The Neriums species are popular and commonly grown as tropical shrubs. These are erect shrubs about 2-4.8 m tall. Leaves mostly in whorls of three's or four's, opposite, lanceolate, 16 to 35 cm long and 1.6 to 3.2 cm broad; leathery, entire, dark green in colour. N. indicum Mill. (= N. odorum Soland) is known as "Indian Oleander". The other species N. oleander Linn. is known as "Oleander" or "Rose Bay". Flowers are double or single, appear in terminal panicles, fragrant, 4.5 to 7.8 cm across and showy. There are several varieties in N. oleander L.

Uses : All parts of the plants are poisonous. Seeds, barks and roots contain cardio-active glycosides. The bark contain small quantities of tannin, a deep red colouring matter and an essential oil. Aqueous extracts of leaves, branches, roots and flowers are toxic to certain insects. The plant is used as a rat poison. The leaves are used in cutaneous eruptions. A decoction of leaves is used to destroy maggots infesting wounds. The paste of the root is applied externally in haemorrhoides and ulcerations. Extracted oil from root bark is used in skin diseases of a scaly nature. Fresh leaf juice is applied in eyes for inducing lachrymation in ophthalmia.

Nicotiana Linn.
Family : Solanaceae

Nicotiana tabacum Linn. :

This is a half hardy annual, may attain a height of 1.8 m. Flowers are rose coloured. There are different varieties in it.

The plant is cultivated in tropical countries and grown as annual crop.

Uses : The ill effect of these plants are well known. Its green leaves are applied to skin to cure scabies, destroy lice and have a salutory effect in cases of neuralgic pains and toothache. As a rectal injection, it can act as a direct stimulant in obstinate constipation and helps to reduce hernia or as a revulsive in apoplexy and asphyxia. It is used for cigar, cigarette and pipe tobacco.

Nymphoea Linn.
Family : Nymphaeaceae

Under this genus there are about fifty species and four are native to India. These are perennial, ornamental aquatic herbs with creeping rhizomes.

Plate 3.

Catheranthus roseus

Carica papaya

Plate 4.

Cassia fistula

Cochlospermum gossypium

Nymphoea alba Linn. **:**

This is a hardy aquatic tuberous rooted perennial. The plants bear stout creeping rhizome, at the tip it is bent up, and bear stipuled leaves and flowers on long stalks. The leaves are large and float on the surface which is nearly circular, entire and leathery, with stomata, cuticle and palisade tissues on the upper side. Flowers are white. Commonly known as Water Lily or Water Rose.

Uses : They were administered against burning and insomnia. They were also said to improve the voice. They are mainly used as emollients and demulcents. The pulp of rhizome when applied externally acts as a rubefacient.

Nymphoea odorata Soland **:**

This is mostly used as an antiseptic and astringent. It is commonly known as "American Pond Lily" and native to America.

Ocimum Linn.
Family : Labiatae

A genus of about one hundred and fifty aromatic species, herbs or shrubs. Five species are native to India.

Ocimum basilicum Linn. **:**

This is sacred in Hindu religion. The plant is native to Tropical Asia. There are several varieties in it like bullatum, minimum and violaceum. The plant is an annual herb.. The leaves have numerous dot-like oil glands, which secrete strongly scented volatile oils. The leaves are simple, petiolate, ovate or subovate. Small, white flowers appear in clusters. Flowers are two-lipped in raceme fashion.

It is native to Punjab and Himachal Pradesh. This crop is cultivated in USA, France, England, Rumania and India.

Mainly its leaves and tender parts of shoots yield 0.4% volatile oil. The oil from, this plant constitute methyl chavicol, methyl cinnamate, eugenol, linalool, camphor and geraniol. The leaves contain protein, carbohydrate, volatile oil, fixed oil, cellulose, mineral matter and vitamins A,B & C. Commonly known as Sweat Basil or Basil.

Uses : This plant possesses stimulating properties. An infusion of leaves is used to treat gonorrhoea, burning urine and nephritic affections. The plant is useful in the treatment of worms, stomach complaints, fever, cough and gout. Leaf juice is employed as a nasal douche, and also used to treat skin diseases.

For relieving constipation and piles seeds are eaten. The juice of leaves is poured into the ear to cure otitis. The seeds are said to have refreshing and sedative properties. The essential oil extracted from this plant is cephalic and antispasmodic. It is used in tonic neurosis. It is also used as food flavourant, perfumery and cosmetics as well as in medicines.

***Ocimum canum* Sims. :**

This is a herbaceous plant. Bracts are petiolate. Flowers are more conspicuous. Seeds are black, ellipsoid in shape and become mucilaginous when wetted.

It is distributed in Jammu & Kashmir, Punjab, Himachal Pradesh, Uttar Pradesh and Delhi.

Uses : The oil obtained from this plant constitute linalool and camphor. Seeds are tonic and diuretic. The oil from its leaves has the property of destroying bacteria and insects.

***Ocimum kilmand-scharicum* Guerke. :**

This is a perennial undershrub. Bracts are petiolate. Flowers are more conspicuous. Seeds are black, ellipsoid and become mucilaginous when wetted. Commonly known as "Camphor Basil".

It is introduced from Kenya in India and it does not have any natural range of distribution. It is commercially cultivated in West Bengal, Jammu & Kashmir, Dehradun and Lucknow for obtaining camphor.

Uses : The essential oil obtained from this plant constitute camphor. It is a source of camphor. This camphor is used in medicine in preparation for local applications on pains and sprains.

***Ocimum pilosum* Linn. :**

This is used to mostly alleviate the pain of child birth.

***Ocimum sanctum* Linn. :**

A much branched erect aromatic herb which may attain a height of 85 cm, entire plant is hairy. It is biennial or triennial in habit. Stems and twigs are usually quadrangular. Leaves are opposite and decussate, upto 6 cm long, hairy, margins entire or toothed, dotted with minute glands. Bracts are sessile. Flowers are small, less conspicuous, purplish or reddish, appear on slender spikes in small compact clusters. Fruits are small, seeds are reddish or yellowish. Seeds are subglobose in shape and do not become mucilaginous when wetted.

It has a wide distribution in the Indian subcontinent. Commonly known as "Sacred Basil".

Uses : The leaves and seeds of this plant have medicinal value. The infusion of juice of leaves is used to treat digestive complaints, bronchitis and catarrh and also administered locally to cure ringworm and skin diseases. The juice of leaves is dropped in ears to relieve earache. To cure common colds decoction of leaves is prescribed. A decoction of root is used in malarial fever. The oil extracted from leaves has the property to destroy insects and bacteria.

Oenanthe Linn.
Family : Umbelliferae

***Oenanthe crocata* Linn. :**

This is a stoloniferous glabrous herb.

The species is native to Temperate Europe. Plants grow in moist and wet places. Commonly known as "Water Dropwort".

Uses : The resin extracted from the root of these plants is poisonous. The plant is also toxic. The plant can be used in caution in cases of dysentery, leucorrhoea and epilepsy. It is also used for scabies, whitlow and piles.

Olea Linn.
Family Oleaceae

***Olea europeae* Linn. :**

This is an evergreen flowering shrub, flowers are white and fragrant. Plants may attain an ultimate height of 15 m.

This species is native to Asia Minor and Syria. It is cultivated in Mediterranean region. Commonly known as "Oleve".

Uses : Ripe fruit of olive is used for its oil. Around 15 to 14% non-drying oil is available which constitute fatty acids, squalene, phytosterol and tocopherol. The oil is obtained by bruising and pressing the fruit.

This oil is used as vehicle for injections. This is also known as emollient and demulcent.

Ononis Linn.
Family : Leguminasae (= Fabaceae)

***Ononis antiquorum* Linn. :**

This is a herbaceous plant viscid. Pods are oblong and turgid. The species is native to Canaries and Mediterranean region.

The root of this plant is used for medicinal purposes. The roots are bitter in taste and is regarded as aperitive, diuretic and astringent. Commonly called "Rest Harrow".

Onosma Linn.
Family Boraginaceae

Onosma echioides (Gaertn.) Linn. :

This is an erect much branched hairy herb. Flowers are purple in colour. It is native to Europe and Asia. It occurs rarely in Tehri Garhwal at Bhai.ro Ghati and also in Alaknanda Valley, Dhodi Ghatie, Kedar Nath and Madmaheshwar of Garhwal Hill between 3500 to 4500 m elevation.

Uses : It is used in the Tibetan system of medicine. The roots are locally applied on wounds and cuts. It is also used as dye.

Onosma emodi Wall. :

This is a perennial hairy herb.

Occur in boulder hewn, scrubby slopes of Alpine region of Central Himalaya between 3500 to 4200 m.

Uses : Roots are used as medicine against skin diseases.

Operculina S. Manso
Family : Convolvulaceae

Operculina turpethum (Linn.) Silva Manoso :

This is a large climbing herb having milky juice. Leaves are ovate, cordate, upto 15 cm long and 8 cm broad. Flowers are upto 6 cm long, funnel shaped, appear in few flowered bunches, white in colour. Roots are long, branched and fleshy. Fruit is enclosed by sepals.

This species is native to tropical climate. It is widely distributed in India upto 1500 m altitude.

Uses : The dried root with intact back has medicinal properties. It contains turpethin. Powdered roots are used as purgative.

Ophioglossum Linn.
Family : Ophioglossaceae

Ophioglossum reticulatum Linn. :

This is a hardy deciduous fern. Fronds are barren, egg shaped, pale green, fertile ones contracted spike like.

The plant is native to America and Tropical Africa, uncommonly occurring at Mt. Abu of Rajasthan. Commonly known as "Bourbon".

Uses : The paste of the plant is applied on burns as cooling agent. The extract of leaf is also used in the preparation of tonic.

Opuntia Mill.
Family : Cactaceae

Opuntia tuna Mill. :

This is a hardy succulent plant. Stems are fleshy, flat and bristly, attain a height upto 8 m. Flowers are reddish orange; fruits are pear shaped, purple wth spines.

It is native to South America.

Uses : Flat branches and fruits constitute the drug. The flat succulent branches are rich in vitamin C. To cure rheumatism a poultice of the boiled stems are used. To treat gonorrhoea, diarrhoea and asthma, fruits are useful.

Orchis Linn.
Family : Orchidaceae

Orchis latifolia Linn. :

A Stem of this plants are usually fistular upto 95 cm long. Tubers are paired, lobed and palmate. Leaves to 1& cm long. Spike is cylindric, dense flowered, to l6 cm long. Flowers are 2 cm long, variable in colour from pink to purple to almost pure white. It is native to Europe, Africa and Himalaya. The plant is found in Western Himalaya, Kashmir between 3200 to 4200 m altitude. Commonly known as "Cuckoo flower".

Uses : The tubers are used as tonic and expectorant. It is used in powder form and diluted in water or milk. It is administered in intestinal irritation, chest irritations, phthisis, diarrhoea and chronic dysentery. It is also used in paste and chocolate and regarded as an emollient and demulcent.

Origanum Linn.
Family : Labiatae

Origanum vulgare Linn. :

This is an erect, pubescent aromatic, perennial herb to 1.5 m tall. Leaf stalkcd, ovatc, cntirc. Flowcrs arc pale white, tinged with pink and sm;ill, appeai in August-September.

It is distributed in Europe, Asia and Africa, commonly found at Thunera of Uttarkashi between 2200-350 m and different places of Himachal Pradesh between 1800-4800 m. It is also common in the tropical zone, growing upto 1500 m in Garhwal. Commonly known as Bastard Majoram or Banchari.

Uses : The leaves are aromatic and contain protein, fat, crude fibre, carbohydrates, minerals and vitamins A,B and C. The herb yield 0.15 to 0.40% volatile oil which contains thymol, carvaerol, alcohols, esters etc.

The oil of this plant possesses diuretic, carminative, stomachic, diaphoretic and emmenagogue properties. In homoeopathy it is administered in hysteric conditions. It is used in gargle and bath. It stimulates growth of the hair. This plant is used for infusion or fumigations, asthma, chronic catarrh, rheumatic pains and stiff necks. It is also regarded as stimulant, stomachic, expectorant, sudorific and emmenagogue. The crushed leaves are smelt in the cold. The leaf juice is applied on cuts and scratches to check the flow of blood. Oregano is an essential ingredient of chilli sauce.

Ostrya (Mich.) Scop.
Family : Betulaceae (= Caprinaceae)

Ostrya virginiana (Mill.) K. Koch. :

This is a tall tree of 25 m height. Leaves are upto 13.5 cm long, slightly fuzzy on both sides, and edges toothed. Flowers are greenish white. The fruit clusters are bladder like, about 6 cm long.

The species is native to North America. Commonly known as Iron wood or Lover Nood.

Uses : Bark of this plant constitute the drug, which is laxative and tonic. It is used to cure fever and indigestion. To reduce swelling of neck, a poultice of bark is applied. It also furnishes a hard wood.

Oxalis Linn.
Family : Geraniaceae (= Oxalidaceae)

A large genus consisting of eight hundred species, native to South Africa and America.

Oxalis acetosella Linn. :

This is a stemless herb, perennial, with rhizomatic creeping root. Flowers are pinkish or whitish in colour. It is found in the damp shady areas of Garhwal upto 3000 m, known as Red Wood Sorrel.

Uses : The leaves of these plants are eaten as salad. A drink is prepared with this plant with water and used to treat fever.

Oxalis corniculata Linn. :

A delicate perennial hairy herb, procumbent, hairy with alternate, stalked, 3 lobed leaves. Flowers are pale pink or yellow, small in size and appear on long stalked inflorescence.

Commonly occurring on different parts of H.P. upto 3000 m elevation, in shady places. It is also found throughout in the damp or moist places upto 2000 m in Garhwal Himalaya, known as clover sorrel or Wood Sorrel.

Uses : The plant is used as a cure for scurvy. It is a good appetiser. Leaf juice is used locally in removing of the warts and also known as useful in cataract of the eyes. Paste of top shoots along with a few fruits of black pepper is applied to boils, absceses, wound and weeping eczima. Leaves and roots are used to treat dysentery and diarrhoea.

Oxalis violacea Linn. :

This is a herbaceous perennial and bulbous rooted plant. Leaves arise from the basal bulb which consists of three rounded leaflets. The flowers are light purple in colour, known as Wood Sorrel or Sheep Sorrel.

Uses : From the above ground part of these plants in full bloom stage is used. To expel intestinal worms a teaspoonful of fresh and dried powdered leaves boiled in a cup of water is given to drink in the morning. Green leaves are eaten and known as blood purifier. The plant is also useful to treat fever and to increase urine flow.

Oxydendrum DC.
Family : Ericaceae

Oxydendrum arboratum (Linn.) DC. :

It is a hardy deciduous tree of 25 m in height. Bark deeply fissured. Leaves are 22 cm long, lance-shaped. Flowers appear in druping clusters, white in colour and small.

The species is native to Eastern North America. Commonly known as Sour wood or sorrel tree.

Uses : Leaves and bark of this tree are useful. Leaves are considered tonic and used to treat fever and urinary troubles. To reduce swelling a poultice of leaves mixed with bark is used.

Palms
(Other Palms)
Family : Palmae

The Palms trees are fascinating plants widely distributed in the tropics of both hemispheres. The family Palmae has nearly 240 genera and comprising of 3600 species, majority of which are natives of Tropical Asia and Tropical America. They may be grouped into two groups, one with pinnate or feather like leaves are other with palmate or fan-like leaves. The inflorescence usually is very large and much branched, mostly axillary, in few cases terminal. The spadix is enclosed in a spathe. Some have bisexual and some have unisexual flowers. Fruit is a berry or drupe, many of which are attractive usually hold one seed, but sometimes two or three.

Uses : Having apart from ornamental value of palms, they furnish food, fibre, shelter, clothing, timber, building material, paper, oil, wine, dyes, wax etc. for the use of humanity. Phoenix dactylifera L. (Date Palm) is the staple food of some parts of North America, which provide fruit, wine, sugar. Cocas nucifera L. (Coconut) is very important as edible fruit and oil. Fruits of Bactris maraja Mart (Maraja palm) and Enterpe edulis are edible from which nutritious drinks are prepared. Bactris gasipaes H.B. ex K.N. Granat (Palm Peach) fruits are used as vegetable. Palm honey is prepared by the extraction of sap from Jubaca spectabilis. The sap of Nypa fruiticans also produce syrup and alcohol. Fruit of Serenoa pepeus is edible. Borassus flabellifer L. (Palmeira) yield sugar and food . Caryota urens L (Bastard Sago Palm) is widely cultivated for palm sugar, sago and wine. Sago is also prepared from the trunks of Mextroxylon sagu Rottb. “Common Sago Palm”. Cocos nucifera L., Coroza oleifera, Elaeis guineensis Jacq. (African Oil Palm) etc. are valuable source of vegetable fats, used for cooking fats, margarine soap etc.

Panax Linn.
Family : Araliaceae

A genus of nine species of shrubs or trees.

Panax ginseng C.A. Mey. :

This is an erect perennial herb upto 80 cm tall. Leaves are compound, palmate. Petiole is longer. Leaflets are usually 6, sometimes 4, upper three layers upto 15 cm long and lower two upto 4 cm long, elliptic or ovate. Inflorescence simple, umbelliform,. Flowers greenish-yellow. Fruits berry-like. Roots are thick, fleshy, short, 1.2 to 2.6 cm in diameter, cylindrical or fusiform.

This is grown for the roots which are employed in medicine.

This species is native to Northern China and Korea. Commercially cultivated in China, Korea and Japan., Known as "Ginseng"

Uses : The root of the plant constitute the drug which contains panaxosides, steroids, vitamins and trace elements. The drug prepared from this plant is said to be tonic, stimulant and aphrodisiac. It is claimed that ginseng increases the natural resistance and recuperative power of the body. It is used for the treatments of amnesia, headache, convulsion, dysentery and cancer. It is also prescribed for neurasthenia, dyspepsia, palpitation and asthma.

Panax quinquefolium Linn. is known as American ginseng which is a dwarf perennial greenhouse herb. The roots contain panaxosides and cultivated in United States to employ in medicine for the same purpose as that of P. ginseng.

Papaver Linn.
Family : Papaveraceae

A genus of about one hundred species of which seven are native to India. These are annual or perennial herbs with milky latex.

Papaver somniferum Linn. :

This is an erect herbaceous plant called "opiumpoppy", assumes great importance because it yields commercially valuable drug opium. It is one of the most important medicinal plants used as a source of drug alkaloids. Plants attain a height of about 125 cm, leaves are large serrated and flowers are attractively coloured. There are red, pink white, purple and multi-coloured flowers in it. The countries in which opium poppy is cultivated in large scale are India, Iran, Turkey, USSR, Yogoslavia, Bulgaria, Australia, Poland, Belgium, France and Spain. Out of these countries, only India is producing opium and rest of the countries obtain alkaloids direct from straw. These plants are grown in India commercially as annual crop during winter in restricted areas to a contiguous tracts in few districts of Madhya Pradesh, Rajasthan and Uttar Pradesh.

They grow well in a well drained fertile soil, neutral to slightly acidic in reaction. Deep clay loam soil rich in organic matter, adequate sunshine, temperature below 20°C during active vegetative phase and 30 to 35°C during reproductive phase is conducive for high yield. Warm and dry weather conditions during February to March is favourable for good flow of latex. Hailstorm and high gusty wind, frost and cloudy weather during lancing period are harmful.

Plate 5.

Coleus forskholii

Daphne mezereum

Plate 6.

Gliricidia maculata

Emblica officinalis

Plate 7.

Uses : The latex of the immature fruit contain several alkaloids like morphine, codeine, thebaine, noscapine, narceine and papaverine. Other minor alkaloids are aporeine, codamine, cryptopine, gnoscopine, hydrocotarnine, lanthopine, laudanidine, laudanine, laudanosine, meconidine, narcotoline, neopine, papaveramine, prophyroxine, protopine, reticuline, rhoeadine and xanthaline. Presence of meconic acid and methyltransferase enzyme has also been reported from the plant.

Jacaranda acutifolia

Hyocyamus niger

Plate 8.

Iris germanica

Medicinal value of opium is due to its presence of numerous alkaloids. Number of drugs are prepared from opium. Opium alone or along with quinine salts, is used in cases of delirium, spasmodic coughing or spasms. It reduces all secretions, except for perspiration. It is prescribed for diarrhoea and sugar diabetes. The plant contain more than two dozen alkaloids, out of which morphine, codeine, papaverine, naseopine and thebaine are generally used as pain killer, cough depressant, sedative, antitussive and smooth muscle relaxant. Except papaverine, which is also obtained from svnthetic sources, all other alkaloids are still obtained either from the latex opium or from direct extraction of capsule called husk or straw. Seeds are used in bakery products and in confectionery. Seeds are important source of oil, valued for culinary purposes.

Papaver bracteatum Linn. :

This is a herbaceous plant, perennial in nature and attain a height of 1.6 m. Flowers are deep red in colour, borne singly

The species is distributed in nature of Russian Caucasus and Northern and Westhern Iran between 1600 to 3200 m.

Uses : Fresh capsule is a potential source of thebaine alcohol and used as analgesic and cough remedy. Seeds are eaten and is a source of oil. Commonly called Great Scarlet Poppy.

Papver rhoeas Linn. :

Uses : It has sedative and demulcent properties. Because of its slightly diaphoretic and sedative action, this can be used in acute phlegmasias of the chest. It is a'so used fur treatment of pulmonary affections, in persistent cough, whooping cough, sore throats and for soothing of a sharp pain.

Parkia R. Br.
Leguminasae (Mimosaceae)

In this genus there are about forty species and native to Tropical region. Two species are indigenous to India.

Parkia roxburghii G. Don. :

These are tall unarmed trees. Flowers are in dense, long-peduncled heads. The pods are large, flat and coriaceous.

Uses : The bark and leaves of this plant are used for preparing lotions forskin disease and ulcers. The bark, root and leaves are reported to be cyanophoric. Seeds and pods are reported to be employed for stomach disorders.

Pods pounded in water is sometimes used for washing head and face. The seeds are eaten. Tender pods are used for flavouring food and are edible. It occurs in Assam. Commonly called "Netta Tree".

P. biglandulosa Wt. & Arn. :

Uses : The ferinacious pulp of the pods is edible. The flower pollens with water makes a refreshing drink.

Passiflora Linn.
Family : Passifloraceae

Passiflora incarnata Linn. :

This is climbing plant with axillary tendril. The plant may reach a height of 15 m. Leaves are 4 to 6 lobed, finely toothed, alternate. Flowers are purple and flesh coloured, attractive, upto 6 cm in diameter. The fruits are berry like, ovate, yellow in colour, smooth, upto 8.5 cm long, enveloped in a fleshy aril.

The species is native to America.

Uses : Leaves, flowering tops and fruits of these plants are useful. It is aphrodisiac. Fruit juice is used for sore eyes. Crushed leaves or plant tops are used as a poultice to treat bruises, burns, skin eruptions and haemorrhoids.

Pedicularis Linn.
Family : Scrophulariaceae

Pedicularis hoffmeisteri Klotz. and Gracke. :

This is a semi-parasite, with loose-pollen. Flowers are fertilised by bumble bees.

The species is distributed at Chyangseel of Uttarkashi between 2200-3200 m.

Uses : The paste of root and leaves is applied on laceration during rainy season.

Peganum Linn.
Family : Rutaceae (= Zygophyllaceae)

Peganum harmala Linn. :

This is a shrub which attain a height upto 95 cm. Leaves are divided into numerous narrow segments, upto 8 cm long. Flowers appear singly from the axils of leaves, white in colour, upto 4 cm across. Fruits are round, deeply lobed and upto 2 cm across. Seeds are with reticulated seed coat, brownish. The plant yield "Turkey-red".

It is native to America, Mexico and Mediterranean region. Plants are widely distributed in Northern and North-Western India and in drier regions of the Deccan peninsula. Commonly known as "Harmal" or "Syrian Rue".

Uses : Seeds contain harmaline, yageine and harmine alkaloids. The plant and dried seeds are used as drug. It has narcotic, emetic and anthelmintic properties. It is used in the treatments of jaundice, fever, painful menstruation, colic pains, gallstones, asthma, hysteria and rheumatism. Higher doses are toxic and cause nervous depressions. The alkaloids act as hallucinogens, it stimulates motor tracts of cerebrum and central nervous systerns.

Penstemon Schmid.
Family : Scrophulariaceae

Penstemon pallidus Small. :

This is a hardy perennial plant of 1.2 m tall. Flowers are white in colour. The species is native to America. Compost consisting of two parts rich loamy. Commonly known as "Pearl Tongui".

Uses : Roots of this species constitute the drug. Roots are used to treat rattlesnake bites and to hasten the movements of the after birth from a woman who has delivered a child. Roots are chewed to get relief from painful tooth cavities.

Pergularia Linn.
Family : Asclepiadaceae

Pergularia daemia (Forsk.) Chiov. :

A hairy climber with milky juice. Leaves are ovate or round, upto 6 cm long, lower surface is hairy. Small flowers appear in short clusters, yellowish green in colour.

Fruits are upto 9 cm long and 1.6 cm broad, covered with spinous outgrowths.

It is native to Madagascar to India. The plant occurs in almost all parts of India upto 1600 m altitude. Commonly known as Pergularia.

Uses : The plant is emetic, purgative, expectorant and used as uterine tonic. Leaf juice is used to treat infantile diarrhoea, asthma, and catarrhal

Plate 9.

Largerstroemia speciosa

Juglans regia

Plate 10.

Mangifera indica

Lavendula officinalis

affection, rheumatism and amenorrhoea. It is also useful in curing gynoecological conditions like excessive bleeding. Squeezed fresh leaves are applied as poultice in carbuncle.

Perilepta Bemek.
Family : Acanthaceae

Perilepta auriculata Nees. :

A shrub with pale blue tubular flowers. The fruiting brownish hairy spike bears oil droplets which emit camphorous odour.

It is native to India to China and Indo-China. Distributed in Kangra and Hamirpur districts of Himachal Pradesh.

The oil droplets from the fruiting spike are mixed with Kajal (Collyrium) and is applied to the patient suffering from cataract and trachoma.

Petroselinum Hill.
Family : Umbelliferae

Petroselinum crispum (Mill.) Nym. ex. A.W. Hill :

This is cultivated as condiment. It is a hardy aromatic biennial herb. Leaves are rosette and finely divided. Flowers are yellow or yellowish green, appear in compound umbels. Fruits are 3-4 mm long, crescent shaped.

It is native to Europe and Mediterranean region and extensively cultivated in America and Mediterranean region. Commonly known as Porsalay.

Uses : The colour of the dried herb is green which is pleasantly fragrant and spicy. The green leaves contain protein, fat, carhobydrate, fibre, minerals, caroteme, thiamine, nicotinic acid and ascorbic acid. The leaves, stems and fruits contain a glucosi.de apiin. All parts of the plant contain essential oil.

Fresh leaver are used for garnishing and seasoning of food. Leaves are eaten frvsh. They are used in culinary. The roots are used as vegetables in soup. The roots of these plants are aperitive, diuretic, stimulant and diaphoretic. It is effective particularly on the urinary system, obstructions and the skin. In blennorrhagoea essential oil of this plant has been used in dose of 1-2 drops. It is said that parsley juice in white wine cured blennorrhoea, leucorrhoea ond congestion of abdominal viscera. The juice is administered for anti-periodic and seminal discharges.

Peucedanum Linn.
Umbelliferae

Peucedanum ostruthium (Linn.) Koch. :

This plant species is native to Eurasia.

The root stock of these plants was used to make theriacal water and quack medicine.

It was used as a tonic and stimulant. Commonly known as "Master wort" or "French Ben Zoin".

Pholidota Linn
Orchidaceae

Pholidota articulata Linn. :

It is an epiphytic orchid. Pseudobulbs of these plants are jointed, 2 leaved, cylindric upto 12 cm long. Leaves are thickly membranous, many nerved, elliptic, acute, short-stalked to 12 cm long and 5 cm broad. Inflorescence upto 16 cm long, drooping, many flowered. Flowers to 1.24 cm across, musk scented and yellowish-white.

It occurs in Kumaon, Sikkim, Manipur and Khasia Hills between 800 to 1500 m altitude.

Uses : The whole plant is used as tonic. Commonly Known "Rattle Snake Orthid".

Pholidota imbricate Linn. :

Its Pseudobulbs are tufled to 8 cm long, broadly conical. Leaves are 40 cm long and 6 cm broad, acute, broadly elliptic lanceolate. Inflorescence is drooping, densely flowered to 26 cm long. Flowers are small upto 8 mm long, musk-scented or odourless, pale pink in colour.

It is distributed in Orissa, Chota Nagpur, Khasia Hills, Andaman & Nicobar Islands, Kumaon and Sikkim.

Uses : The plant is used as tonic.

Phyllanthus Linn.
Family : Euphorbiaceae

A genus of about over six hundred species, distributed in warm countries. About forty species are native to India.

Phyllanthus niruri Linn. :

A small annual herb. Leaves are small, pale yellow in colour. Flowers are greenish yellow.

It is a weed commonly found in different parts of India.

Uses : Fresh leaves are used for treatment of jaundice and intermittent fever. The plant possesses diuretic properties. It is useful in treating the diseases of genito-urinary tract.

Phyllanthus urinaria Linn. :

It is an erect annual herb with greenish yellow flowers.

This species is occurring in the hill slopes of Garhwal and also in other parts of India.

Uses : The plant is used to treat liver diseases and jaundice. The whole plant is diuretic.

Physostigma Balf.
Family : Leguminasae (= Fabaceae)

Physostigma venenosum Balf :

This is a woody climber which may attain an ultimate length of 18 m., branches are twining. Leaves are trifoliate to 16 cm long and 10 cm wide. Leaflets rounded at base and pointed at the apex. Inflorescence axillary appear in pendent clusters. Flowers are kidney shaped, purple. Fruit pods with horizontal veins, yellow brown in colour. Seeds are hard, smooth, 3 cm long and 1.6 cm wide.

It is native to West Tropical Africa. It can be grown in the coastal areas of Southern India and Nilgiri hills.

Uses : Seeds contain 48% starch, 23% albuminoids and several alkaloids like physostigmine, eseridine, calabarine, physovenine etc.

Seeds are used to treat skin diseases, rheumatism and the liquid extract of seeds is used as a sedative. Physostigmine salicylate is administered orally, subcutaneously or intramuscularly as an antispasmodic in rheumatoid arthritis, bursitis and fibrostitis. Physostigmine is also used in ophthalmology, it also acts as a strong sedative on the spinal cord. Eseridine is a purgative. Commonly known as Ordeal Bean of Calabor.

Phytolacca Linn.
Family : Phytolaccaceae

Phytolacca americana Linn. : P. decandra :

This is a perenial plant of 2.5 m tall with extensive root system. Leaves are dark green, broad, ovate. Flowers are borne in terminal clusters, white to pinkish in colour. The plants bear dark purple berries.

The species is native to North America. Commonly known as "Virginian Poke".

Uses : Immature green leaves, berries and roots of these plants have medicinal properties. A poultice from dried leaves is applied on ulcers, wounds and swellings. A poultice of fresh leaves is applied on old scabs.

Juice of the berries is used to treat skin eruptions and cancerous skin ulcers. A poultice of powdered roots is used to treat cancer.

Picea D. Dietr.
Family : Pinaceae

A genus of about fifty five species of large evergreen trees. Cones of which are cylindrical, pendulous or erect when young.

Picea abies (Linn.) Karst. Linn. :

It is a hardy evergreen coniferous tree upto 35 m tall. Leaves needle shaped, spirally arranged. Branches are drooping, brown and fuzzy. Cones are rounded, short to 2.6 cm long, purple in colour.

It is native to Northern Europe. Comonly known as sprouce or "Spruce Fix".

Uses : Twigs, cones and resin of these plants have medicinal properties. Powdered resin is applied on wounds to hasten healing. Twigs and cones are rich in vitamin C. A tea prepared from inner bark is used to treat kidney stones, rheumatism and stomach trouble. This plant furnishes valuable wood, resin and turpentine.

Picea rubens Sarg. (= P. rubra Linn.) :

This is an evergreen coniferous tree of 30 m tall. The needles are quadrangular. Bark is reddish in colour. Cones are reddish brown.

This species is native to North America.

It is generally grows well in mountaineous region and commonly known as "Red sprouce".

Uses : Pitch of the plant is useful. It is used to treat stomach or chest pain and rheumatic joints pain.

Picrasma Blume.
Simaroubaceae

Picrasma excelsa Planch. (= Pincraena excelsa Linn.) :

This is a tree with pithy branches and glabous.

The species is native to Himalaya. Commonly known as Bitter Wood.

Uses : Wood of this plant constitute the drug which contains picrasmin, neoquassin and thiamine.

It is known as bitter tonic and employed as enema to expel thread worms. The bitter wood and bark are used as a substitute for quassia.

Picrorhiza Royle ex. Benth.
Family : Scrophulariaceae

It is a genus of two Indian species. which are perennial low herbs.

Picrorhiza kurooa Royle ex. Benth. :

A prostrate, perennial herb with bitter rhizome and leaves. The leaves turn black on drying. Flowers are light blue in colour with long filaments.

It is native to western Himalayan Region. Distributed along stream borders and moist rocks of Alpine region in Central Himalaya between 3500-4000 m elevation.

Uses : Root and stem are used in medicine. Rhizomes of the plant is administered against fever and gastralgia. The powdered rootstock is used in abdominal pain and it is also used as laxative. It is administered in liver complaints, anaemia and jaundice.

Picrorhiza scrophulariflora Pennell. :

A small herb with spathula shaped leaves. Rhizomes to 28 cm long and woody. Flowers in cylindric spike, borne on leafless erect stems .

Rarely found in Chyangseel of Uttar Kashi above 3000 m, and also in Kashmir and Sikkim.

Plate 11.

Uses : The extract of root is given to drink in fever. The dried rhizomes improve appetite and stimulates gastric secretion. It is a bitter tonic and is useful as antiperiodic and promotes secretion of bile.

Narium

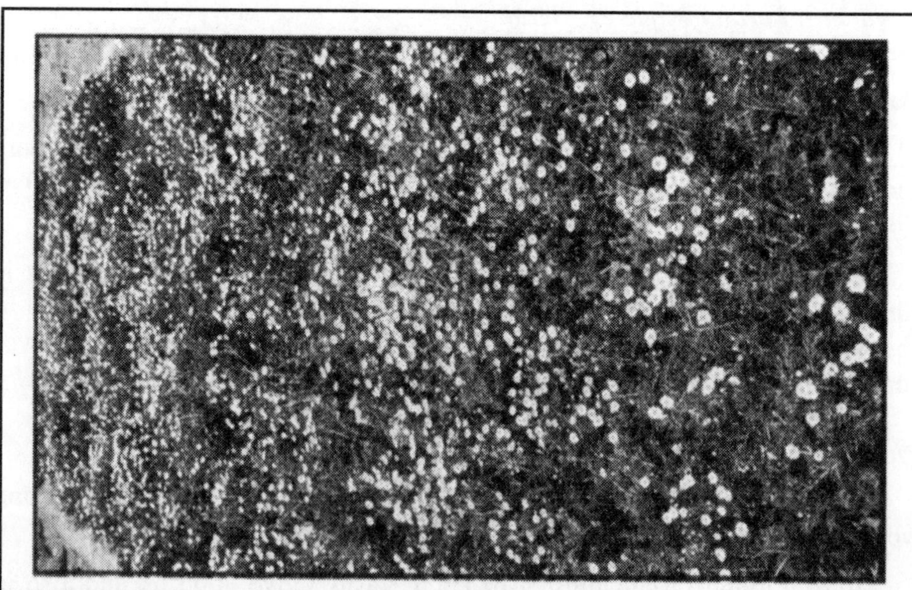

Matricaria chamomilla

Plate 12.

Ocimum sanctum

Opuntia

Pilocarpus Vah.
Family : Rutaceae

Under this genus there are about twenty species. These are shrubs or small trees, sometimes attaining a 4 m height. Leaves are pinnately compound and flowers appear in elongated racemes.

This genus is native to Tropical America, mostly are from Brazil.

Uses : Several species of Pilocarpus have medicinal properties. These are shruby plant and commonly known as "Jaborandi". These plants are distributed in Brazil and Rio Acre region. The important species are Pilocarpus jaborandi Holmes, P. microphyllus Stapf., P. pennatifolium Lem. and P. trachylophus Holmes. Leaves are aromatic. Abundant flowers are produced in axillary or terminal racemes or spikes. Flowers are stem like, white or purple and fragrant. Fruits are one seeded and carpels with 1 to 5 ridges.

Leaves contain essential oil which constitute methyl ketone and various hydrocarbons. It also contains alkaloids mainly pilocarpine, pilocarpidine, pilosine, isopilosine and isopilocarpine. From the leaves, flowers and fruits of P. pennatifolium, the alkaloid jaborandine has been isolated.

Pilocarpus jaborandi Holmes (= P. Pinnatus Martius, P. selloanus Engler) :

The leaves are shorter than P. pinnatifolium. Branchlets and stems are stiffy, pubescent, more openly flowered racemes.

Pilocarpus pinnatifolium Lam. :

The branchlets are glabrous or pubescent. Leaves are alternate upto 60 cm long. Leaflets are two or three in pairs, besides the terminal one upto 2.6 cm long, oblong; apex rounded or emarginate, coriaceous and yellowish green in colour. Inflorescence is spike-like and many flowered. Flowers are reddish brown.

There is another species Pilocarpus microphyllus, leaves of which possess medicinal properties.

Uses : The alkaloid pilocarpine in derived from the leaflets of the above threespecies. Pilocarpine is the standard drug used for initial and maintenance therapy in certain kinds of glaucoma in eye.

Pimenta Linn.
Family : Myrtaceae

Pimenta officinalis Linn. :

This is an evergreen aromatic tree which may attain a height of 14 m. Leaves are aromatic. Flowers are white in colour. Fruits are berries, globular,

6-8 mm across, reddish brown in colour and with a somewhat rough surface. It is native to West India, Tropical Central America and Jamaica. This plant is primarily cultivated in Jamaica and Guatemala. It is grown in the gardens of Bangalore, Bihar, Orissa, Kerala and West Bengal. Commonly known as Allspice.

Uses : Allspice possesses the combined characteristic flavour and aroma of black peppers, clove, cinnamon and nutmeg. Leaves and unripe dried fruits have medicinal properties. The essential oil of fruits contains eugenol, phellandrene, caryophyllene, methyleugenol and cineole. The berries constitute protein, fat, fibre, carbohydrates, minerals, vitamins A, B and C.

The essential oil is used as a carminative and flavour. The berries of Allspice are used as condiments for flavouring various preparations, and also valued in liquor industry as flavouring ingredient. It possesses stimulant, tonic, purgative, digestive and carminative properties. The powdered fruit is used to treat flatulence, diarrhoea and dyspepsia. It is also used against neuralgia and rheumatism.

Pimpinella Linn.
Family : Umbelliferae

***Pimpinella anisum* Linn. :**

This is a hardy annual, pubescent, aromatic herb of 70 cm tall. The basal leaves are long and dissected with long petioles. The leaves on stem are pinnately decompound. Flowers appear in umbels, small, white in colour and borne centrepetally. Fruit is small, ovoid, laterally compressed, greyish green in colour. After drying, the fruits become greyish brown.

It is native of Greece to Egypt. The crop is commercially cultivated in South and South-East European countries as an aromatic spice crop. It is also cultivated in Egypt, Syria, Iran, Mexico and India. Commonly known as "Anise".

Uses : Anise seeeds contain protein, fatty oil, essential oil, sugar, starch, crude fibre, minerals and choline. It contains 4 to 5% essential oil which constitute 95% anethole and also nethylchavicol, p-nethoxyphenyl-acetone and acetic aldehyde.

Fruits and essential oils are mainly used for medicinal purposes. Seeds are used as condiments. It has carminative, diuretic and diaphoretic properties. This plant is used externally in the form of fomentations, lotions and poultices for lecteal ecchymoses and obstructions. Fruits are regarded as a stimulant, carminative, stomachic, diuretic, expectorant and emmenagogue. It is used in the debility of digestive organs, gastralgia, flatulence, windy and spasmodic colics, dyspepsia, giddiness, dizziness and nervous headache. Seeds and oil is also used for flavouring baked food, bread, biscuits, soups, cordial drinks and candies. The oil also possesses fungicidal and insecticidal properties.

Pinus Linn.
Family : Pinaceae

It is a genus of over one hundred ten species of which six are native to India. Which are evergreen and resinous trees.

P. edulis Engelm. (= P. cembroides Zncc. :

It is a hardy evergreen coniferous tree. A slow growing round tree with edible seeds upto 12 m tall. Branchlets are orange coloured and then turned to grey brown and reddish tinge with age. The rigid, triangular stout needles grow in clusters of two or three.

It is native to Mexico. Commonly known as Nut Pine.

Uses : Needles, nuts, wood and resin of these plants are useful. Gum or resin is chewed for sore throat. A poultice of hot resin is used to treat muscular pains, soreness and pneumonia. A preparation from needles, used for treating syphilis. Nuts are eaten and nutritive.

Pinus elloittii Engelm. :

A quick growing erect plam attain a height of 35 m, and 1.5 m in diameter. Outer bark is grey and inner bark red-brown. Needles are in two's or three's and upto 36 cm long. Cones are red brown. Seeds are avoid upto 8 mm in length.

It occurs naturally in South-Eastern United States. It is cultivated in Brazil and Texas.

Uses : Resin obtained from slash pine contain 66% resin acids, 25% turpentine, 7% nonvolatiles and 2% water. The turpentine contains 1- -pinene. The green needles from young pine leaves yield 0.27% balsam-scented oil.

Various products of these plants are used in pharmaceutical preparations for the treatment of gonorrhoea, leucorrhae, catarrh, rheumatism, chronic inflammations of bowels, skin diseases. Talc oil rosin contains sterols, which is potential precurser of cortisone.

Plate 13.

Papaver somniferum

Plantago ovata

Plate 14.

Plumeria

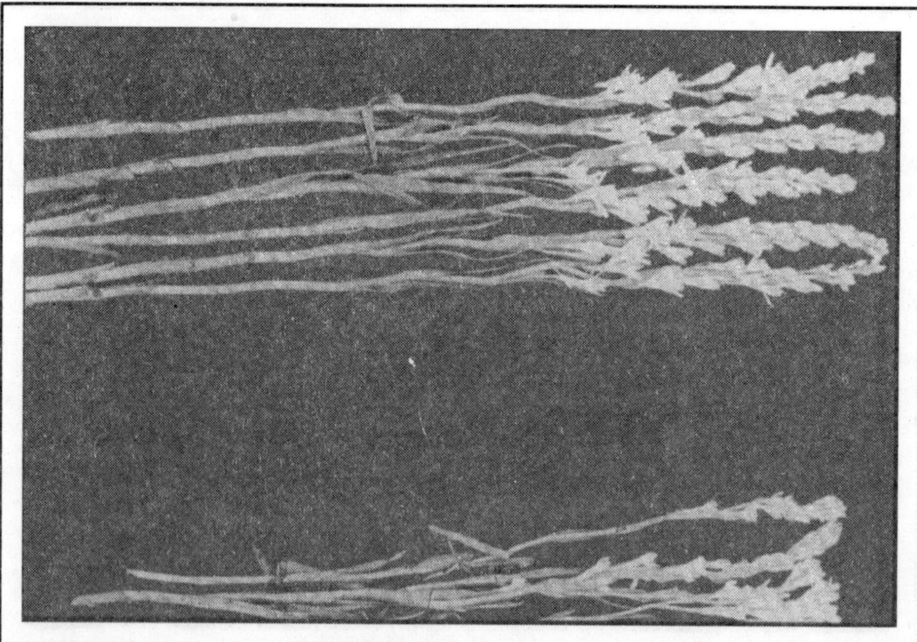

Polianthes tuberosa

Pinus palustris Mill. :

This is a tall tree of 32 m height with 0.78% m trunk thickness. The bark is coarse and scaly. Outer bark is dark and inner bark is reddish. The needles are drooping to 60 cm long, in clusters of three. Cones are stalkless to 30 cm long, cylindrical. Seeds are dark brown to 6 cm long . It occurs naturally in the South-Eastern United States. It is widely cultivated in America. Commonly known as long leaf pine or pitch pine.

Uses : Sap, turpine and gum of these plants are used. Natural oleoresins obtained from this plant primarily contains 66% resin acids, 25% turpentine, 7% nonvolatiles and 2% water. The turpentine contains some d-pinene. Resin consists mainly of diterpene resin acids of the abietic type and pimaric type. Crude oil contains 40 to 60% resin acids, 40 to 55% fatty acids and 5 to 10% neutral properties. Green needles yield 0.42% of balsam-scented oil which constitute camphene, beta-pinene and borneol cadinene.

Turpene oil is used to treat chronic diarrhoea, tape worm, colic, obstinate constipation and also applied on tooth socket to stop bleeding. Gum is used to treat tuberculosis and kidney ailments and for increasing menstrual flow. Pine-tar was an ingredient in expectorant syrup for the relief of bronchitis. Vitamin A and E is produced from pine needles. Pine tar is stimulating and antiseptic.

Pinus roxburghii Sargent :

This is a tall tree. Needles are in bundles of three, cones are upto 25 cm long.

It is found in Himalaya region upto 2500 m elevation. Commonly known as chir.

Uses : In the preparation of Indian medicine resin and wood oil is used. It is also used in the manufacture of lubricating compound, linoleum sealing wax and oil cloth. The crushed leaves with cow dung is massaged in backache. The oil extracted from this plant has medicinal value used for the cure of chronic bronchitis, constipation.

Pinus silvestris Linn. :

It is known as Pine.

Uses : Decoction of the pine needles is beneficial for rheumatism and gout. Medicines are prepared from buds for chronic catarrh in the bronchia. It has diuretic properties. The scales contained terpintine which is active principle.

Pinus strobus Linn. :

This is a long leaved pine of 42 m tall. The needles are soft and bluish green and arranged in five whorls. Cones are upto 30 cm long with rounded scales.

It is native to North-East America. Commonly known as White Pine.

Uses : Inner bark, buds, gum and tar of these plants are used. Bark is used to treat cold and cough. Water soaked soft bark is used to treat wounds. The inner bark is boiled in water and the liquid is drank in dysentery. Gum is used for muscular and rheumatic pains. Tar is used to treat cough, burns and itching.

Finns taeda Linn. :

It is a tall pine of about 42 m height. Bark is covered with thin scales. Needles are arranged in whorls of three, twisted and 22.2 cm long. Cones are stemless to 16 cm long. They are grown on wet clay soils or sandy soils. Plants are propagated by seeds.

Uses : Resin and turpine oil extracted from crude turpine of this pine is used for medicinal purposes. Oil of turpine is a stimulant and laxative and used to treat intestinal worms and increase urine flow. In pulmonary diseases the tar vapour is inhaled, it is also used to treat blemishes and skin diseases.

Pinus wallichiana Jakson. Linn. :

It is a hardy evergreen tree of 52 m tall. Leaves are in bundles of 5. Cone woody upto 32 cm long. It is distributed in Himalayan region upto 2500 m elevation.

Uses : The resin or colophony is used for making soap, for sizing paper and cloth. Wood is used as timber and for making match boxes.

Piper Linn.
Family : Piperaceae

This is a very large genus consisting of about two thousand species and nearly fifty species are distributed in India. These are mostly climbing shrubs. Flowers appear in sympodial spikes and bracts are closely oppressed to the axis.

Piper betl Linn. :

This is a perennial bisexual vine. Branches arises from the stem nodes. Nodes are thick and smooth. Flowers are in sympodial spikes, the bracts closely oppressed to the axis.

The plant is native to Central and East Malavsia and India. Commonly known as Betel leaf or Betel Pepper or Betelvine.

Uses : The characteristic smell and hot taste of the leaves is due to an essential oil. The leaves contain B and C vitamins. It stimulates brain, lungs and heart.

***Piper brachystachyum* Wall. :**

This is a trailing herb with greenish spike. The species occurs in Ukhinath and Ghomti of Garhwal Himalaya between 950 to 1600 m altitude in moist places of tropical zones.

Uses : Dried fruits are carminative and tonic. It is used to treat diarrhoea, cold, cough and bronchitis. Roots are used as piplamul.

***Piper cubeba* :**

This species is distributed in East Indies and Ceylon and cultivated in West Indies.

Commonly known as Failed Pepper or Cubeb.

Uses : Dried seeds of unripe fruit is used in medicine. It is administered for the treatment of chronic bronchitis, and as an urinary antiseptic. It also promotes healing of mucous membranes.

***Piper longum* Linn. :**

This is a creeping aromatic herb. Lower leaves are broadly ovate, deeply cordate with big lobes at base, upto 12 cm long, upper leaves are dark green, oblong-ovate, cordate. Flowers are in spikes, bracts or male spikes narrow and female spikes circular. Fruits are small, ovoid, shining, blackish green in colour. Commonly known as Long Pepper.

It grows in tropical zones and distributed in central Himalaya, Khasi and Mikir Hills, Western Ghat, Lower hills of W. Bengal and Assam. It is cultivated in Assam, Tamil Nadu and Cherrapunji.

The fruit of long pepper contains piperine and piplartine alkaloid; starch, resin, gum and fat. The dried fruits on steam distillation yield 0.7% essential oil with spicy odour.

Uses : The dried unripe fruits are useful in cold, cough, chronic bronchitis and diarrhoea. It is also used in liniments for rheumatic pains and paralysis. Fruits and roots are used in Ayurvedic and Unani systems of drug. It is used to treat insomnia, epilepsy, obstruction of bile duct and gall bladder, dysentery and leprosy. It possesses anthelmintic and carminative properties. Fruits are used as spice, pickles and preserves.

Piper nigrum Linn. :

This is a tropical perennial climbing plant and requires support to climb. Leaves are large, broad, ovate and round. Flowers appear in long spikes.

Fruit, a berry gathered from ripe and dried, forms a black pepper corn, which are 6 mm across. Fruits are turning red when ripe. The dried mature but unripe berries are known as black pepper. It occurs in hot and moist parts of Southern India. The species is cultivated in Sri Lanka, Thailand, Malaysia, Indonesia and India. There are large number of varieties in this species. Commonly known as Black Pepper or Kali Mirch.

Uses : It contains starch, resin, a volatile essence, an acrid oil, a colouring matter, cellulose, albumin, malic acid, mineral salts, a crystallizable principle and piperine. The alkaloid piperine is responsible for biting taste of black pepper. On steam distillation the black pepper yield 1.0 to 2.5% volatile oil which has characteristic odour of pepper and also of phellandrene.

It is said that it has anti-blennorrliagic properties. It also possess stomachic, stimulant and carminative properties. It is used to treat dyspepsia, malaria, haemorrhoids, tremors and delirium. It increases appetite and fortifies the digestive system. It is used as an aphrodisiac. Pepper mixed with cinnamon is used for treating migraine. It possesses preservative properties and is used to preserve perishable foods.

Plantago Linn.
Family Plantaginaceae

A genus of over two hundred and fifty species and ten species are found in India, Which are scapigerous herbs. Psyllium husk, which is commonly known as "Isopgul", is obtained from ripe seeds of Plantago ovata Forsk and Plantago psyllium Decne.

Plantago ovata Forsk. :

This is a small stemless plant covered with dense or soft hairy growth. Leaves are narrow, 12-22 cm long. Flowers are small, in oval or cylindric spikes. Fruit 10 mm long, its upper half separates like a lid. Seeds are boat shaped, pinkish gray brown in colour. Each seed is encased in a thin, white, translucent membrane called husk. The husk is odorless and tasteless. It is distributed in

Canary Islands, Mediterranean regions of the Southern Europe and North Africa to Pakistan. It is cultivated in large scale in India and Pakistan. Jn India the crop, is commercially cultivated in Banaskantha and Mehsana districts of Gujarat and adjoining areas of Southern Rajasthan.

Uses : The husk of the seed yields a colloidal mucilage which primarily consists of xylose, arabinose and galacturonic acid. Presence of rhamnose and galactose has also been reported. Two polysaccharide fractions have been isolated from the mucilage. The first fraction is soluble in cold water contains 20% uronic acid .and on hydrolysis it yields 46% d-xylose, 40% aldobiouronic acid, 7% 1-arabinose and 2% soluble residue. The second fraction, having 3% uronic acid, dissolves in hot water and when cool, forms a viscous gel yielding on hydrolysis, 80% d-xylose, 14% arabinose, 3% aldobiouronic acid and a trace of D-galactose. The mucilage is over 30% of the whole seed. The husked seeds contains semi-drying oil, glycoside aucubin, tannin and an active principle resembling acetylcholine.

Seeds are mostly used in medicine because of its mucilage, which is colloidal in nature and serves as a safe bulk-laxative which promotes regular bowel movement. For chronic dysenteries of amoebic and basillary origin, this is administered. The seeds and husk are used to cure inflammation of the mucous membrane of gastro-intestinal and genitourinary tracts, dueodenal ulcer, gonorrhoea and piles. It can also be used as cervical dilator for termination of pregnancy.

Plantago psyllium Linn. :

This is a much branched, erect, glabrous annual herb, upto 60 cm tall. Flowers are rose-pink or yellowish in colour. Capsules are two celled, seeds are smooth and dark brown in colour.

The species is native to Europe and North Africa. Commercially cultivated in Spain, France and Cuba.

Uses : The mucilage is over 10% by weight of the whole seed. It is used for the same purposes as that of P. ovatq.

Pleurospermum Hoffm.
Family : Umbelliferae (= Apiaceae)

It is a genus of perennial or biennial herbs, consisting of over eighty species. About fourteen species are native to India.

Pleurospermum angelicoides DC. :

This is a robust perennial herb and a temperate plant and istributed in inner moist pockets of alpine region in Central Himalaya from 3500-4000 m elevation.

Uses : Preparation of rhizomes of these plants are used in dyspepsia.

Pleurospermum brunonis (DC.) Benth. :

This is a small tufled herb to 25 cm tall, highly scented. Leaves are finely dissected. Flowers are dark-brown in compound umbels, closed by the creamy white bracts

The species sporadically grows on rock boulders, rocky slopes and rock crevices in sub-alpine and alpine zones between 3600-4000 elevation in Himachal Pradesh.

Uses : The powder of flowering shoots is mixed in cow's fresh butter and massaged over the entire body to allay fevers. The dried herb is kept in the boxes containing clothes as a preservative against attack of moths.

Plumbago Linn.
Family : Plumbaginaceae

Plumbago zeylanica Linn. :

This is an evergreen flowering shrub upto 60 cm tall with white flowers. Plant is spreading with rambling branches. Inflorescence is racemose.

It is native to tropics and occurs at Gelduba, Chhindwara District, M.P. and sporadically distributed in the hills of Garhwal upto 1600 m altitude. Commonly known as Leadwort or Chitawour.

Uses : The rogis contain plumbagin. The root paste about 20 gm is applied thrice a day for seven days on abscess. The root is used as a masticating for toothache. The plant is acrid, caustic, produces irritation and even rubefaction when applied to skin. The roots are used to promote appetite and improve digestion. It is also used to treat piles, diarrhoea and anascasa.

Plumeria Tourn. ex. Linn.
Family : Apocynaceae

Plumeria is called "Pagoda Tree" or "Temple Tree", is a genus of laticiferous trees and shrubs. It comprises of about fifty species. The trees are with stout but weak and fleshy branches and milky juice. The leaves are large, lanceolate, narrow and spirally arranged near the ends of branches, 22-42 cm long. They have waxy funnel shaped highly fragrant flowers, grow in clusters at the tips of the branches.

It is native to frofical warm America.

Uses : Different plant parts are edible and are also used as medicine. The dried flowers of P. acuminata Roxb. are used as substitute of tea. The flower buds are eaten with betel leaves as a febrifuge. The fruits of P. alba L and P. rubra.'L. are also edible. P. acuminata fresh flowers yield an essential oil on steam distillation ranging from 0.04 to 0.07% which possesses antifungal activity. The leaves contain 5-6% pectin having low jellying quality. A decoction of bark is reported to be purgative, emmenagogue and febrifuge. The latex of the plant contain calcium salts of plumieric acid, cerotic acid etc. and is applied for itches, rheumatism and gum troubles. The root is cathartic. P. rubra L. also possess same medicinal property as that of P. acuminata and flowers are used in pectoral syrup.

Podophyllum Linn.
Family : Berberidaceae (= Podophyllaceae)

They are monthly scapigerous herbs with creeping rootstock, consisting of ten species and one species is native to India.

Podophyllum emodi Wall.

It is a succulent erect herb, glabrous, upto 32 cm tall, with creeping long knotty rhizomes. Stem one or two, simple, leafy without top. Leaves are alternate, palmate, to 30 cm in diameter, deeply divided in 3-5 lobes, toothed, purple spotted. Flowers white to pinkish, to 5 cm across, appear in the fork of the stem; sepals 3, petaloid; stamens usually 6. Fruits are ovoid, pulpy, to 6 cm long, scarlet when ripe, many seeded.

It is native to Himalayan region in India, Pakistan and Afganistan. Rarely occur above 3500 m at Chyangseel of Uttar Kashi and also other ranges of Himalayas in India. Commonly known as May Apple or American Mondrake the plants can be cultivated in Himalayan mountain regions at .an altitude of 1600-3500 m in Kashmir, H.P., U.P. and Sikkim. It prefers shady, moist forested areas but will grow in open glades.

Uses : The rhizomes and root contain resinous mixture called podophyllum resin or podophyllin. The primary constituents are lignin glyosides, podophyllotoxin, podophyllic acid and picropodophyllin.-peltalin and B-peltalin. The rhizomes also contain gum, starch, albumin, gallic acid, calcium oxalate, lignin, flavones. The root paste is applied on ulcers, cuts and wounds. It is also used for curring skin diseases and tumerous growth.

The drug of rhizome is employed as a vermifuge and in treating constipation, typhoid fever, jaundice, dysentery, chronic hepatitis, scofula, rheumatism; kidney and bladder and prostrate problems; gonorrhoea and syphilis. The podophyllum is used as a purgative and also for treatment of vaginal warts. Two derivatives of podophyllotoxin, called eloposide and teniposideare employed for treatment of cancers.

Podophyllum hexandrum Royle. :

A succulent erect herb with creeping rootstock. Flowers are white or pinkish. Rarely occur above 3200 m at Chyangseeel of Uttar Kashi and also in other inner ranges of Himalayas.

Uses : The rhizomes contain a resin podophyllin. The root paste is applied on ulcers, cuts and wounds. It is also used as purgative, for curing skin diseases and tumerous growth.

Podophyllum pelts turn Linn. :

A perennial herbaceous plant of 50 cm tall, erect in habits. It has creeping, jointed, fibrous branching rhizomes of considerable length and 6 mm thickness. It has single stalk with two umbrella like leaves at the top of the plant. Leaf is dark green and smooth upto 30 cm in diameter. Flowers are white to 4 cm in diameter which appear at the axils of the petioles. Fruits are yellow, plum like, disagreeably scented when immature, contains mucilaginous pulp. After fruits, leaves fall and the plants die.

It is native to Canada and America.

The rhizomes for medicinal purposes are mainly collected from wild. Rhizomes alongwith roots are washed and divided into pieces of 10 to 20 cm and dried for shipment. Large quantity of the rhizome is consumed in U.S.A. and also in other countries. Rhizomes with 5% resin ar accepted commercially.

Uses : Rootstocks and resins are used as medicine. The rhizomes and roots contain 3.5 to 6% to resinous mixture called podophyllum resin or podophyllin. It primarily contains lignin glycosides and also constitute podophyllic acid, picropodophyllin and peltatin. The ointments containing podophyllin are applied on hypertrophic and hyperkeratotic lesions.

The tea preapared from roots is effective in controlling constipation. Roots are also used in jaundice, syphilis, fever, liver ailments and cancer. The resin from this plant exhibits antitumour properties and to treat veneral warts. Podophyllin, a yellow brown powder is used as a cathartic. Podophyllum extract is an ingredient of various liver pills.

Polianthes Linn.
Family : Amaryllidaceae

Polianthes tuberosa Linn. :

This plant has tuberous root stock from which grows an erect flower stem 95 to 130 cm high. It produces mass of grass like foliage, arching in shape, pale green, long, narrow and very dense. It is approximately the top third of the stem which bears 14 to 20 pairs of pure white, waxy textured raceme of blooms. Flowers are funnel shaped.

The plant is native to Central America and is widely cultivated in India and also in other countries.

Uses : The flower yields valuable essential oil. Extraction of essential oil from flowers is carried out extensively by enfleurage or by use of volatile solvents. The tuberose oil contains methyl benzoate, methyl anthrinilate, benzyl alcohol, butyric acid, phenyl acetic acid, methyl salicylate, eugenol, geraniol, nerol and farnesol.

The oil is used in items like non-alcoholic beverages 0.25 ppm,. icecream, ices etc. 0.45 ppm,. candy 1.5 ppm and baked goods 1.7 ppm. From tuberose bulbs sapogenins like hecogenin and tigogenin are extracted. The plants are considered to be hot, dry, diuretic and emetic. Dried and powdered bulbs are used as a remedy for gonorrhoea. To remove small red pimples they are rubbed up with turmeric and butter.

Polyalthia Blume.
Family : Anonaceae

Under this genus there are about one hundred twenty species distributed in palaeotropical region. Twelve species are found in India. It is an erect trees or climbing shrubs.

Uses : The aqueous extracts of the bark of P. longifolia Benth. et Hook.f. stimulates the isolated ileum and uterous. This extract lowers blood pressure, stimulates respiration and depresses the heart. The fruits are eaten in times of scarcity. The leaves are used for decorative purposes.

A decoction of fresh roots of P. suberosa is used as abortifacient. The bark of P. simialum yield fibre from which rope is made. The wood of P. cerasoides

Bedd. P. coffeoides and P. fragrans are used for construction and box making etc.

Polygala Linn.
Family : Polygalaceae

Polygala senega Linn. :

This is a hardy herbaceous perennial of 95 cm in length, much branched. Leaves are narrow and arranged alternately on stems. Flower spikes are crowded, arises at the top of the stem, pink to white with green tinges.

It is native to Canada and U.S.A. Commonly known as Rattle Snake Root or Senega Snake root.

Uses : Root of this plant constitute the drug. It contains triterpinoid saponins and polygalitol. It is laxative and emetic. It was used to treat gout, rheumatism, pleurisy, heart disease, pulmonary infections, pneumonia and uterine disorders. It is also used in veterinary medicine.

Polygonatum Mill.
Family : Liliaceae

A genus of fifty species and fifteen are native to India. These are mostly herbs with creeping rootstocks and the stems are leafy above.

Polygonatum biflorum (Walt.) Ell. :

This is a hardy herbaceous perennial plant upto 1-^m tall. There is a sympodial fleshy rhizomes, upon which the annual shoots leave curious seal-like marks when they die away. Leaves are 15 cm long and hairy below. Flowers hang from the axils of the leaves, homogamous, green or white in colour. Fruits are globular and blue.

The species is native to North America. Commonly known as Solomon's Seal.

Uses : Roots of these plants constitute the drug. Roots are used to treat arthritis and rheumatism. Root preparation with boiled milk is used as mild laxative, for increasing perspiration and for treating haemorrhoids. To treat poison ivy rash and skin irritations, a decoction of roots.is prescribed.

Polygonatum multiflorum (Linn.) All. :

This is a hardy herbaceous perennial of 1.5 m tall, native to Europe. In India, commonly found at Bhatanu of Uttar Kashi between 2200-3500 m.

Commonly known as solomon's seal or David's Hemb.

Uses : The paste of seeds and leaves is applied in itching and chopped skin.

Polygonum Linn.
Family : Polygonaceae

Under this genus there are about three hundred species and eighty species are native to India. These are herbs or shrubs.

Polygonum hydropiper Linn. :

This is an annual plant which grows erect upto 80 cm height. Leaves are with smooth margins, narrow, alternate upto 15 cm long. Stems are reddish in colour. Flowers are produced on the ends of terminal spikes, green in colour and small.

Comonly known as Smart Weed.

Uses : Upper part of the plant in full bloom stage is harvested. It is used to treat internal bleeding, uterine disorder, to promote menstrual flow, as mouth wash for sore mouth in nursing mother.

Polygonum recumbens Royle ex Bab :

This is a much branched decumbent herb with small pale white flowers in axillary whorls. Leaves are lanceolate and elliptic. Plants are obtained in different parts of H.P. between 1600 m 3000 m elevation.

Commonly known as knot weed.

Uses : The paste of leaves, and apical parts is used in weeping eczima and wounds between fingers in a toe. It makes emollient poultice.

Pongamia Vent.
Leguminasae (Papilionaceae)

A monotypic genus of Indo-Malayan region.

Pongamia pinnata (Linn.) Pierre.:

It occurs throughout the greater part of India and Andaman Islands. Leaves are imparipinnate. Flowers are pinkish white in colour.

Uses : The juice of this plant is used to treat diarrhoea, cough, dyspepsia, flatulence, leprosy, gonorrhoea. The juice of the roots is used to clean teeth, strengthening gums, gonorrhoea and cleaning foul ulcers. A decoction of dried flowers is given in diabetes. The oil extracted from seeds is used for medicinal uses and soap making. The seed paste is also used to treat painful rheumatic joints, skin diseases and leprous sores. Flowers are edible and the leaves are used as manure.

Populus Linn.
Family : Salicaceae

Under this genus there are about forty species and nine are found in India. They occur in North Temperate region. Trees are with scaly or often resinous buds.

Populus angustifolia James. :

This is a pyramidial shaped tree, growing to a height of 25 m. Leaves are with wavy margins and narrow.The tree bark is greenish in colour.

Commonly known as Narrow Leaf Poplar.

Uses : Flowers and ripe fruits have medicinal properties. The decoction of fruits and flowers are used to treat infected gum, toothache. To stop internal haemorrhage, a decoction of fresh flower is administered. Tea of fresh flower is utilised for drinking to purify the blood.

Populus balsamifera Linn. :

A beautiful tall tree grows tall upto 38 m and the tree trunk reaches a diameter of 2.5 m. Leaves are pale green below and shiny dark green above, upto 16 cm long and 15 cm wide, finely toothed, rounded or heart-shaped at the base. The buds are aromatic. Commonly known as Balsam Poplar.

Uses : Mature leaves, bark, flower and flower buds have medicinal value. The buds and fresh flowers of this tree are steeped in cold water and stained off, the liquid is drunk for purification of blood. To treat snake bite, the leaves and bark are boiled in water and vapour inhaled. For treating pulmonary ailments, kidney disorders, stomach disorder, scurvy, rheumatism and gout the tincture prepared from the tree bark is used.

Populus nigra Linn. :

It is a hardy deciduous tree and attain a height of 35 m. Commonly known as Black Poplar.

Uses : The plant yield essential oil which contains glycosides like salicin, chrysin, populin, salicortin and nigracin. Presence of gallic acid and tannin has also been reported.

The dried bark is most important plant part. The bark in this plant contain salicin and a sugary tasting principle, populin. Salicin has been used in treatment of rheumatism. The buds are bitter, covered with resinous, aromatic substance. Buds are diuretic, sudorific and expectorant in a dose of 35-45 gm in a litre of water. The bark of P. euphoratica Oliver from N.W. India and Punjab plains is a vermifuge and the twigs of these plant are chewed for cleaning teeth.

The bark of P. alba L. (White Poplar) is used as tonic, diuretic and antipyretic and the plants are found in W. Himalaya. The bark of P. ciliata n Kat) possesses tonic, stimulant and blood purifier properties.

Potentilla Linn.
Family : Rosaceae

A genus of about five hundred species and forty species are distributed in India. These are herbaceous plants with creeping stems rooting at nodes. Leaves are pinnately compound and leaflets are usually lobed.

Potentilla anserina Linn. :

This is a herbaceous plant with creeping stems, rooting at nodes. Flowers with epicalyx of small green leaves. Flowers are homogamous and visited by flies.

It is native to temperate region. Commonly known as "Agremone" or "Silver Weed".

Uses : For preparation of homoeopathic medicine this plant is used which is very effective as an antispasmodic.

Potentilla fulgens Wall. :

This is a herbaceous perennial upto 30 cm tall with crimson flowers, native to Himalaya.

It is commonly occuring at Kaliz of Uttar Kashi between 1600-3500 m elevation. Commonly known as "Bajradanti".

Uses : The paste of root is filled in hollow cavity of teeth to check the infection.

Primula Linn.
Family : Primulaceae

Primula veris Linn. P. Officinalis Jacq. :

This is an erect herbaceous perennial plant .native to Asia Minor. Commonly called cowslip or coslip.

Uses : The infusion of leaves and flowers of these plants were said to be antispasmodic, cordial and sedative. The rhizomes are bitter in taste and is applied in cases of rheumatism.

Prinsepia Royle.
Family : Rosaceae

Prinsepia utilis Royle. :

This is a hardy deciduous flowering shrub with spines. The species is a glabrous undershrub with white flowers.

It is native to Temperate Himalaya and commonly distributed at Thunera of Utter Kashi between 1500-2500 m. It is also found wild in Nilgiris. Commonly known as Bhenkal.

The paste of root is used for healing the wounds. From seeds, oil is obtained which is used in rheumatism.

Prosopis Linn.
Leguminasae (= Fabaceae)

Prosopis glandulosa Torr. :

This is a thorny shrub may attain a height of 10 m. Flowers are greenish yellow. Fruits are upto 19 cm long and 1.6 cm broad pod. Commonly known as Mesquite.

Uses : Roots, new leaves, inner bark and gum of these plants are useful. A tea made from root is given in umbilical hernia in children. Inner bark boiled in salt water is drunk for indigestion. Gums are dissolved in water and used to treat sore throat. Powdered leaves are soaked in water and the liquid is squeeged in sore eyes.

Prunella Linn.
Family : Labiatae

Prunella vulgaris Linn. :

This is a herbaceous perennial of 35 cm tall, highly aromatic. Plants are tufted. Leaves are finely cut, round and bract-like. Flowers are purple, arises from the axils of the leaves.

The species is native to Britain. Commonly known as self heal.

Uses : Leaves, flowers and flowering tops are useful. An infusion of leaves and flowers are used as a gargle, sore throat and throat irritations. It is also used to treat diarrhoea, haemorrhages, for relieving gas and colic.

Prunus (Tourn.) Linn.
Family : Rosaceae

Under this genus there are about forty species and nineteen are found in India.

They are distributed in North Temperate region and few species are from tropical region.

Uses : These are shrubs or trees. Many species are cultivated for their fruits.

Prunus amygdalus Batoch. :

This is a tree of 15 m tall with pale pink flowers.

The species is native to South Europe and Algeria. Commonly known as Almond.

Uses : Bitter almond is poisonous, contain amygdaline and emulsin, which by fermentation give hydrocyanic acid. In minute doses it has anti-spasmodic effects and prescribed in an emulsion for asthma, whooping cough, and all convulsive coughs. Sweet almond are not poisonous and do not contain amygdaline. Seed almonds are used among other things to make emulsion.

Prunus armeniaca Linn. :

It is a small tree of 8 m tall. Flowers are white or pale pink.

The species is native to North China, California, Washington, France, Turkey and Australia. Commonly known as Apricot or Chuloo.

Uses : The kernels contain the cyanogenetic glycoside, amygdaline.

The oil obtained from seeds is massaged after delivery and also in general body pain. The bark of the tree is applied in burns for cooling effect. This is said to be effective in controlling cancer. It is used as a substitute of almond oil.

Prunus dulcis D.A. Webb. :

This plant species is native to California, France, Italy, Spain, Morocco, Portugal and Iran.

Seed carnal oil is used as a vehicle for injections, demulcent and lubricant. Commonly known as "Almond".

Prunus laurocerasus Linn. :

This is a quick growing tree upto 8 m tall, with dull white flowers and black fruit. There are several varieties in this species like "Caucasica", a fine large leaved from and "magnoliaefolia", leaves upto 35 cm long.

It is native to East Europe and Asia Minor. Commonly known as "Cherry Laurel".

Uses : The leaves and flowers are used to flavour milk or custard. Distilled water made from leaves are narcotic. This plant is often used in homoeopathic medicine.

Prunus Padus Linn. :

It is a good woodland tree upto 16 m tall, flowers are white, fragrant. It is native to North Europe, North Asia to Japan. There are several varieties in this species. Commonly known as Bird cherry.

Uses : The bark of this plant is bitter in taste and is recommended in intermittent fevers, rheumatism and arthritic complaints.

Prunus persica Stokes. :

The plant is native to USA, NewZealand, Australia and Temperate Asia. Commonly known as Peach.

Uses : It contains fixed oil in the seed kernel which is known as "Persic oil". The oil from seed kernel contain B-sitosterol and squalene. The oil can be used to make emulsion. It is used as a vehicle for injection, and also used as demulcent and lubricant.

Prunus serotina Ehrh. :

A very tall tree with a straight trunk and may attain a height of 36 m. The tree bark is dark black in colour and is aromatic. Leaves with finely serrated edges, shiny, smooth upto 12.6 cm long. Flowers are white, fruits are black, round and bitter in taste. Commonly known as wild black cherry.

The bark is used for medicinal purpose, which contains cyanogenetic glycoside prunasin and enzyme prunase. The powdered bark is administered for remedy of cough, cold and measles. The bark is also administered against fever, tuberculosis, intestinal worms and indigestion. The fruit is recommended in fevers which have tendency to be putrid. An infusion of dried and powdered berries has been used to treat diarrhoea.

Prunus virginiana Linn. :

This is a tree of 12 m height. Upper surface of leaves are dark green and shiny and lower surface pale green in colour. Fruit is small, round, black in colour. The tree bark has a disagreeable smell. Commonly known as "Common Choke Cherry".

Uses : Leaves, fruits, bark of stem and root have medicinal properties. The dried powdered bark is smoked to relieve headache and cold attack. The warm infusion of bark is used to ease the pains of child birth. The tea prepared from stem bark is used in measles and colds, and tea of root bark is said to be sedative. In early days, root bark was used for treatment of tuberculosis, fever, indigestion, worms and malaria. To cure dysentery, fruits were used. The tea prepared from leaves, bark and dried roots has been used to cure lung ailments.

Psoralea Linn.
Papilionaceae (Leguminasae)

Psoralea corylifolia Linn. :

A much branced annual herb, erect in habit and grow to about a maximum height of 1.6 m. The stems and branches are covered with conspicuous glands and white hairs. Leaves are round, dotted With black glands on both surfaces. Flowers appear from the axils of leaves in bunches. Individual flowers are small, bluish purple in colour. Fruits are roundish and small. Each fruit bear one smooth seed.

The species is native to India and Arab. In India, it is distribted in M.P., U.P., Rajasthan, Bihar, Gujarat and A.P. and found as weed in waste places. Commonly known as Babchi or Bawachi.

Uses : The seeds of this plant constitute the drug. The seed are covered with a stickly, oily pericarp which contains coumarins. The seed oil is very effective against certain bacteria causing skin diseases. Various preparations of seeds are used for leucoderma, leprosy and psoriasis. The drug is useci in the form of ointment as well as taken internally. This has also antimicrobial, antifungal and antiinsecticidal activities. For promoting urination and as anthelmintic the drug is administered. Psoralen and isopsoralen obtained from the seeds are being investigated against several diseases including AIDS.

Pteridium Scop.
Family : Polypodiaceae

Pteridium aquilinum (Linn.) Kuhn. (= Pteris aquilinum Linn.) :

This is hardy fern with dull green fronds, upto 1.28 cm long. Fronds are stiff, reddish at base and triangular in shape. On the undersurface of the fronds, brown fruiting bodies are found.

It is a cosmopolitan species. Commonly known as Blacken Ferns.

Uses : Entire mature plant is useful. To expel intestinal worms and treat diarrhoea an infusion of plant is used. It is also used to relieve stomach cramps and increase urine flow. A tea made from the plant is prescribed to women with caked breasts.

Pterocarpus (Linn.) Jacq.
Leguminasae (Papilionaceae)

This genus have about one hundred species and four are native to India, Which are mostly trees and flowers are yellowish.

Pterocarpus marsupium Roxb. :

This is a large tree commonly knwon Indian kino or Amboyna kino tree. Leaves are compound with 5 to 7 leaflets. Leaflets are oblong or elliptic upto 12 cm long. Flowers are scented, yellow, upto 1.5 cm long, appear in very large dense bunches. Fruits are roundish, winged, to 5 cm long, each fruit possess one seed.

The species is occurring in Central and Penisular India in a mixed deciduous forest.

Uses : Leaves, flowers and gum constitute the drug. The tree is useful in treating diabetic patients. Bruised leaves are applied on skin diseases, sores and boils. Gum of this tree is astringent and used to cure diarrhoea, fever and urinary discharges.

Pterocladia J. Ag.
Gelidiaceae
(Algae)

Pterocladia capillacea J. Agardh. :

This is a red algae distributed in Australia, New Zealand, Egypt and Japan. Agar is used, which constitute a polysaccharides. A colourless, odorless, tasteless, transparent mucilaginous substance.

Uses : In pharmaceutical industries, it is used as suspending agent, suppositories, emulsions, lubricants and used as ingredient in capsules. It serves as a bulk laxative.

Pulmonaria Linn.
Family : Boraginaceae

Pulmonaria officinalis Linn. :

This plant species is native to Eurasia.

It is an astringent. It is said that the plant preparation is effective in lung diseases. commonly known as Linngwort.

Punica Linn.
Punicaceae

Punica granatum Linn.:

This is a deciduous small tree, may attain a maximum height of 10 m with reddish scarlet flowers. Fruits are edible. Both flowers and fruits are attractive. The fruit is large globose, shiny red, yellowish green or whitish when ripe. The fruit contains large number of hard seeds covered with a juicy red, pink or yellowish white, sweet-astringent acid pulp.

The species is originated in Persia, Afganistan and Baluchistan. It is also found wild in the warm valleys of Himalays. This fruit tree is commercially cultivated in tropics. In India, commercial pomegranate orchards are confined to the arid and semi-arid regions of Western India.

Commonly known as Pomegranti or Dalim.

Uses : Bark of trunk or roots possess medicinal properties which contains 20 to 22% tannin; 0.5 to 1% alkaloids. The seeds contain steroidal estrogens. The fruit pulp contain protein, fat, fibre, carbohydrate, minerals, oxalic acid, vitamins A, B and C.

Bark of this plant is anthelmintic, and for expelling tapeworm is said to be very effective. The seeds and the fruit pulp are stomachic. The bark and the rind of the fruits and seeds are commonly used in dysentery and diarrhoea. The fruit pulp is also said to be cardiac. Fruit juice is valued for its medicinal properties. Leprosy patients get benefit from fruit juice.

Putranjiva Wall.
Family : Euphorbiaceae

There are about four species distributed in India and Burma and one is native to India.

Putranjiva roxburghii Wall, occurs throughout India, wild and cultivated. The plant has a typical appearance due to the colour and structure of leaves.

Uses : A decoction of leaves of P. roxburghii Wall. "Wild Olive" is used to treat cold, ıever and rheumatism. Leaves are applied to swollen throats of cattle. Seeds of this species yield fatty oil which is used for burning.

Pyrethrum Zinn.
Family : Compositae (= Asteraceae)

Pyrethrum carneum Linn. :

The plant is native to Eurasia. The plant is used as a febrifuge, sternutatory and stimulant of the nervous system.

Pyrethrum powder is used as insecticide. Commonly known as fever few.

Quercus (Tourn.) Linn.
Family : Fagaceae

About three hundred species distributed in North Temperate region, Indo-Malayan region and Pacific coasts. Around twenty species are native to India.

It is deciduous or evergreen trees, rarely shrubs. Leaves are short-petioled, serate, lobed or pinnetifid, rarely entire. Flowers appear with or before leaves.

Quercus alba Linn. :

This is tree of 30 m tall with light gray coloured bark. The leaves are divided into 8 to 10 rounded lobes, upto 25 cm long and smooth.

The species is native to America. Commonly known white oak or Quebee Oak.

Uses : Bark and roots of this plant are useful. A tea made from bark is used to treat sore mouth, burns and bleeding haemmorrhoids. A tea from root bark is used to treat gonorrhoea and to expel phlegm from the lungs.

Quercus gambelii Nutt. :

It is a tree of 16 m tall. Leaves are 5 to 9 lobed upto 10 cm long.

The species is native to America. Commonly known as Gambal Oak.

Uses : Bark and branches are useful. Powdered bark or lotiorr made from boiled branches are used to treat skin cancers and skin sores. To treat malaria and diarrhoea a tea prepared from the bark is drunk.

Quercus infectoria Oliv. :

It is native of Greece and Iran. Through the attack of insect species Adleria gallae-tuictoriae Oliv., galls are formed on the tree which constitute the drug. It contain 60-70% gallotanni£ acid, Tannic acid is used as an astringent and styptic in case of anal fissures and haemorrhoids.

It is also useful in dvsentery and diarrhoea. Commonly known "Dyer's Oak".

Quercus velutina Lam. :

It is deciduous tree of 30 m tall and the trunk attain a diameter of 1.5 m. Leaves are seven lobed, oblong, dark green ir. colour. Bark is permeated yellow.

The species is occurring in Eastern and Central America. Commonly known as "Black Oak".

Uses : Bark is used for medicinal purpose. A decoction of bark is used to treat sore eyes. The bark is used to treat rheumatism, tuberculosis, diarrhoea, nervous tension and it is also known as laxative and tonic.

Quercus virginiana Mill. :

It is an evergreen tree of 25 m tall with horizontal and spreading crown.

It is a sea-coast plant and require moist atmosphere. Commonly known as live Oak.

Uses : Bark has medicinal properties. A decoction of bark is used to treat sore eye. Bark is also useful for treatment of dysentery.

Rauvolfia Linn.
Family : Apocynaceae

This genus have around one hundred species and five are native to India. T&es-e are herbaceous or woody shrubs and glabrous. Ripe carples are drupaceous, distinct or connate. Different species of Rauvolfia have been used as a source of hypotensive alkaloids. Important species are Rauvolfia serpentina Benth. ex Kurz and R.vomitoria Afzel. An important medicinal plant in this genus is R.serpentina. R. vomitoria is exploited in West African countries like Zaire.

R. serpentina Benth ex Kurz.

This is an erect glabrous perennial shrub, attain a maxium height of 80 cm. Leaves are arranged in whorls of 3-4 upto 10 cm long and 5 cm broad, gradually tapering into a short petiole. Leaves are bright green above and pale

Plate 15.

beneath. Flowers are white or pinkish, peduncle deep red, about 1.5 cm long, appear in small, clusters. Fruits are round shaped, to 5 cm in diameter, dark, purple or blackish when ripe.

This species is native to Bangladesh, Burma, Sri Lanka, Malaya, India and Indonesia. In India, the plant is occurring in sub-Himalayan tracts in Punjab, in lower ranges of Eastern and Western Ghats, Assam, Chhotanagpur, Orissa, Bihar, U.P., M.P., Maharashtra and Andaman Islands.

Ricinus communis

Rosemarinus officinalis

Plate 16.

Saraca indica

Terminalia arjuna

Uses : Roots are important source of drug in this crop, and the crop is ready for harvest at 18 months. In winter months, when the plants are completely defoliated, roots are dug. Before digging a light irrigation may be given. The roots goes upto 40 cm depth. All roots including secondary and tertiary are dug throughly with much care. The collected roots are cleaned properly and washed if necessary. Roots are air-dried completely, until they become brittle. In 18 months period the yield of air dried root will be around 2000 kg/ha. About 90% of the total alkaloids remain in the root bark. Although roots of Rauvolfia contain more than three dozen alkaloids, but the most common alkaloids are reserpine, rescinnarnine and deserpidine. From R. vornitoria reserpine and ajmaline is obtained. It has antihypertensine sedative properties used for treatments of high blood pressure as well as tranquilizers. R. vomitoria contain higher amount of alkaloids as compared to R.serpentina. The roots of R. serpentina contains 1 to 1.9% total alkaloids depending on the quality of roots. The drug is suitable for mild anxiety cases or patients of chronic mental illness. The roots of the plant are also useful in diseases of bowels and in fever.

Reseda Linn.
Family : Resedaceae

Reseda odorata Linn. :

A herbaceous species is native to Europe and Central Asia. regarded as a diaphoretic and used to treat poisonous bites. only one species is R. lutea Linn. "Dyer's Weld", yields a yellow dye.

Rhamnus Linn.
Family : Rhamnaceae

It has about one hundred and fifty species and nine are native to India, T-h-ese are deciduous or evergreen shrubs or trees.

Rhamnus purshiana DC. :

A dwarf deciduous tree which may attain a maximum height of 15 m and a trunk about 46 cm. Leaves are upto 17 cm long and 6 cm wide, thin broad, elliptic with short point at apex and conspicuous veins, hairy beneath and veins on upper surface. Flowers are bell-shaped, greenish yellow in colour, arises in axillary clusters. Fruits are small, round, juicy, black in colour when ripe with 2 to 3 seeds in each fruit. Tree bark is bitter in taste, reddish grey colour externally and yellow coloured inside. It is native to Western-North America. Commonly known as "Cascara Buckthorn".

Uses : The bark and ripe fruit is used as medicine. It contains anthracene compounds, anthraquinones, emodin, oxanthrone, aloe-emodin and chrysophenol in the free state and their dianthrones. It also constitute tannin, resins, glucose and starch.

The bark is also used for medicinal purposes. It is cathartic. The preparation from Dark is administered as laxative in cases of chronic constipation. It is also employed for the treatment of dyspepsia and haemorrhoids. From flowers, honey is prepared which is also mildly laxative. Fruits are eaten raw or cooked. Debittered extract of bark serves as flavouring in baked food, ice cream and soft drinks.

Rhamnus cathartica Linn. :

A thorny shrub, native to England and North Africa.

Fruits and bark have medicinal properties.

Uses : It is used as a vermifuge, and for the diseases of skin. Veterinary medicine is prepared out of this plant, particularly for treating illness of young dogs. It was said that the berries are used for treating fits and gout. Berries are bitter in taste. Roasted and powdered berries are also used as purgative.

Rheum Linn.
Family : Polygonaceae

This genus has fifty species of which about ten species are found in India. Which stout perennial herbs are mostly from Himalayan region.

Rheum australe D. Don. :

Flowers are like Rumex, but coloured and entomophilous. Rarely found at Chyangseel above 3200 m elevation. Commonly known as "Rhubarb".

It thrives well in the hills and cool climate. Pteats-are propagated-by

It thrives well in the hills and cool climate.

Uses : The paste of the root is taken orally with Curcuma in internal injury. The paste of the root is applied around the forehead to get rid of headache.

Rheum emodi Wall ex. Meissen. :

It is a stout herb upto 4 m tall with dark brown, thick, long branching ^ roots. Leaves are large, to 50 cm long, broadly ovate. Flowers are dark I purplish, minute, appear in a very large bunch.

Sporadically occurring between 3000-4000 m in Kullu, Simla and other parts of H.P., also distributed in Kashmir, Punjab, and Uttar Pradesh. The plant is native to Himalaya.

Uses : The watery extract of roots is given orally in stomach pains, constipation, dysentery, swelling of throat and tonsilitis. Lotion is dropped in ears in earache.

Rheum officinale Baill :

This is a hardy herbaceous plant upto 4 m in height and native to Tibet. Ordinary rich soil and sunny position are preferred. Plants are propagated by seeds and division of roots.

If is used as a medicine in the form of powder, tea, extract or syrup.

Rheum palmatum Linn. :

It is a hardy herbaceous perennial plant upto 2.5 rn tall. The species is native to North-East Asia. There is one variety in it known as "tanguticum".

Uses : It contains a sweet smelling resinous principle, cellulose, starch, tannin and various acids. It has culinary uses.

Rhododendron Linn.
Family Ericaceae

It has about five hundred species of North Temperate region, about eighty species are native to India, mostly Himalayan. Which are mostly evergreen or deciduous trees or shrubs.

Rhododendron anthopogon D.D. :

An aromatic dwarf shrub, leaves ovate, lower surface covered with red brown scales. Flowers are sulphur coloured. The species is occurring in sub-alpine area of H.P. between 3000-4000 m elevation.

Uses : The powder of the dried flowers mixed in bland oil is used as massage over the entire body in post-delivery complications like fevers, cough and cold. The decortion of the leaves is used in cold, cough and chronic bronchitis.

Rhododendron arboreum Sm. :

It is a tree of 14 m tall with red flowers. The species is native to Himalaya. This plant is commonly found at Glumari Dhar of Uttar Kashi, between 1500-2500 m elevation. Commonly known as Flame Tree.

Uses : The dried flowers are fried with ghee and is taken to check blood dysentery.

***Rhododendron hypenanthum* Balf. f. :**

This plant species is rarely found at Chyangseel of Uttar Kashi at 3000-3800 m elevation.

The dried flower is eaten with butter in leucorrhoea and gonorrhoea.

***Rhododendron lepidotum* Wall, ex G. Don. :**

It is a shrub of 1.5 m tall, produce yellow, pink or purple flower.

The species is native to Himalaya.

Uses ; The tea made from bark is supposed to be purgative. Other species of importance are given below : The chloroform extract of leaves and shoots of R. barbatum is a respiratory depressent. The leaves of R. campanulatum D. Don is used to treat rheumatism, syphilis and sciatica. The leaves of R. falconeri are employed as platters and for lining baskets. The leaves of R. cinnabarinum are poisonous to live stock, the smoke of leaves and wood causes inflammation of the eyes and face.

Rhus (Tourn.) Linn.
Family : Anacardiaceae

Under this genus there are about two hundred fifty species and thirteen are native to India. Which are trees or shrubs with mikly or resinous juice.

***Rhus glabra* Linn. :**

This is a hardy deciduous flowering shrub of 3.5 m tall with close erect panicles of small fruits, densely covered with crimson hairs. Flowers are greenish yellow produced in bunches at the end of branches.

The species is native to America. Commonly known as smooth sumach or caroling sumach.

Uses : To treat infections of lymph glands, diarrhoea, skin ulcers and gonorrhoea, a decoction of bark or roots are used. Leaves ripe fruits, stem bark and roots of this plant are useful. Leaves and smoked in asthma. A refreshing drink is made from fruits used as a gargle in sore throats. Infusion of fruit is given in fever. The bark boiled in milk is used to treat burns.

Rhynchostylis Bl.
Family : Orchidaceae

Rhynchostylis retusa Bl. :

It is an epiphytic herb. Stems are stout and woody, clothed with sheaths of fallen leaves, upto 65 cm long. Leaves are leathery, strap shaped, linear, deeply channelled, gracefully arching, upto 30 cm long and 2.5 cm broad. Inflorescences are pendulous, compact, many flowered, cylindrical, to 60 cm long. Flowers are 1 cm across, fragrant, long lasting, white, spotted with bluish purple.

It occurs in Chhota Nagpur, West Bengal, Orissa, W.Ghats and Malabar Coast. Commonly known as Cats tail orchind.

Uses : They grow in large chunks of tree fern fibre or in a media of chunks of hard wood charcoal. Plants grow well in high humidity and moderate temperature. Plants are benefitted by as much light as possible with the free movement of air. Leaves of these plants are used to treat rheumatic diseases.

Ribes Linn.
Family : Grossulariaceae (= Saxifragaceae)

Ribes nigrum Linn. :

It is commonly known Block Currant. The fruits of this plant contain malic acid and citric acid.

Uses : It stimulates the appetite, it nourishes as well as greatly refreshes. The fruit is recommended for gastric troubles, spasmodic vomiting and dysentery. The persons suffering from lung diseases accompanied by a cough are not advised to take these fruits. Children, old people, weak and nervous women must not be allowed to eat too manv fruits.

Ricinus Linn.
Family Euphorbiaceae

This is a tall perennial herb, many plants become tree like with stout fibrous roots, thick and soft woody trunk and attain a maximum height upto 10 m. Leaves are large, broad, roundish in outline with 5 to 9 deep lobes which are coarsely toothed and pointed at the apex. Young leaves are purple- bronze and silky, mature leaves are grey-green or dark purplish-red. Flowers are large, in big terminal bunches. Female flowers are uppermost, bristly, greenish with bright-red stigma. Male flowers are found below, have green or red, yellowish stamens. Fruit is a prickly capsule when immature. Green, bluish or brilliant red, turning brown when mature and dry. Seeds are oblong, seed coat crustaceous and the colour of seed coat varies.

This plant is native to Africa. It is, however, naturalised in all subtropical and tropical countries of the world. The crop is commercially cultivated in Brazil, India, Russia, Texas in U.S. , Mexico, Peru, Paraguay, Thailand, Ethiopia, South Africa, Tanzania, Rumania and Yugoslavia. In India, it is also found wild in agricultural fields, gardens near habitations in waste places, Commonly known as Arandior Caster Seed tree.

Uses : Seed consist of 20% coat and 80% kernel. Castor seeds constitute glycoprotein, ricin, ricinnie and lipase. Small seeds are richer in oil than the larger seeds. The oil is nearly odourless and has a mild flavour.

The oil leaf and root are used against various ailments. The oil is useful for skin diseases. Castor oil seeds have tonic effect. The oil is emollient, laxative and it can be used in cases of inflammation of the intestine or dysentery. The oil is made into contraceptive jellies, foams and cream. The oil is used in ointment as a soothing agent and as an oil vehicle in cyc drops. Undecyclenic acid, derivative from castor oil, is an antifungal agent. The fresh leaf may be simply worn under a hat or bound around the head to relieve headache and fever.Fresh leaves are laid on the abodomen to relieve internal complaints. Heated and oiled leaves are placed on rheumatic joints, swellings and inflammed muscles. The root bark, oil and pounded young leaves are used as purgative.

The root paste about 15 gm of at least three years old plant is taken in the morning for three days for abortion. These plants are mosquito repellant. Castor oil is also used in food and beverage industry. The seeds are a commercial source of lipase and is used by soap manufacturers. In some areas this plant is cultivated for its leaves which are fed to a special silkworm. Paper can be made from the fibre of the stem. In Indonesia, inflorescence and tender fruits are cooked as vegetables and mature roasted seeds are eaten.

Robinia Linn.

Family : Leguminasae (= Fabaceae)

***Robinia neo-mexicana* Gray. :**

It is a hardy deciduous flowering tree of 10 m tall. Leaflets are numerous. Flowers appear in dense clusters which are purple in colour and fragrant.

The plant occurs in coniferous forests at an elevation of 1600-3000 m in America. Commonly known as New Mexico Lowest or Rose Acacia.

Uses : Foliage and flowers arc useful for treating rheumatism and to induce vomiting.

Rosa Linn.
Family : Rosaceae

This genus has about two hundred and fifty species which are prickly shrub, sometimes climbing or trailing. About 10 species are distributed in India.

Rosa canina Linn. :

This is a vigorous and very prickly shrub with stout stems. Leaves are toothed all round the margin. Flowers are scented, white or pink, single, upto 5 cm across and appear in clusters. Hips are oval and scarlet in colour. It is distributed throughout Britain, North Europe and West Asia. Commonly known as Dog Rose.

Uses : A syrup made from the rose hips are given to children for extra vitamin C. Flowers are purgative. The decoction of the receptacles is used for diarrhoea.

Rosa moschata Mill. ex. Herrmann. :

It is a large thorny climber with reddish, sparsely prickly stems. Foliage dark green, with purplish flush and upto 4-7 leaflets, greyish below. Flowers are white, 5.0 cm across appear in terminal cluster, sweet scented. Hips ovate, orange red or dark brown.

It is found in Middle East, the Himalayas and Europe. In India, it is distributed in Garhwal Himalayas upto 2000 m altitude.

Uses : Rose water prepared from flowers are used in medicine. Perfume is also made out from the flowers.

Rosa sericea Linn. (= R. omeiensis Linn.) :

It is a thorny shrub upto 3.5 m tall. Leaflets ranging from 6-18 which look like those of fern. Flowers are white, single with four petals and 5 cm in diameter. Hips are small, pear-shaped and bright red.

It is distributed in nature in Garhwal hills betwen 2400-3600 m altitude particularly at Kedar Nath Ponwali, Badri Nath. It is also found in North Bengal, Manīpur, Sikkim, Bhutan and Assam.

Uses : Syrup prepared from flower is used to treat epilepsy and insomnia.

Rosemarinus Linn.
Family : Labiatae

Rosemarinus officinalis Linn. :

This is a hardy evergreen fragrant shrub of 2.6 m tall. Leaves are rolled back and stomata in hairy grooves on lower side. Leaves are narrow upto 2.6 cm long. Flowers are violet blue in colour.

The species is native to South Europe. It is cultivated in USA, France, Tunisia, Morocco, Yugoslavia, Portugal and Spain and also in some other temperate and subtropical parts. In India, it is grown in Nilgiri Hills and Temperate Himalaya. Commonly known as Rosenary.

Uses : The essential oil contains d and dl- -pinene, B-pinene, dipentene, P-cymene, d-linalool aldehyde C_{10} geraniol and acetic acid. The essential oil also contains camphor, bornyl alcohol, cineole, and comphene. The essential oil content varies from 0.4 to 2.4%.

Oil and flowers are used for medicinal purposes. Flowers are used in medicine as stimulant, diuretic, aperient, emmenagogue. The essential oil obtained from the plant is also stimulant. The oil is employed as a flavour and carminative. A balm is made from it for frictions, to stimulate the circulation and productive revulsive effects. Fresh tender tops of these plants are used for garnishing and for flavouring food, cold drinks, soaps etc. Dried leaves are used as condiments.

Rubus Linn.
Family : Rosaceae

Rubus idaeus Linn. :

It is a hardy evergreen, fruit bearing flowering perennial upto 2 m tall. Flowers are white and the berries are red or yellow.

The species is native to Europe. Commonly known as 'Raspberry'.

Uses : Fruits of these plants are medicinal. Raspberry is used to stake the thirst in the heat of fever, it promotes perspiration and urination. The fruit is full of insoluble fibre and absorb fewer of calories.

The fruit juice is given to children to control dysentery. Tea made from root bark is used to treat dysentery, while tea made from roots is used to dry up running noses.

Rudbeckia Linn.
Family : Compositae (= Asteraceae)

***Rudbeckia laciniata* Linn. :**

This is a much branched herbaceous perennial of 2.6 m tall. Stems are short with stiff hairs. Flowers are yellow.

It is native to North America.

Uses : Entire plant in bloom is useful. Leaf preparations are used to treat gonorrhoea and to induce menstruation. To treat kidney trouble and to increase urine flow these plants are said to be useful.

Rumex Linn.
Family : Polygonaceae

It is a genus of about two hundred species and thirteen species are distributed in India. Which are herbaceous plants mostly perennial.

***Rumex hymenosephalus* Torr. :**

This is a hardy perennial herb growing to lJm in height. The leaves are large with sheaths at base of each leaf. The flowers are rose coloured, appear in clusters.The tubers resemble sweet potatoes.

It is native to America. Commonly known as 'Canaige', "Red Dock" or "Wild Pie Plant".

Uses : Older roots have medicinal properties. Grounded powdered dried roots are applied as a poultice for sore joints. The extract of the root of this plant is used as a tonic, depurative and slight purgative. It is also used to treat sore throat and diarrhoea.

***Rumex obtusifolia* Linn. :**

This is a perennial herb upto 1.6 m tall. Leaves are medium. Flowers are greenish or brownish.

It is native to Europe, Asia and Africa. The plant is a weed found in waste areas, empty lots and roadsides.

Uses : Decoction of root is used externally to treat skin diseases. This decoction is taken internally as laxative. The poultice of roots is applied to skin itch, ulcers and skin diseases. Decoction of root is also administered for curing veneral disease. Commonly known as "Batter Dock" or Blunk Sorrel".

***Rumex nepalensis* Spreng.:**

This is a hardy perennial herb. It is found abundantly at Thunera of Uttar Kashi. Commonly known as "Dock" or "Sorrel".

Uses : The sap of the leaves and stem is applied on cuts for its astringent properties.

Ruta Linn.
Family Rutaceae

***Ruta graveolens* Linn. :**

This is a hardy evergreen aromatic shrub upto 1.5 m tall. The etheral oil is present in the leaves. Flowers are terminal, yellowish green with toothed petals.

The species is native to South Europe. Commonly known as "Herb of Grace" or "Rue Plant".

Uses : Plant in full bloom and mature leaves are harvested for medicinal purpose. Rue is narcotic and stimulant. It is a poisonous plant. The plant contains an acid and a colouring matter, rutin. In minute doses it is stimulant and antispasmodic. Infusion of plant is used to treat windy colics, epilespy, and nervous distress. It is said to promote fertility, promote delayed menstruation. Tea prepared from leaves is used to treat indigestion. To stop convulsions and fits leaf juice is useful. The leaf poultice is used to treat rheumatism.

Salix Linn.
Family : Salicaceae

It is a large genus consisting of about five hundred species. Which are mostly deciduous trees or shrubs and about thirty five species are distributed in India.

***Salix alba* Linn. :**

It is an ornamental tree of 16 m tall with spreading and drooping branches. Young branches are silky white in colour and with maturity they turn olive-brown . Leaves are glossy green above and silky white below, narrow, pointed with finely toothed edges upto 15 cm long. The flowers are catkins.

It is mostly cultivated in North-West Himalaya. Commonly known as White Yellow.

Uses : Young leaves and flowers of this plant are used as medicine. Tea made from the leaves and bark is used in fever. The bark preparation is known as a tonic in convalescence and also used for treatment of menstrual irregularities and diarrhoea.

Salix nigra Marsh. :

This is tall tree of 15 m height with reddish brown leaf and twigs. The bark is rough and scaly. Leaves are narrow, finely toothed and sharply pointed upto 16 cm long, 3 cm wide. The flowers are catkins upto 8 cm long. It is native to North America. Commonly known as Black Willow.

Uses : Barks of two to three years old branches and catkins of this plant are used to treat some ailments. The catkins are said to be aphrodisiac. The infusion of bark is prescribed for treatment of colds, asthma and increased perspiration. Tea pepared from bark is used to treat gangrene, ulcers, indigestion and diarrhoea.

Salix purpurea Linn. :

It is a shrub or small tree of about 6 m tall. Shoots are reddish purple.

It is native to Europe, Northern Africa, Northern and Central Asia. Commercially cultivated in Belgium and Poland. Commonly known as "Purple Osier".

Uses : The dried bark contain phenolic glycosides, salicin, populin, fragilin, glycosmin, grandidentatin, picein, salicyloyl tremuliodin.The bark contains a bitter astringent substance salicin, it is used in the treatment of rheumatism.

Salsola Linn.
Family : Chenopodiaceae

Salsola Kali Linn. :

This is an annual plant of Km tall, round and bushy in habit. It is a very fleshy plant with leaves ending in spines. Flowers are small, solitary, arises from the apex of the branches, whitish pink in colour.

The plant is native to North America. Commonly known as "Russian Thistle" or "Glass Wort".

Uses : Plants in bloom are used for medicinal purposes. It is useful for increasing urine flow and menstrual flow and decreasing water retention in body.

Salvia Linn.
Family : Labiatae

Under this genus there are about seven hundred species and twenty four species are distributed in India. Which are herbs or shrubs, in which leaves are opposite and often whorled below. Flowers are showy, racemose or spicate, usually in whorls.

Salvia lanata Roxb. :

This plant species is distributed commonly at Donpur of Uttar Kashi.

Uses : The crushed leaves are applied in laceration of does during rainy season.

Salvia officinalis L. :

This is a hardy herbaceous perennial upto 1.4 m tall. Leaves are with toothed edges. Flowers grow in terminal spike, blue in colour.There are several varieties in it like auriculata, major and minor.

The species is native to South Europe. Commonly known as "Sage".

Uses : Leaves of these plants are useful during flowering. It is used to treat colds, coughs, night sweats, sore throat and worms. A tea made from leaves is a mild laxative and used to treat stomach gas. Plant is also used as a tonic, antiseptic and to relieve cuts and skin sores. It is used to prepare many alcoholate and other drugs. It is a stomachic and tonic and prescribed for digestive complaints. Infusion in water have been used externally for bruises, ecchymoses, oedema, ulcerous or articularly obstructions in gums and aphtha.

Sambucus Linn.
Family : Caprifoliaceae (= Sambucaceae)

It is a genus of forty two species and four are native to India. Which are mostly shrubs or trees.

Drupes are crowned by the calyx teeth.

Sambucus canadensis Linn. :

This is a shrubby tree of about 4.5 m tall. The branches are yellowish-grey. The leaves are one-toothed. Flowers are white. Fruit is purplish black.

It is native to Eastern North America. Commonly known as "American Elder"

Uses : The Bark, roots, flowers and fruits have medicinal value. Poultice made from flowers and fruits are used to treat sores, burns and rheumatism. The mashed leaves are used as poultice for headaches. Bark preparations are used to treat abrasions and ulcers. Medicinal beverages and wines are prepared from berries.

Sambucus edullus Linn. :

This is a herbaceous perennial plant, upto 1.4 m tall. Flowers are white tinged with pink.

It is native to Himalaya. It grows well in dry banks and propagated by division. Commonly known as "Dwart Eden".

Uses : Almost all parts of this plant is used in medicine. The flowers in infusion are effective in rheumatic affections. Leaves are used as poultice, to dispense tumours in the joints caused by contusions. The juice of the root is purgative and diuretic. Seeds yield a mucilaginous oil which is purgative, The berries juice is a gentle purgative.

Sambucus nigra Linn. :

This is a small tree upto 6.5 m tall and produce white flowers. The bark of this plant has purgative, vomitive and narcotic properties. The leaves are laxative, purgative and diuretic

Dried flowers are used in an infusion as an emollient, sedative and diaphoretic. Berries are laxative. The fresh leaves in a poultice are said to alleviate the pain of piles. Locally known as "Common Elder" or "European Elder".

Sanguinaria Linn.
Family : Papaveraceae

Sanguinaria canadensis Linn. :

This is a hardy perennial, low growing herb with a solitary leaf stem. The leaves are palmately lobed around the edge. Flowers are white. It has a thick red rhizome and with red juice. Seeds are arillate.

It occurs naturally in Canada, Nebraska, Arkansas and Georgia. Commonly known as "Blood Root", "Sweet Slumber" or "Tetter Wort".

Uses : Dried rhizomes and roots are useful. It primarily contains sanguinarine alkaloids. Apart from this it also constitute protopine, oxysanguinarine, berberine, chelilutine, chelirubine, coptisine, sanguilutine and sanguirubine.

This is known as emetic and expectorant. A tea prepared from roots are used to treat fever and rheumatism. It is externally used for eczema and skin cancers.

Sanguisorba Linn.
Family : Rosaceae

Sanguisorba officinalis Linn. :

This is a hardy perennial herb and native to Arnerica.lt grows well in sunny or shady borders or by water side. Plants appreciate abundant moisture during growing seasons. Propagated by seeds or by division of roots. Commonly known as "Greater Burnet".

Uses : This plant has astringent, vulnerary and haemostatic properties.

Santalum Linn.
Family : Santalaceae

Santalum album Linn. :

It is an evergreen tree with drooping branches. Bark is dark, rough, with vertical cracks, mature wood scented. Leaves are opposite, shining on upper surface upto 8 cm long. Flowers are small, purplish, appear in small bunches. Fruits are roundish upto 0.6 cm across, purple black in colour.

The species is native to India and distributed in Deccan Peninsula. Commonly known as "Sandal Wood" or "Sandal Tree".

Uses : The oil extracted from the heartwood of the tree is used to promote urination, inflammation of bladder, gonorrhoea and cough. The wood, ground up with water into a paste and this paste is applied on skin diseases, local inflammations and on the forehead in fever. Seed oil is also used in skin diseases. The drug of this plant is used to treat tuberculosis of gall bladder. The oil is used in perfumery industry. The wood powder is used to make incesne powder or agarbati. The wood has pleasan[f] perfume and is used to make valuable handicraft items.

Saponaria Linn.
Family : Caryophyllaceae

Saponaria Officinalis Linn. :

This is a hardy perennial plant upto 70 cm in height. The leaves of these plants lather if rubbed with water. Flowers are in dense clusters, pink in colour. Flowers are protendrous, butterfly visited.

The species is native to Europe. Commonly known as Bouncing Bet, Boston Pink or Soap Wort.

Uses : Roots of these plants are medicinal and used to treat gout, rheumatism, skin diseases and syphilis. The plant has been used in washing clothes because of high saponin content. The plant acts as a tonic, depurative, diaphoretic and softening agent. It is also used in jaundice and abdominal congestion in the form of tea, extract or syrup.

Saraca Linn.
Family : Caesalpiniaceae (Leguminasae) = Fabaceae

Saraca indica Linn. (= Saraca asoca (Roxb.) Dewilde :

This is a medium sized evergreen tree, reaching to a height of 15 m with spreading branches. Bark is dark brown. Leaves are alternate, compound upto 32 cm long with 6-12 leaflets which are oblong or oblong lanceolate, acute to 20 cm long, entire, short petioled. Young shoots are pendent. Flowers are yellow or orange and gradually turn vermillion. Flowers are scented at night. Pods upto 15 cm long and contain 4-8 seeds which are grey in colour.

It is widely distributed in West Coast of Bombay, in Khasi Hills, Chittagong, Arakan, Tenasserium, Upper Burma, Ceylon and Malaya. Commonly known as "Asoka".

Uses : The flowers, seeds and dried bark of the tree is medicinal. Pounded flowers mixed with water is given for haemorrhagic dysentery. Seeds are used to cure urinary discharges. Dried bark is astringent and used to treat excessive menstruation as a uterine sedative.

Sarcococca Linn.
Family : Buxaceae

Sarcococca saligna (D.Don.) Muell. Arg. :

This is a hardy evergreen shrub of about 1.8 m tall. Flowers are scentless,unisexual without petals, greenish in colour.

Plants are native to Northern India and Himalaya, commonly occurring at Donpur of Uttar Kashi. Commonly known as "Sweet Box" or "Tilyor".

Uses : The sarbat prepared from root is given in phosphoteuria and gonorrhoea.

Sassafras Trew.
Family : Lauraceae

Sassafras albidum (Nutt). Nees. :

This a hardy deciduous aromatic tree of 16 m tall. Leaves are of three different shapes like 3-lobed, 2-lobed and mitten-shaped and unlobed. Flowers are borne in clusters, fragrant, male and female flowers appear on different plants, greenish yellow in colour. Fruits are dark blue, one seeded berry.

It is native to Eastern Europe and USA. Commonly known as Sassafras.

Uses : A volatile oil is extracted from the root which contains 80% safrole and little quantity of eugenol, pinene, phellendrene, sesquiterpene and d-camphor. Sassafras oil is antiseptic which destory lice and allay insect stings and bites. The oil is carminative and stimulant. A tea of bark is used to treat bronchitis, dysentery, kidney trouble and used to increase urine flow, relieve gas and upset stomach. The bark is used as poultice for sore eyes. The wood of this plant is regarded as a sudorific.

Satyrium SW.
Orchidaceae

Satyrium nepalense D. Don. :

This is a terrestrial orchid. Plants attain a height of 80 cm. Tubers are oval, producing stolon each with a small tuber at its end.Stem glabrous, sheathed near the base. Leaves are 2-3 , mostly basal, fleshy, narrowly elliptic, upto 30 cm long and 9 cm broad. Inflorescences are densely many flowered, to 25 cm long, bracts are much longer than the flowers. Flowers are licm in diameter, fragrant, pink or white.

In India, it occurs in Kashmir, Khasia Hills, Darjeeling, Manipur and Hilly districts of Southern India betwen 2800 to 3200 m altitude. Commonly known as “Satyrion”.

Uses : They are planted in shallow earthen pots filled with equal parts of white sand, well rotten dried leaf mould, rich loamy soil, crushed and dusted morter. Roots of these plants are used to treat malaria and dysentery. It is considered as tonic.

Saussurea DC.
Family : Compositae (= Asteraceae)

It is a large genus, consisting of about four hundred species and fortysix are distributed in India. These are glabrous or tomentose herbs.

Saussurea gossypiphora Don. :

It is a dwarf shrub to 16 cm tall. Flower heads are in the form of brown spots at the top. From a distance the plant resembles a dove.

It is rarely occurring in glaciated rocky slopes in Thamsar jol (Bara Bhangal), Dainasar (Chhota Bhangal) , Chansal Peak in Simla and Sangla Pass in Kinnaur district. Commonly known as "Saussuree".

Uses : The wool of the herb is applied to fresh cuts, where it sticks easily, seals the wound, checks bleeding and enhances healing.

Saussurea lappa C. B. Clarke :

A robust perennial herb of about to 2.5 m tall with bitter roots. Leaves are very large, the lower basal ones upto 1 m long, borne of winged stalks. Flowers bluish purple, borne on round flower heads.

It is native to Himalaya, occurs in Kashmir between 3000-4500 m and also in Garhwal of Uttaranchal. Commonly known as Costus.

Uses : The dried root of ths plant is antiseptic and disinfectant. Administered in bronchitis, asthma, flatulence and certain cardiac complaints.

Saussurea obvallata (DC.) Edgew. :

An erect tufled herb of about 80 cm tall.

The species is distributed in Rohal Kanda, Chansal Peak and Chandramohan of Shimla district; Sangla and Chitkul Passes of Kinnaur district of Himachal Pradesh.

Uses : The bracts of these plants are boiled in water and used as hot fomentation to cure hydrosele. The roots are often applied as a paste in cuts and bruises.

Scabiosa Linn.
Family : Dipsacaceae (= Compositae)

Scabiosa arvensis Linn. :

It is a herbaceous plant, woody at base.

The species is distributed in temperate Eurasia.

Uses : This plant is said to be useful formerlv as medicine for scurvy complaints, tinea, leprosy and scabies. Commonly known as "Corn Scabious".

Schinus Linn.
Family : Anacardiaceae

About eighteen species, native to Mexico to Argentina.

Schinus molle Linn. :

Trees are resinous. Flowers are small, white, in axillary or terminal bracteate panicles.

Uses : The leaves of these plants are applied on ulcers and sores. The leaves are also chewed to harden the gum. To treat amenorrhoea and dymenorrhoea, the aqueous extracts of leaves are used. The fruits of this species is edible and a syrup is prepared of the fruits.

Schleichera Willd.
Family : Sapindaceae

Schleichera Oleosa (Lour.) Oken. :

It is a large tree. Leaves are red when young. Mature leaves are uniformly green. Flowers are greenish yellow in colour, often arises on short branches.

It is native to Indo-Malayan region. Found at Geidubba of Chhindwara District, Madhya Pradesh.

Uses : The aril of seed is edible. The oil extract of seeds is used in leg swelling. It is a useful timber plant. It also furnishes the best lac (Mirzapore Lac). The seed oil is used as illuminant, in perfumery and for soap making.

Scilla Linn.
Family : Liliaceae

Scilla amoena Linn. :

It is a herb with unicate bulbs. Leaves are radical, linear, lorate or lanceate. Flowers appear in racemes, on a simple, leafless scape.

The species is native to temperate Eurasia. Commonly known as "Star Hower squill".

Uses : This plant is known as a bitter narcotic poison. It is used medically to stimulate urination, and is one of the best diuretics known. To activate its property, the squill is sometimes associated with other drugs. Tincture of this plant is externally used either alone or with digitalis in frictions.

Scrofularia Linn.
Family : Scrophulariaceae

A genus of about three hundred species and nearly twenty three species are found in India. Which are herbs or undershrubs.

***Scrofularia aquatica* Linn. :**

This is a perennial herb with opposite leaves. Flowers appear in tall inflorescence, whose primary branching is racemose.

The species is native to Europe. Commonly known as "Bishop's Leaves", "Figwort" or "Pilewort".

Uses : This plant is recommended to treat scrofula piles. In small doses it is purgative and in higher doses it acts as vomitive.

***Scrophularia marilandica* Linn.**

This is a perennial plant of 4 m tall. Stems are four angled. Leaves are with toothed margins, thin and tapering to a point upto 32 cm long. Flowers are small reddish brown in colour. Commonly known as "Mary and Figwort".

Uses : All parts of plants are medicinal. The plant is useful to treat haemorrhides, tuberculosis, scabies, wounds and also to increase menstrual flow. To treat anxiety, restlessness and insomnia in pregnant women, the infusion of fresh roots in water is used. The plant is also considered a tonic. It increases urine flow, cures intestinal worms and breaks fever by increasing perspiration.

Scutellaria Linn.
Family : Labiatae

***Scutellaria laterifolia* Linn. :**

This is a herbaceous perennial of 95 cm tall and erect in habit. Leaves are coarsely toothed, pointed upto 12 cm long. Flowers are blue.

It is a cosmopolitan species. Commonly known as "Skull Cap".

Uses : This plant at the stage of blooming is useful. It is used to treat heart disease, diarrhoea, to promote menstruation, to help eliminate a mother's after birth, to increase perspiration and also used as sedative.

Sedum Linn.
Family : Crassulaceae

Sedum ewersii Ledeb. :

This is a succulent plant. Fleshy leaved xerophyte.

It is Commonly occurring at Donpur of Uttar Kashi between 1800-3500 m.

Uses : The paste of root and leaf is applied on moles. Leaves are emollient, refrigerant and are prescribed as a remedy for sores, burns and piles. There is another species Sedum acre which is used for worts and piles.

Sempervivum Linn.
Family : Crassulaceae

Sempervivum tectorum Linn. :

This is a hardy succulent leaved perennial, and rosette type of plant upto 12 cm across. Pinkish flowers appear on hairly branches. This plant is planted on cottages to keep slates in position.

The species is native to Asia and Europe.

Uses : The juice of this plant is externally used for treatment of corns, gangrenous sores, dirty ulcers and burns. From this plant, ointments are prepared to treat burns and piles.

Senecio Linn.
Family : Compositae (= Asteraceae)

This is a very large genus consisting of about two thousand species and seventy five are native to India. Which are herbs, undershrubs or shrubs. Heads are solitary or clustered, variously coloured.

Senecio aureus Linn. :

This is a herbaceous perennial upto 95 cm tall, erect in habit. Flowrs are bright yellow.

It is a cosmopolitan species. Commonly known as "Golden Ragwort".

Uses : Roots and above ground portions in full bloom constitute the drug. An infusion of the plant in water is used to cure coughs, weakness due to prolonged fever and nervous disorder. A tea prepared from flowering tops is used to relieve pains of child birth, to hasten delayed child birth and also for kidney problems.

Senecio graciloflorus DC. :

This is a herbaceous plant upto 2.5 m tall, glabrous. Flowers are yellow in colour.

It occurs in the temperate zones of Garhwal Himalaya.

Uses : The leaves of these plants are used to treat abdominal pain and fever.

Senecio nudicaulis Buch-Ham. ex. Don. :

This is a hardy evergreen herb occurring at Kaream of Chhindwara District, Madhya Pradesh. Commonly known as "Grand Sil" or "Ragwort".

Uses : The whole plant of Senecio nudicaulis (about 2p gms.) is boiled in water and the extract is used three times a day for two weeks in cough and cold.

Senecio aurbavilla Pers. is an antisyphilitic, Senecio Candida is very poisonous. Senecio jacoboca Linn. is regarded as an aperitive. Senecio vulgaris Linn. is an emollient and resolvent.

Serenoa Hook. f.
Family :L Palmae

Serenoa serrulata (Michx.) Hook. :

This is a low growing palm. Leaves are fan-shaped, toothed, green.

The species is native to South-East United States. Commonly known as "Saw Palmetto".

Uses : Leaves, ripe fruits, inner bark and roots of this plant are used. An infusion of leaves and roots is used to treat dysentery and stomach pain. A poultice made from inner bark is applied on insect bites, snake bites and skin ulcers. Dried ripe fruits are said to be sedative and are also used for improving digestion and respiratory infections.

Sesamum Linn.
Family : Pedaliaceae

Sesamum indicum Linn. (= S. orientale Linn.) :

This is an erect herb.

The species is native to Africa. It is largely cultivated in India, China, Burma, Mexico and South America, for the oil from seeds. In India, it is mainly grown in M.P., U.P., A.P., Rjasthan, T.N. and Maharashtra. This species is often found wild on roadsides and waste lands. Commonly known as "Sesame", "Gingeti".

Uses : Seeds of these plants contain 50 to 56% oil. This oil is a semi-drying, polyunsaturated oil having the fatty acids like linoleic, oleic, palmitic and stearic.It possesses antioxidant and synergistic properties. This oil is employed as a vehicle for injections and also used as emollient and demulcent. The seeds are used in confectionery. Seed oil is edible and is also used for cosmetics, manufacture of soap etc.

Sesbania (Scob.) Adans
Family : (Papilionaceae)

It has about fifty species native to tropical and subtropical region. In India, six species are found. Which mostly are herbs, shrubs or small softwood trees, often prickly. Flowers are in lax, axillary recemes.

Sesbania bispinosa (Jacq.) W.F. Wight. :

This is native to W. Himalaya and southwards to Peninsular India.

Uses : The stem fibre is used as a substitute for hemp and for making ropes.

Sesbania grandiflora Poir. :

A decoction of bark of this plant in small doses is given against dysentery and diarrhoea. The root juice is used as an expectorant. Paste is applied on swellings.

This species is native to India.

Uses : S. sesban Merr. leaves are eaten and the bark of this plant is used to make ropes. This species is found wild throughout India. Seeds of S. speciosa are used as famine food. Leaves and tender branches of S. cannabina and S. speciosa are used as fodder for cattle.

Shorea Roxb. ex. Gaertn.
Family : Dipterocarpaceae

Shorea robusta Roxb. :

It is a large sized resinous tree. Flowers in axillary or terminal, are cymose panicles.

It is native to India, found at Rater of Chhindwara District, M.P. and also in Punjab, Bihar, West Bengal, Assam and Uttar Pradesh. Commonly known as "Sal" or "Saul".

Uses : The resin powder is taken three times a day for 20 days in dysentery. The wood of this tree ranks next to teak and used chiefly for railway sleepers and for construction work.

Sida Linn.
Family : Malvaceae

Sida cordifolia Linn. :

This is a small shrub, minute star-shaped hairs are present all over the plant. Leaves are 6 cm long, ovate or roundish. Flowers are small yellow. It is distributed throughout India in waste places and open scrub forest. Commonly known as "Mahot Balais".

Uses : The seeds make general tonic for improving sexual strength. Decoction of roots is administered in fever.

Silybum Vaill ex. Adans
Family : Compositae (= Asteraceae)

Silybum marianum (Linn.) Gaertn. :

This is an erect, glabrous, thistle-like annual or biennial herb upto 1 m tall. Leaves are large, variegated with spinose-toothed lobes, conspicuously white-spotted above. Flowers are rose purple in colour, achenes transversely wrinkled or smooth, black or grey.

The species is occurring frorn Western Himalayas, Kashmir and Punjab. Commonly known as "Holy" or "Mary's Blessed".

Uses : Ordinary rich soil and open situation are preferred for cultivation

of this species. Plants are propagated from seeds and seedlings are transplanted 65 cm apart. The species is cultivated for the edible heads, roots and leaves.

Sinapis Linn.
Family : Cruciferae (= Brassicaceae)

Sinapis alba Linn. :

The stems this plants are hairy. Leaves are irregularly pinnate. Flowers are large yellow and spreading. Pods are hairly and few seeded. The seeds are large, white. It is a self fertile species and native to Southern Europe and Western Asia. A winter season crop and suitable for temperate region. Commonly known as "White Mustard", "Safed Rai" or "Yellow Mustard".

Uses : The seeds contain 29% crude fat and yield 0.16% volatile oil. The seeds also contain the glucoside, sinablin and the enzyme ryrosin.

Young leaves and tender shoots are used as a pot-herb. It is a rubefacient in the form of powder, mustard plaster or baths. White mustard seed is a depuratine, it is effective for treatment of dyspeptic troubles, constipation, flatulence, apepsia and digestive disorder.

Smilacina Desf.
Family : Liliaceae

Smilacina amplexicaulis Nutt. :

This is a hardy parennial of 95 cm in height. Leaves are with clasping base upto 12.6 cm long. Flowers are small and whitish in colour. Fruits are berry, and red in colour.

The species is native to Himalaya and America. Commonly known as Fat solomon Plume.

Uses : Mature leaves are roots are useful. A tea prepared from leaves was thought to prevent conception. Smoke of the burning roots is used to revive a fainting person.

Smilax Linn.
Family : Smilacaceae (= Liliaceae)

Under this genus there are about three hundred species and twenty four are native to India. Which are shrubby oir herbaceous, climbing or straggling plants.

Smilax aristolochiaefolia Mill. :

This is a herbaceous plant. At the base of leaves spring two tendrils, one on each side, usually regarded as modified stipules. Stems are furnished with recurved hooks.

The species is native to Southern Mexico, Guatemala and British Hondurus. Commonly known as Mexican Sarsaparilla.

Uses : The roots of this species contains steroidal sapogenins. The other species of this genus like S. febrifuga Kunth "Ecuadorean Sarasparilla" and S. regelii Killip & Morton (Syn. S. ornata Hook) "Honduras Sarsaparilla " also contain steroidal sapogenins. This sapogenins mainly constitute sarsapogenin, smilagenin and also sitosterol and stigmasterol. For pharmaceutical purpose Smilax species are utilised for partial synthesis of cortisone.

Smilax aspera Linn. :

This is a climber, shruby in nature, with white flowres. Root stock thick and creeping type. It is a distinct species, having the Dioscorea like leaves and spiaite infloivscence.

The species is native to Europe and India. It occurs in the Garhwa] Himalaya upto 2000 m altitude. It is also distributed throughout the greater part of India from Kashmir, Sikkim, Assam and Deccan Plateau. Commonly known as "Prickly Ivy" or "Raugh Bird Weed"

Uses : Roots of these plants are used as blood purifier and to treat skin diseases.

Smilax ovalifolia Roxb. ex. D. Don. :

It is commonly known as "Saraspatrilla Plant" or "Greanbraiar" and it is occurring at Rater of Chhindwara District, Madhya Pradesh Compost consisting of light loam, leaf mould and sand, well drained position and warm situation are good. It is propagated by division of roots.

The root decoction is taken twice a day in spermatorrhoea and weakness.

Solanum Linn.
Family : Solanaceae

A large genus of about one thousand ten hundred species and forty five species are native to India. Which are herbs, shrubs or small trees under this genus.

Solanum viarum Dunal Linn. :

This is a very important species under this genus for steroid drug industry. Its berries are important source of steroid hormone solasodine. It is a stout, much-branched spiny herb upto 1.6 m tall. The plant produces berry like green fruits with white streaks which turn yellow after ripening. Numerous small brown coloured seeds are produced which remain embedded in a sticky mucilage of fruits.

These plants are found wild in Eastern to North-Western Himalayas, Nilgiri Hills and in the plains of Andhra Pradesh, Karnataka, Kerala, Maharashtra, Tamil Nadu, Andaman Islands, West Bengal, Jharkhand and Orissa.

Uses : The dry berries containing 10% moisture and more than 2% solasodme are preferred by pharamaceutical farms. The economc potential of this crop as a raw material for steroidal drugs is very limited. The steroidal hormones are used as antiinflammatory, antiarthritic and hormonal.

***S. dulcamara* Linn. :**

Uses : This useful for gout rheumatism, cutaneous and syphilitic affections. Generally the infusion, the decoction, syrup and alcoholic and aqueous extracts are used in pharmaceutical prepartions. This also cause abundant sweating and stimulates renal secreation. Commonly known as “Better Sweet”.

***S. indicum* Linn. :**

This is a spiny shrub, flowers are blue; fruits are round and smooth, light green with dark green stripes.

Uses : The plant is carminative. The expectorant used in cough, fever and asthma.

***Solanum nigrum* Linn. :**

This is an erect annual herb. Flowers are white, berries are greenish to black and red when ripe. Commonly known as “Ghamai” or “Gurkamai”.

Uses : The juice of the plant is useful on ulcers and other skin diseases. The infusion of plant is useful is dysentery and fever. The drug made from this plant act as laxative, improve appetite and this is administered against asthma.

***Solanum surattense* Burm. F. :**

A prickly spreading herb. Leaves are with sharp yellow prickles. Flowers are purple. Berries are yellow when ripe.

It is distributed throughout India in waste places and roadside, common at Isvalee of Uttar Kashi between 1200-2200 m and at Harvalachar of Chindwara District, Madhya Pradesh, Commonly known as Kateri, Katsareya or Bhafur.

Uses : The fruit is tied as garland to relieve jaundice. In swelling of any part “of the body, the garland prepared from the fruits is tied to the organ to reduce the swelling. The dried root is used in cough, asthma, pain in chest and fevers. The plant is carminative and diuretic.

***Solanum trilobum* Linn. :**

This is a perennial thorny straggling undershrub.

The roots and leaves of these plants are prescribed in bronchitis and consumptive cases. The species contain 3% total alkaloid.

S. tuberosum Linn. :

Uses : The juice of the tuber is an antacid and it is very rich in vitamins. It is native to South America.The starch is used in some skin ointments and tablets. Commonly known as Potato.

Sophora Linn.
Family : Leguminasae (= Fabaceae)

Sophora japonica Linn. :

This is an ornamental plant which may attain a height of 30 m. Trees are with attractive pinnate foliage and ornamental creamy white flowers. The variety "variegata" possess leaves margins with creamy white and "violacea" bear rosy violet flowers.

Its natural distribution is in China, Japan and Korea. Commonly known as "Japanese Pagoda Tree".

Uses : Well drained loamy soil and sunny situation are preferred. Propagated by seeds or by cuttings of half ripened young shoots. Flower buds constitude rutin.

Sorbus (Tourn.) Linn.
Family : Rosaceae

Under this genus there are about one hundred species and seven are native to India. Which are deciduous trees or shrubs.

Sorbus americana Marsh. :

This is a hardy deciduous tree of 10 m tall. Bark is smooth and grey. Leaves are compound. Leaflets pointed with toothed edges. Flowers appear in clusters, white in colour. Berries are bright red.

It is native to America. Commonly known as American Maintain Ast.

Uses : Berries and bark are useful. A tea made from bark is used to treat nausea and to cleanse the blood. Bark preparation is also used for heart diseases. Berries are rich in vitamin C and used to cure scurvy.

The ripe fruit of S. aucuparia L. is used in the preparation of extracts, syrups and juices. The extract of the fruit is used in digestive disorders and gall-bladder irritation. It is native to Kashmir and Kumaon. The fruits of S.cuspidata and S. lanata are edible.

Spathodea Beauv.
Family : Bignoniaceae

Under this genus there are two species native to tropical Africa, which are handsome evergreen Trees with large showy flowers.

Spathodea campanulata Beauv. :

This is a fast growing, tall tree with light grey and rough bark. Leaves are deep green, oval, coarse with deeply marked veins. Flower buds are velvety olive green which appear in huge bunches at the end of branches. Flower is orange and bright red in colour. Fruits are lance shaped, smooth, woodlike and dark brown containing winged seeds.

It is an East African tree widely grown in gardens of India. Commonly known as Tulip Tree.

Uses : The bark extracts of this tree is used for dressing of ulcers and also used to treat skin diseases. The seeds of this species is edible.

Sphaeranthus Linn.
Family : Asteraceae (= Compositae)

Sphaeranthus indicus Linn. :

It is a hairy herb, glandular with many branches, upto 80 cm tall. Flowers are pink or purple in colour.

The species is occurring at Rater of Ghhindwara M.P. District. It is also found in the moist damp places of tropical zones of Garhwal Himalaya.

Uses : The plant contain essential oil 0.22% and possess bitter alkaloid.

The stem, leaves and flower paste are used six to seven times a day for a fortnight in small pox. The plant possesses anthelmintic, aphrodisiac and stomachic properties.

Spigelia Linn.
Family : Loganiaceae (= Spigeliaceae)

Spigelia marilandica Linn. :

A hardy herbaceous perennial plant of about 70 cm tall. Leaves are lance shaped. Flowers are funnel shaped, yellow on the inside and red on the outside.

The species is native to North America. Commonly known as Caroling Pink.

Uses : Leaves, plant in full bloom and the roots have medicinal properties. A tea prepared from leaves, act as digestive, while tea made from roots are used to treat malaria. To treat intestinal worms entire plant along with roots is useful.

Spiraea Linn.
Family : Rosaceae

A genus have about one hundred species and ten are native to India. Thos-j are deciduous shrubs with simple leaves, sometimes pumately lobed.

Spiraea tomentosa Linn. :

This is a shrub of 130 cm tall. The leaves are toothed, woolly beneath, upto 7.5 cm long. The branches are woolly and reddish brown in colour. The flowers are pinkish or white in colour.

lt is native to North temperate region and Himalaya. Commonly known as "Hardhack".

Uses : Leaves, flowers, stems and root bark of this plant are useful. A tea "made from leaves and bark is used to treat diarrhoea. Tea made from flower is given to pregnant women to ease child birth.

Spiraea ulmaria Linn. :

Uses : The plant is commonly known as "Meadow Sweet" has tonic properties. The leaves and roots are used for metrorrhagia. It is administered in a tea like infusion whose taste is fairly pleasant and useful for coughs and inflammatory complaints.

Stellaria Linn.
Family : Caryophyllaceae

This genus have of about one hundred and twenty species of which twenty are distributed in India. Which are usually diffuse, tuffed herbs. Flowers are white, appear in dichotomous cymes or rarely solitary and terminal.

Stellaria media (Linn.) Vill. :

This is a hardy herbaceous annual of 40 cm tall. Leaves are ovate in shape. Flowers are white. It has small homogamous white flowers, that fertilise themselves in absence of insects.

The species is native to America and also found throughout the temperate regions of India. Commonly known as Chick Weed.

Uses : Plant in full bloom are useful. It is used for eye infections and haemorrhoids. A poultice made from the plant is used to treat ulcers and skin sores.

***Stellaria pubera* Michx. :**

This is a hardy perennial plant upto 60 cm tall. Leaves are upto 5 cm long, short-stalked. Flowers are 15cm long, white in colour. Commonly known as "Giant Chick Weed".

Uses : Plant in full bloom and juice of mature plant have medicinal properties. Plant juice is rich in vitamin C and useful for treatment of scurvy. An infusion of plant is used to treat weakness after illness, lung congestion and tuberculosis.

Sterculia Linn.
Family : Sterculiaceae

This genus have of two hundred species and twelve are native to India. Which are trees and shrubs under this genus.

***Sterculia urens* Roxb. :**

This is a tropical medium sized deciduous tree upto 15 m tall with erect straight trunk and broad top. The bark is smooth, glossy, papery, geyish-white, easily peeled off. All parts of tree, when injured, exude a soft gum. Leaves are crowded at the end of branches, alternate, palmate, shallowly fine lobed, very densely hairy on the under surface. Flowers appear in large erect terminal panicles, yellow brown, and small. Male and hermaphrodite blooms are intermingled. There is no corolla, the calyx is bell shaped, leathery, fine toohed. Fruit is star shaped, densely covered with stinging hairs.

The species is native to tropical Himalayas and Southern India. Some commercial plantations of this tree is found in tropical Africa. Commonly known as "Kuttera Gum Tree" or "Karaya".

Uses : The gum exudate from the stem of the tree has laxative properties. It is used in throat infection and in dental fixture powders. The gum is eaten in curries and is also used for making sweet meats. Roasted seeds are eaten and grounded seed is used as a substitute for coffee. It is also used in toiletries and cosmetics. From wood of this tree packing cases, toys, boats and doors are made. From the bark of this tree, fibre is extracted which is used for making coarse cloth and ropes.

***Sterculia villosa* Roxb. :**

This is a deciduous tree with grey bark. Flowers appear in bunches, yellow in colour. Fruits are follicular and seeds are black.

The species is occurring almost throughout India. Commonly known as "Oadal Tree".

Uses : The white clear gum exudes of this tree is used to treat throat infection and in dental fixture powders, lozenges and pastes.

***Sterculia foetida* Linn. :**

It is commonly known as Bastard Poon and occurs in Southern India. The oil extracted form this plant is laxative.

Stillingia Garden ex. Linn.
Family : Euphorbiaceae

***Stillingia sylvatica* Linn. :**

This is a perennial plant of 95 cm tall which excrete milky juice. The leaves are leathery and fit-shy. Flowers are yellow in Colour.

Uses : Routs are useful for medicinal purposes, ft is used to treat pulmonary diseases, liver ailments, constipation and syphilis. It helps to correct menstrual irregularities and induce vomiting.

Strophanthus DC.
Family : Apocyanaceae

***Strophanthus gratus* Franch. :**

This is a woody, stout and evergreen climbing plant which attain an ultimate height of 10 m. The leaves are elliptic, leathery, thick, pointed at the apex and rounded at the base upto 16 cm in length and 8.5 cm in breadth. Petiole is purple, and midrib of leaves are reddish. Flowers appear in terminal clusters which are fragrant, bell shaped, rose-purple in colour. Seed pods are in divergent pairs upto 16 cm long and 8.5 cm wide, which takes nearly twelve months to mature. Seed pods split open on maturity and release numerous sharp edged seeds. Seeds are odourless when dry and bitter in taste.

The species is native to West tropical Africa. Commonly known as "Smooth Strophanthus".

Uses : Seeds are the principal source of ouabain. It also contains small amount of other cardenolide glycosides and the alkaloid trigonelline.

Crushed leaves, stems and seeds are used for medicinal purposes. Crushed leaves are rubbed on the body to treat fever. Leaf decoction is used for gonorrhoea. Crushed stems are given for extreme debility. The leaf juice is applied on ulcers, wounds and Guinea worm. Ouabain extracted from seeds is cardiac stimulant which is administered intravenously or intramuscularly. For acute heart failure and pulmonary edema, ouabain is used in emergency.

Styrax Linn.
Family : Styraceae

This genus have about one hundred and twenty five species, are deciduous or evergreen trees or shrubs. Flowers are while in colour.

Styrax benzoin Dryanl. Linn. :

It is quick growing trees and graceful. A fragrant acrid resin is secreted when wounded. Leaves are ovate-oblong to lanceolate, broadly rounded at base and pointed apex, alternate, short stemmed. Inflorescence and leaves are coated with white or brownish hairs. Flowers are white, fragrant, cup like or bell shaped, calyx-5, narrow, toothed, petals are silky and stamens are 10. Flowers appear in compound, pendent racemes. Fruit is globose, drupaceous, dry, dehiscent. Seeds are round, 1 or 2 in number.

S. benzoin occurs in Java, Malacca and Malaya. S. paralleloneurum is distributed in Malaya Peninsula and Sumatra. S. tonkinense occurs in Laos, Tonkin, Annam and Thailand. These are cultivated in Sumatra and Java. Commonly known as "Benzoin Tree", "Storax Tree" or "Benjamin Tree".

Uses : Fresh resin of S. tonkinense contains 79% crystalline coniferyl benzoate, 12% benzoic acid etc. While S. benzoin and S. paralleloneurum contains less benzoic acid. These resins are completely soluble in ether and alcohol. Benzoin is carminative, expectorant and diuretic. Benzoin is added to boiling water and its vapour inhaled in case of bronchitis or laryngitis. Benzoin is an important ingredient in compoind benzoin tincture. Thistincture acts as antiseptic and protective coating on abraded, cracked and friction-blistered skin, on indolent ulcers and facial cold sores. Diluted with water, alcoholic solution of benzoin becomes milky and is used to heal cracked nipples and also employed in feminine hygiene.

Swertia Linn.
Family : Gentianaceae

Under this genus there are about one hundred species and thirty species are native to India. Which are annual or perennial herbs with opposite leaves.

Swertia alata Royle ex D. Don. :

This is an erect herb upto 70 cm tall. Flowers are yellowish green.

It is found in the temperate zones of Garhwal Himalaya between 2700 to 3300 m altitude.

Uses : The whole plant is medicinal and is used in fever.

Swertia angustifolia Buch-Ham ex. D. Don. :

This is an erect herb upto l$m tall. Flowers are white or pale blue in colour. It is native to Himalaya and occurs in Garhwal Himalaya at an altitude upto 3500 m.

Uses : The whole plant has medicinal properties and used as blood purifier and febrifuge.

Swertia cordata Wall ex. Clarke :

This is a much branched herbaceous plant upto 1.5 m tall. Flowers are yellowish white.

It is distributed in Uttarakhand Garhwal region between 2800-3500 m altitude.

Uses : The whole plant is medicinal and is used as a bitter tonic.

Swertia chirayita (Roxb.) Karst. :

This is an erect annual herb known as "Chirayta" and can attain 1.6 m height, much branched and robust. Leaves are opposite in pairs, without stalks, pointed at tips upto 12 cm long. Flowers are greenish yellow tinged with purple. Fruits are ovoid 10 mm long.

The species is distributed in temperate Himalaya between 1800 to 3200 m from Kashmir to Assam. It can be cultivated in the hilly regions.

Uses : During flowering, the whole plant is collected and dried which has "medicinal value. It is bitter, stomachic, febrifuge and anthelmintic. It is used to treat diarrhoea, malarial fever and weakness etc.

Swertia ciliata Buch-Ham :

This is a perennial herb known as "Cheretta" or "Kala Chirata". The corolla segmnets each bear two nectaries on the upper side, consisting of little pits covered with hairs.

It is native to Himalaya. Commonly occurring throughout temperate and alpine slopes of Central Himalaya near forest edges and moist slopes upto 4200 m.

Uses : Leaves are used for medicinal purpose for fever.

Symphytum Linn.
Family : Boraginaceae

Symphytum officinalis Linn. :

This is a hardy herbaceous perennial, attain a height of 130 cm and produce white or dull purple flowers.

The species is native to Europe. Commonly known as "Common Comfrey" or "Black Wort".

Uses : The plant is demulcent, emollient and mucilaginous. The root is astringent. It is used to treat haemoptysis, haematuria, diarrhoea, dysentery, urine discharges and renal inflammation.

Symplocarpus Salisb.
Family : Araceae

Symplocarpus foetidus (Linn.) Salisb. :

This is a herbaceous perennial. The odour of this plant is skunky. Leaves are upto 80 cm long, heart-shaped. Flowers appear in spikes, spotted and striped, purple and yellow green in colour.

It occurs in America, North-East Asia and Japan. Commonly known as "Skunk Cabbage".

Uses : Leaves and roots of these plants possess medicinal properties. Preparations from dried powdered roots are used to treat skin itching, asthma, chest ailments, ring worm and rheumatism. A poultice made from leaves and roots is used on swellings and sores. The tubers of these plants are used to increase menstrual flow, expel intestinal worms and also to treat epilepsy, hysteria and coughs.

Symplocos Jacq.
Family : Symplocaceae (= Styraceae)

This genus have about three hundred species and forty are distributed in India. Which are trees or shrubs. Flowers are axillary or terminal or in simple or compound spikes or racemes.

Symplocos chinensis (Lour.) Durce. :

This is a small tree. Leaves are ovate and acute, turning yellow on drying. Flowers are fragrant white.

It is distributed in the oak forest in the Garhwal Himalaya between 1500 to 2500 m altitude.

Uses : The leaves are astringent and used to treat diarrhoea. The bark is used to cure menstrual disorder. Powdered bark is applied on face for medicinal purpose.

Symplocos paniculata Wall :

This is a small tree.

The species is distributed in Himalayan region from Punjab to Assam.

Uses : It has the same medicinal properties as that of S.racemosa, and is considered a tonic.

Symplocos racemosa Roxb. :

This is a tree which attain a maximum height of 8 m. Leaves are dark green, leathery, usually pointed at tip, margins are entire or toothed, upto 25 cm long. Leafstalk is small upto 3 cm long. Flowers appear in small axillary clusters white or pale yellow in colour. Fruits are purplish black upto 1.5 cm long.

It is native to China and India, distributed in the plains and lower hills of Central and Eastern India. Commonly known as "Lodh or Lodhra".

Uses : The fresh air dried bark has medicinal value. The decoction of the bark is used as gargle in bleeding gum. The bark is an astringent and is used to treat excessive bleeding during menstruation. It is also used to cure disgestive disorders, ulcers and eye diseases.

Syzygium Gaertn.
Family : Myrtaceae

This genus have about five hundred species of Paleotropical region and over fifty species are native to India. Which are trees or shrubs. The seed-coat adhering more or less loosely to the pericarp and the two distinct cotyledons attached chiefly near the middle of the opposing faces concealing the hypocotyl within. The inflorescence is often paniculate-cymose.

Syzygium cumini (L.) Skeels :

A tall evergreen tree. Leaves are leathery, smooth, shining to 30 cm long. Flowers are small, dull white, appear in large bunches. Fruits upto 5 cm long, ovoid, black when ripe, juicy and edible.

This species is found at Geldubba of Chhindwara District, M.P. and generally grow in moist deciduous forest. Commonly known as Jammu, Jam.

Uses : Fresh bark juice mixed wtih milk is used in diarrhoea. The seed powder about 10 gm is administered orally thrice a day for 4 to 5 months in diabetes. The bark is used in sore throats, bronchitis, asthma, ulcers and dysentery.

Syzygium jambos (Lam.) Alston. :

This is a tall evergreen tree with greenish sub-sessile flowers. Berries are deep pink when ripe.

The species is distributed in different parts of India.

Uses : The tree bark, seeds and fruits of these plants have medicinal properties. The bark is used to treat sore throat, ulcers, dysentery and bronchitis. The seeds are used to treat diabetes. The roots of S. malacense are diuretic and used in application of itch.

The fruits of S.aqueum, S. bracteatum, 5. cuminii Skeels, S. jarnbos, S. sainarangense and S.zeylanicum, are eaten. The bark of S. ceresoideum is given in bilousness, dysentery and bronchitis. The wood of S. alternifoliumis rich in tannin. The bark of S. gardneri is a source of black dye.

Tabebuia Gomez. ex DC.
Family : Bignoniaceae

This genus have about one hundred species native to Tropical America and West Indies. Which are evergreen trees or shrubs.

Tabebuia pentaphylla Hemsl. :

This plant is commonly known as “Bamata” or “Poyer”.

The flowers of these plants are rose coloured, pink or white.

Uses : The fruit and bark are diuretic.

Tamarindus Linn.
Family : Leguminasae (= Fabaceae)

Tamarindus indica Linn. :

This is a perennial slow growing tree and more or less evergreen in nature. But leaves fall after the harvest of fruits. Leaflets measure upto 15 cm long and 10 to 20 in pairs. It has spreading branches and grows to a height of 18 m or even taller. The bark of the trunk is scaly. It flowers in clusters and takes 8-10 months to harvest fruits after flowering. There are African and Indian varieties in it. Variety with fruits having pink coloured pulp in common. The pulp is sour. Red coloured pulp is sweet and sour. The tender fruits are bent, flat, knotty up to 20 cm long and 2.5 cm broad. The skin looks rusty and breaks if pressed.

The species is native to Africa and grows in all tropical countries. It grows well in India on a commercial scale. Commonly known as "Tamarind".

Uses : The ripe fruit contain 55% tamarind pulp, 33% seed and 12% fibre. The pulp contains tartaric acid, citric acid, malic acid, sugar, protein, minerals, carbohydrate, fibre, vitamin A and niacin.

The leaves of these plants are applied to reduce inflammatory swellings and ringworm. Flowers are appetising, sweet and sour. The seeds are astringent, aphrodisiac and useful in giddiness and vertigo. The fruit is slightly laxative and refrigerant. Drinks made from it are used in some inflammatory illness. Lozenges are made from the pulp, mixed with Senna and chocolate and used as a cathartic in infantile diseases. It is used in culinary preparations. Fruits are appetising, tonic to the heart and anthelmintic. Fruits are digestive and useful in diseases caused by deranged bile, such as intoxication from liquors and burning of body.

Tamus Linn.
Family : Dioscoreaceae

Tamus communis Linn. :

This is a climbing plant, hibernating by tubers formed by a lateral outgrowth of the first two internodes of the stem.

It is native to Europe, Persia and Africa. Commonly known as "Beaten Women's Herb" or "Blank Bryony".

Uses : Fruits of this plant are poisonous and fatal. The rhizome of plants are used as a poultice on bruises.

Tanacetum Linn.
Family : Asteraceae (= Compositae)

This genus have of about seventy species, and nine species are found in India. Which are odorous, annual and perennial herbs. Leaves are usually pinnatisect and flower heads are yellowish in colour.

Tanacetum longifolium Wall ex DC. :

This is a hardy herbaceous perennial plant, distributed at Chota Changseel between 2500-3500 elevation.Commonly known as "Tansy" or "Kuberu".

Uses : The root powder is given with milk or tea in extreme stomach pain.

Tanacetum vulgare Linn. :

This is a hardy herbaceous aromatic perennial upto 95 cm tall. The leaflets are toohed. Flowers are inconspicuous and yellow. There is onc variety in it crispum, which is finer with more crisped leaves.

The species is native to Europe and Asia. Commonly known as "Tansy" or "Gingor Plant".

Uses : Leaves, flowering tops and seeds are useful. Leavcs are aromatic, used for flavouring pudding etc., and for garnishing. Crushed leaves are used as poultice for sprains and bruises, it is also useful for stomachache. It is said that aromatic oil of this plant can induce abortion. Dried leaves and flowering tops are used to treat intestinal worms. Powdered seeds have insecticidal properties. A tea made from flowering tops is used to promote menstruation.

Taraxacum Weber
Family : Compositae (= Asteraceae)

Taraxacum officinale Weber. :

This is a hardy perennial herb, growing upto 40 cm tall. This is a weed in most parts of the world and cultivated for its blanched leaves for saladings. Leaves are deeply serrated. Flowers are yellow in few-flowered heads terminating in naked, hollow scape.

It is native to North temperate region, throughout the Himalayas, and also found as weedj.n South Indian hill Stations. Commonly known as "Dandelion" or "Crow Parsriep".

Uses : Cooked young leaves are eaten to purify blood. Leaves are used as salad. Leaves, flowers and roots of these plants have medicinal value. This is an important plant in homoeopathic medicine. It is said that the juice of the fresh plant is effective against liver diseases, chronic hepatitis, visceral congestion, intermittent fever and hypochondria. Roots are used to increase urine flow, as a laxative and tonic, to treat liver and spleen ailments and to stimulate appetite. Tea made by boiling flowers is used to treat heart trouble.

Taxus Linn.
Family : Taxaceae

In this genus there are about ten species and one is native to India. Which are evergreen deciduous trees or shrubs and woods are very tough.

Taxus baccata Linn. :

This is a hardy evergreen tree. Estimated average age of this plant is 1500 to 2200 years. This plant attains a height of 18 m. Leaves are dark green, coriaceous. Flowers are dioecious.

The species is native to Europe and N.Asia, temperate Himalaya and Khasi Hills. Commonly known as "Yew".

Uses : The leaves are very poisonous, but the aril is harmless. The leaves and bark of the plant are bitter, acrid and nauseous.

It is emmenagogue and antispasmodic. This is also used in asthma and bronchitis. The wood of this plant is valuable. In the Middle Ages it was the chief material used in making bows.

Taxus brevifolia Nutt. :

This is an evergreen shrub of 4 m height, with spreading horizontal branches. Bark is dark purple or brown. Leaf needles are dark green and smooth, long and flat, tapering to a point. The leaf stalks are rudimentary. Fruits are red.

The species is native to America. They require a cool damp place for proper growth. Commonly known as Pacific Yew or Wester Yew.

Uses : Leaves are medicinal and are used to induce menstruation.

Tecoma Juss.
Family : Bignoniaceae

It is a woody perennial shrub or climber. These plants bear terminal clusters of campanulate brightly coloured flowers. Plants are propagated by stem cutting.

Uses : A decoction of leaves of T. capensis Ldl. "West Indian Honey suckle" is given for the treatment of diarrhoea and enteritis. The roots of T. radicans Juss. "Scarlet Trumpet Flower" is used to induce sweating and for healing of wounds. The roots of T. stans Juss. is diuretic, vermifuge and tonic. The bark of young branches of T.undulata is employed for the treatment of eczema and syphilis.

Tectaria Cav.
Family : Aspidiaceae (= Tectaariaceae)

Tectaria macrodonta (Fee.) C. Chr. :

This plant occurs frequently at Mt. Abu of Rajasthan.

Uses : The plant is used in asthma and bronchitis. It is also used as anthelimintic.

Tephrosia Pers.
Leguminasae (= Fabaceae)

Tephrosia virginia (Linn.) Pers. :

This is a perennial plant of 80 cm height. Stems are erect and hairy. Flowers appear in terminal spike, yellow white in colour. Fruit pods are fuzzy upto 6 cm long.

It is native to tropical and subtropical region. Commonly known as "Verginia Tephrosia".

Uses : Roots of the plants or plants in bloom are useful for medicinal purpose. A decoction of the plant is applied to the limbs for strengthening them. Plant as well as roots are considered tonic, stimulant and laxative and used to treat worms and syphilis. A decoction of roots is used to treat chronic coughing, bladder trouble and intestinal worms.

Terminialia Linn.
Family : Combretaceae

This genus have two hundred and fifty species from tropical regions. Twelve species are native to India. It is large trees. Leaves are alternate or subopposite, often with glands on the petiole or on the lower part of the midrib beneath. Terminalias are mostly deciduous trees with horizontally spreading branches, and native to tropical region. Plants arc propogated by seeds.

Terminalia alata Heyne ex Roth. :

This tree is occurring at Malni in Madhya Pradesh District.

Uses : The stem bark powder (about 30 gms) is used twice a day for two weeks as cardiac tonic and as diuretic.

Terminalia bellirica Roxb. :

A tall tree often with buttresses, upto 25 m tall. Leaves clustered near ends of branches, to 25 cm long. Flowers small, pale green, ill smelling, arise in simple spikes. Fruits to 3 cm long, ovoid, brownish and densely covered with hairs.

It is native to India and Malava. Found at Baheda of Madhya Pradesh District. It also occurs in different parts of India excepting the dry region of Western India. Commonly known as "Bahera".

Uses : The fruit contain 5 to 17% tannin. The unripe fruit is used as astringent, but ripe fruit is purgative. The fruit powder is used as tonic and laxative. It is also used as brain tonic, useful in piles, leprosy, dropsy and fever.

Terminalia chebula Retz. :

A medium sized deciduous tree. Leaves ovate, acute to 25 cm long with two small glands, near leaf base. Flowers appear in spikes at the end of branches, dull white in colour. Fruit is 5 ribbed to 4 cm long.

It is native to tropical Asia. Occur at Sidhauli of Madhya Pradesh, District, the tree is also common in mixed deciduous forests. Commonly known as Black Myrebolam or Harra or Harada.

Uses : The fruit is astringent and laxative. It is used as a gargle in the inflammation of mucous membrane of mouth. The fruit decoction is used in bleeding and ulceration of gums. The fruit is roasted and taken orally three times a day for one week in cough. The fruit is also used externally to treat chronic ulcer, wounds and scalds. It is used in dyeing and tanning.

Differentones useful species are given below :

The powdered bark of T. arjuna Weight et Arn "Arjuna" is used in hypertension and a decoction of bark is used as wash in ulcers. The bark of T. bislata Steud is also fairly potent medicinally. The juice of young leaves of T. catappa L. "Country almond" is used to prepare ointment for scabies, leprosy and other skin diseases. T. paniculata bark is diuretic. The fruits of T. procera are used for dyeing.

Teucrium Linn.
Family : Labiatae

Teucrium chamaedrys Linn. :

This is a slightly tender evergreen shrub. Flowers with small upper lip protandrous, with movement of style and stamens. The plant, when brushed, gives off a strong unpleasant smell. Flowers are rosy purple in colour.

It is native to Europe and Asia. Commonly known as "Germander".

Uses : It has a bitter and acrid taste and is regarded as an aperitive, sudorific and vulnerary. This is also said to be tonic, stomachic and vermifuge.

Thalictrum Linn.
Family : Ranunculaceae

This genus have about one hundred and fifty species of herbs and shrubs and twenty are native to India.

Thelictrum alpinum Linn. :

This is a hardy herbaceous perennial upto 16 cm tall. Foliage is like Maiden hair fern. Flowers are yellowish green.

It is native to northern and arctic regions. In India, it occurs at Donpur of Utter Kashi between 2500-3000 m elevation. Commonly known as "Meadow Rue" or "Peeli Jari".

Uses : The paste of root is applied around the forehead in headache and hotness.

Thalictrum fendleri Engelm. :

This is a hardy herbaceous perennial with a foliage similar to Maiden hair fern attain a height of 95 cm. The flowers are yellowish-white.

The species is native to North America. Commonly known as "Meadow Rue".

Uses : Plants shen in bloom are dried, and the smoke of that dried plant cures headache.

Thalictrum foliolosum DC. :

This is a hardy herbaceous perennial with a foliage similar to maiden hair fern. Leaves are pinnately compound, three lobed and sharply decompound. Flowers are white palegreen. Roots are yellow in colour, nodular.

The species is occurring at Rater of Madhya Pradesh District. It is also distributed in the temperate and subalpine region at Kanatal, Chamba, Ghutu in Garhwal.

Uses : The root is used for medicinal purpose. The root powder is used in jaundice. It is also used as blood purifier. The root is considered tonic and ophthalmic, purgative, febrifuge, diuretic and is also used to treat atonic dyspepsia.

***Thalictrum minus* Linn. :**

Thalictrum minus

This is an erect herbaceous plant upto 1.5 m tall. Flowers and roots are yellowish green in colour.

This species is distributed in temperate zones of Garhwal Himalaya at Rudar-Ganga Valley, Tali in Garhwal Hills.

Uses : Roots are used for ophthalmia and abdominal pain.

Theobroma Linn.
Family : Sterculiaceae

***Theobroma cacao* Linn. :**

This is an evergreen small tree upto 8 m tall. Young leaves are red and pendant. Flowers appear on old wood. Petals are cap-like at base. Flowers are in fasicles. Flowers are of rose and yellow colour. Fruits are oval, large, tough, indehiscent, yellow or reddish. From the seeds cocoa is obtained.

This species is native to Southern Mexico and tropical South America. It is widely cultivated in West Africa. This species is occasionally cultivated in Southern India. Commonly known as “Cocoa Tree” or “Chocolate Tree”.

Uses : Seeds are useful for various purposes. Cocoa butter or Theobroma oil obtained from seeds, which contains 2.5% fully saturated glycerides. The powdered product of the partially defatted seeds contains 0.85 to 2.7% theobromine and 0.09 to 0.19% caffeine. For pharmaceutical purpose, cocoa butter is utilised as a suppository or as an emollient. Powdered cocoa is used for flavour or for preparation of cocoa syrup and used in confectionery. Cocoa butter obtained from seeds is also used in cosmetics and perfumery.

Thespesia Soland
Family : Malvaceae

This genus have about fifteen species and only one species is native to India. Which species are distributed in warmer region,

Plants are propagated bv seeds. These are woody perennial shrubs or trees.

Thespesia populnea Soland. :

This is tree upto 14 m tall, a medium sized, quick growing evergreen. The large bell shaped flowers are yellow in colour with reddish blotches at the base. Commonly known as Bhendi Tree.

Uses : The leaves, flower, fruit and bark of T. populnea are used to treat eczema, guinea worm, ring worm, psoriasis and scabies.

Thespesia lampens Datz. :

This is a dwarf straggling shrub upto 80 cm tall, flowers are solitary with free petals, yellow with a glossy brown centre about 8.5 cm in diameter.

Uses : The roots and fruits of this tree are used to treat syphilis and gonorrhoea.

Thevetia (L.) Juss. ex. Endl.
Family : Rutaceae (= Apocynaceae)

Thevetia peruviana Merr. (= T. nerifolia Juss.) :

This is small tree native to West India and is widely grown in India. Leaves are linear-lanceolate, acute at both ends upto 14 cm long, shining green. Flowers are large, bright yellow, white or pinkish, funnel shaped. Commonly known as "Yellow oleander".

Uses : The root of this plant are made into a paste and applied to tumours. Seeds are used to treat rheumatism, dropsy. Seeds possess abortifacient and purgative properties.

Thalaspi Linn.
Family : Cruciferae (= Brassicaceae)

Thlaspi bursa pastoris :

This is a half hardy perennial plant. Flowers are small, white, in racemes. Pods are orbicular, obovate or obcordate, laterally compressed.

Commonly known as Bastard Cress or "Besom Weed" or "Sphephord's Purse".

Uses : This plant is used in popular medicine as a vulnerary and astringent.

Thuja Linn.
Coniferae (= Cupressaceae = Pinaceae)

Under this genus there are five species. Which are aromatic, resinous, evergreen trees with scale-like leaves.

Thuja occidentalis Linn. :

This is a hardy evergreen coniferous tree upto-25 m of height. The leaves are small, scale-like. The needles are bright green above and yellow-green beneath. The cones are brownish yellow, upto 1-2 cm long. Seeds are n winged.

The species occurs in North-east America. Commonly known as "American Arborvitae".

Uses : Cones and leafy young twigs are useful. A decoction of fresh leaves is used to treat gout, fever and coughs. Grounded cones and fresh leaves are also used to treat rheumatism. Leaves are useful for treating scurvy, warts, cough and fever.

Different Other useful species are given below :

The leaves of T.orientalis L. "Chinese Arbor Vitae" yield an essential oil which is used as diuretic, antipyretic and tonic. The leaves of T. plicata Donn. "Nee's Arbor Vitae" also yield essential oil on steam distillation.

Thymus Linn.
Labiatae

This genus have about three hundred species and four species are distributed in India. Which are strongly aromatic perennial herbs or undershrubs.

Thymus linearis Benth. :

This is a hardy evergreen aromatic shrub, rarely found at Kaleech of Uttarkashi at 2800-3200 m elevation. Commonly known as "Thyme".

Uses : The crushed leaves are smelled in cold. The decoction of leaves is given for dysentery. The extract of leaves are taken in stomachache.

Thymus serpyllum Linn. :

It is an aromatic prostrate herb, with spreading habit and pungent taste. Flowers are pink, in globose terminal whorls.

The species is native to Europe. It also commonly occurs in open dry slopes all over H.P. between 2000-2500 m, and from Kashmir to Kumaon in the temperate zones. It is cultivated in England, France, Germany, Spain and Italy. There are several varieties in it like "albus" with white flower, "coccineus", carmine flowers; 'languinosus" are woolly leaved and "vulgaris" are lemon scented. Commonly known as "Wild Thyme".

Uses : The herb contains protein, fat, crude fibre, carbohydrate, minerals, vitamins A, B and C. The plant contain 2.5% volatile oil and main constituent is thymol.

This plant yields an essential oil with pleasant flavour and possesses stimulating and diuretic properties. An antiseptic is extracted from it. For treatment of nervous disorder this is effective. They are regarded as vermifuge. Infusion of plant is used in skin diseases. Powder of the flowers is mixed in gur and is given as a vermicide. The decoction of the plant is used as an efficacious home remedy in colds, cough, fever and stomach problem. It is used as tonic and spasmodic carminative. It is also used as tonic, anti-spasmodic, carminative. The leaves are good for kidney and eye and are regarded as blood purifiers.

The two most important species of Thymus are Thymus vulgaris L. and Thymus zygis L.

Thymus vulgaris Linn. :

This is a perennial subshrub and attain a maximum height of 60 cm with hairy branches. Leaves are aromatic and small. Flowers appear in clusters, tubular, lavender and small. Seeds are round and tiny.

It is native to Southern Europe and is widely cultivated in France, Germany and Greece. It is also grown in Nothern India upto 1000 m altitude. Commonly known as "Common Thyme".

Thymus zygis Linn. :

It is a smaller plants, flowers are borne on elongated inflorescence in clusters, white or purple in colour. This species is naturally distributed in Portugal, Spain and Balearic Islands and extensively cultivated in Spain. There are two varieties in it "gracilis" and "floribunda". Commonly known as "Linear Leaved Thyme".

Uses : These plants contain saponins, thiamine, ursolic acid, caffeic acid and resin. About 1 to 2.6% volatile oil can be obtained by distillation of flowering tops. Thymol for medicinal and other purposes is derived from thyme oil. Thymol is employed as an antiseptic on wounds. It is used for treating the fungus diseases of skin. Thyme was valued as an antiseptic and fumigating herb. A decoction of the plant is taken as stomachic, sudorific and remedy of ordinary coughs and whooping cough. Thyme oil possesses germicidal properties and is used in mouth wash, gargles, dentifrices, cough drops and various disinfectant preparation. Thyme is also utilised as culinary herb, it is used in food industry.

Tiarella Linn.
Family : Saxifragaceae

Tiarella cordifolia Linn. :

This is a hardy perennial herb upto 50 cm tall. Leaves are toothed, heart-shaped upto l2 cm in diameter. Flowers appear on a spike, white in colour. The species is native to Eastern North America. Commonly known as "Foam Flower".

Uses : Roots and plants in bloom are useful for medicinal purposes. The roots are useful for increasing urine flow and to loosen phlegm in the chest. A tea made from the plants also increase urine flow and help to pass kidney stone and other urinary ailments.

Tilia Linn.
Family : Tiliaceae

Tilia parviflora Ehrh. :

This is a hardy deciduous tree. The upper surface of the leaves are usually covered with honey dew. Flowers in little cymes, arising from the axils of the leaves of the current year. Flowers are protandrous. Fruit is a nut. Edosperm is very oily. Commonly known as "Lime Tree".

Uses : Flowers of these plants distilled in water act as anti-spasmodic and slightly sudorific. It is used in nervous affections such as cardialgia, nervous vomiting and hypochondria. Tea prepared from it is effective in cases of indigestion. Prolonged bath with infusion of flowers gives good effect in hysteria.

Tinospora Miers.
Family : Menispermaceae

Tinospora cordifolia (Willd.) Hook. f. & Th. :

This is a vigorous climber. Stems are succulent and produce aerial roots. Stems and branches speaked with white glands. Leaves are ovate or roundish upto 10 cm long. Flowers are very small, male and female flowers are separate. Female flowers are solitary, while male flowers are grouped in axils of bracts.

The species is distributed in the tropical parts of India. Commonly known as Gilo.

Uses : Dried stem and bark of these plants are medicinal. It is used as tonic, antiperiodic and aphrodisiac. The starch of roots and stems are nutritious and are used to cure diarrhoea.

Trachelospermum Lam.
Family : Apocynaceae (= Asclepiadaceae)

Trachelospermum lucidum (D. Don) K. Schum. :

This is an evergreen climbing woody species with fragrant flowers. The species is native to India and rarely found at Chiangseel between 1800-2500 m elevation.

Uses : The roots powder is given in vomiting and dysentery, in case of food poison.

Trianthema (Sauv.) Linn.
Family : Ficoideae (= Aizoaceae)

This genus have about twenty species which are diffuse and prostrate branched herbs.

Trianthema decandra Linn. :

This herbaceous plant has medicinal many properties.

Uses : It is laxative and useful in suppression of menses and in inflammation of testicles.

Trianthema portulacastrum Linn. :

This is a spreading, much branched, succulent herb. Leaves are broader towards tip. Flowers and fruits are minute and concealed in the base of the leaf-stalk. Seedf are black, covered with minute outgrowth, kidney shaped. The species is distributed throughout India, in cultivated fields and riverbeds. Commonly known as "Horse Parslane".

Uses : The leaves have medicinal properties which contain punarnavine alkaloid. It is usesd to promote urination and useful in dropsy and kidney disease.

Tribulus Linn.
Family : Zygophyllaceae

Tribulus terrestris Linn. :

This is a trailing and spreading herb, densely covered with minute hair. Leaves compound, in opposite pairs, leaflets 3-6 pair, upto 8 cm long. Flowers are usually silky, white or yellow, solitary, arises from the axils of leaves. Ovary bristly, style short and stout. Fruits are globose, spinous or tuberculate; consisting of fine hairy or nearly glabrous, often muriculate, woodi cocci, each with two pairs of hard sharp spines, one pair longer than the other. Fruit often cling to clothes and bodies of animals. Seeds are many in each of woodi cocci.

Plant is widely distributed in different parts of India upto 3500 m altitude. Commonly known as "Caltrope" or "Chota Gokeru".

Uses : Steroidal saponin and diosgenin is isolated from this plant. It is very rich in proteins and calcium. Dried fruit contain semi-drying oil, peroxides, diastase, traces of glucosides, resins, protein and large amount of inorganic matters. From the roots, stem and leaves, -sitosterol and stigmasterol were also isloated.

This plant is a most important ingredient of an Ayurvedic preparation. The drug is diuretic, tonic, aphrodisiac. The decoction of leaves is useful as a gargle for mouth trouble, painful gum and to reduce inflammation. The leaves increases the menstrual flow, cure gonorrhoea. The fruits are useful in urinary complaints, painful micturition and impotence. Fruits are also used to treat coughs, scabies and anexemia. The roots are said to be stomachic, appetiser, diuretic and carminative.

Trichodesma R. Br.
Family : Boraginaceae

Trichodesma Zeylanicum R. Br. :

This is a hardy perennial plant upto 70 cm tall. Leaves are with three oval shaped leaflets. Flowers are globular, rose-purple in colour.

It is native to-Europe and temperate Asia. It has been introduced and has run wild in the Nilgiris and Western Himalayas. Commonly known as "Red clover" or "Beebred".

Uses : Plant in bloom constitute the drug. Flowers are used as sedative and to treat coughs. The plant is used to cure liver ailments, digestive disorders and to increase appetite. An infusion of the plant is used to treat whooping cough, skin sores and ulcers.

Trigonella Linn.
Family : Leguminasae (= Fabaceae)

Trifolium pratense Linn. :

This is a hardy annual plant. Leaves are feather shaped. Flowers are whitish, in colour.

The species is native to Eastern Europe and Ethiopia. It grows wild in parts of North India. The plant is grown all over India for its green leaves and young pods.

Uses : Seeds are broadcasted uniformly in the flat beds. Light irrigation is required after sowing. Insects like aphids and leaf eating caterpillars and diseases like downy mildew, leaf spot, powdery mildew and rust cause damage to the crop which require proper control in time.

The seeds are used as condiment and dried leaves for flavouring. The vegetative parts are rich in vitamin A, B and iron. The mucilaginous seeds are used in medicine. It is emollient, resolvent and slightly astringent. The decoction of seeds is administered in drinks, rectal injections and lotions. The flour is used in poultice. The seed is also used to resolve inflammatory tumours, such as carbuncles, phlegmons and whitlows.

Trilisa Cass.
Family : Compositae (= Asteraceae)

Trilisa odoratissima Cass. :

This is a herbaceous perennial upto 1.30 cm tall. Stem grows from cluster of leaves. Leaves are pale green and large. Flowers are purple.

It is native to North America. Commonly known as “Carolina vanilla” or “Vanila Plant”.

Uses : Mature leaves are made into tonic to treat malaria. Leaves are rich in coumarin, a vanilla-like flavouring and used to flavour tobacco.

Trillium Linn.
Family : Liliaceae (= Trilliaceae)

This genus have about thirty spec ies and two are native to India. Plants are erect, unbranched herbs with short rootstocks.

Trillium erectum Linn. :

This is a tuberous rooted perennial which can grow upto 70 cm tall. Three broad leaves are arranged in a whorl at the top of the stem. From the centre of the whorl flower arises which is purplish-maroon in colour. The plant is native to North America. Commonly known as "Berth Root" or Purple Trilleum.

Uses : Roots are useful for medicinal purpose and it is said to be aphrodisiac. Roots are used to treat heart palpiation, haemorrhages, skin infections and also to ease pains of child birth.

Trillium govanianum Woll. :

It is a bulbous rooted perennial plant.

Occasionally occurring in Alpine regions of Central Himalaya betwen 2600-3800 m elevation. Commonly known as "Satwa".

Uses : Roots are used for medicinal purpose against dysentery.

Tristania R. Br.
Family : Myrtaceae

This genus have about fifty species of trees or tall shrubs, native to Malaya and Australia.

Flowers are yellow or white and produced in axillary cymes.

Uses : The leaves and bark of T.conferta R. Br. "Brush Box" contain triterpenes. The bark is used for tanning. The wood is used for ship building. T.rnerguiensis Griff., timber is dark and reddish and suitable for house building. This species occurs in Andaman Islands.

Tropaeolum Linn.
Family : Tropaeolaceae

Tropaeolum majus Linn. :

It can be trained as a climber or a dwarf compact bush. The tall climbing type trail or climb to a height of 1.85 m to 2.0 m. While dwarf compact type is upto 30 cm tall. It has wiry stems, large attractive rounded leaved and spurred flowers of 6 cm across. The flowers may be single, semi-double or double and colours are yellow, orange, scarlet or mahogany.

The species is native to Mexico and South America. Commonly known as "Nasturtium".

Uses : This plant is regarded as tonic and stimulant. It is recommended in scurvy, scrofula, cachexy and serous infiltrations. Dried and ripe fruits of these plants are purgative.

Tropidia Bl. (Linn.)
Family : Orchidaceae

***Tropidia curculigoides* Linn. :**

This is a terrestrial orchid with 85 cm long stems, about 6 mm thick near its base, often branched. Leaves are more than six, scattered, upto 26 cm long and 6 cm broad, elliptic lanceolate and plaited. Inflorescences are terminal, leaf opposed, few flowered, upto 2 cm long. Flowers are 9 mm across, white faintly tinged with yellow.

The species is occurring in Assam, Manipur and Sikkim.

Uses : Roots of these plants are used to treat malaria and diarrhoea.

Tsuga Carr.
Family : Pinaceae

***Tsuga canadensis* (Linn.) :**

This is a hardy evergreen coniferous tree with a handsome and elegant habit of growth which attain a height of 35 m. The bark is dark reddish brown and deeply furrowed. The needles are 3 cm long, flattened, tapering to a point. The cones are upto 3 cm long.

It is native to Eastern-North America. Commonly known as "Canadian or Common Hemlock".

Uses : Resin, bark and needles of these plants are useful for medicinal purposes. A tea prepared from the needles are used to treat cold, coughs, flue and make a sick person sweat. Resin has been used to form skin blisters. Inner bark is boiled and pounded and is applied on sores and swellings.

Tussilago Linn.
Family : Compositae (= Asteraceae)

***Tussilago farfara* Linn. :**

The flowers appear in this plant in the spring before the leaves. The flower head is monoecious.

The species is native to Europe.

The plant multiplies and hibernates by underground offshoots. The flowers and leaves of this plant contain mucilage. Commonly known as "Allhouse or Cotts Foot".

Uses : The infusion of leaves is employed in cases of bronchitis. Leaves and flowers of this plant is a component in many herbal for coughs.

Tylophora R. Br.
Family : Asclepiadaceae

Tylophora indica (Burm. f.) Merr. :

This is a twining plant with fleshy roots. Leaves are upto 12 cm long, arranged in opposite pairs. Flowers are large, dull yellow.

It is native topalaeotropical region. Distributed in the plains of Southern, Central and eastern India.

Uses : Dried root of the plant is useful in the treatment of dysentery, asthma and bronchitis.

Typha Linn.
Family : Typhaceae

Typha latifolia Linn. :

This is a hardy aquatic perennial monocotyledonous plant upto 2.5 m tall. The lower part of the stem is a thick rhizome, the upper part projects high out of water. Leaves arise from the base of the stem and grow to a height of 1.25 m. The flowers are clustered in a dense spike, brown in colour.

It is native to temperate region of Britain. Commonly known as "Cat Tail".

Uses : Roots and seeds are useful. Roasted roots and seeds are eaten. Crushed roots are applied on burns and wounds. The roots are used to treat worm, diarrhoea and gonorrhoea.

Typhonium Schott.
Araceae

Typhonium diversifolium Wall. ex Schott. :

This is a small rhizomatous herb and rhizomes are globose. Spathe is upto 20 cm long, erect, green and stripped purple. The species is native to Asia and Indomalayan region.

It is Commonry occurring in shady field borders in the tribal area of Shimla district of H.P. between 2200-3000 m elevation. Commonly known as "Indian Kale" or "Chamush".

Uses : Paste of tubers are effective remedy for chronic dysentery, gastro intestinal problems and piles.

Ulmus Linn.
Family : Ulmaceae

Under this genus there are about fortyfive species which are deciduous trees with stipulate, alternate leaves.

Ulmus americana Linn. :

This is a hardy deciduous tree of graceful habit upto 45 m tall with spreading crown. The leaves are roughened above and fuzzy below, upto 15 cm long. Leaves are asymmetrical, one side larger than the other. Flowers are borne in hanging clusters. Growth of the plant is sympodial, the terminal being suppressed. Fruits are with notched margins, hairly, round, about 1.5 cm long.

It is native to Central North America. Commonly known as "White Elm" or "Ameri can White Elmi".

Uses : Bark is useful for medicinal purpose. Poultice made from bark is used for abrasions and skin inflammations. The inner bark is used as a tonic and to increase urine flow. An infusion of the bark is used to treat kidney ailments and dysentery. Timber of this tree is fine, hard grain and brown in colour, used for making coffins, carts and wagons and furniture.

Ulmus fulva Michx. (= U.rubra Muhl.) :

This is a hardy deciduous tree of 25 rn tall. Leaves are large oval, rough above and fuzzy below. The inner bark is sticky and aromatic.

The species is native to Eastern Canada and Eastern and Central America. Commonly known as "Sleppery Elm" or "Moose Elm".

Uses : Dried inner bark is useful, which contains mucilage. It has demulcent and emollient properties. A tea made from bark is laxative. Poultice of the bark is used to soften the skin and protect it from chapping, it heals the skin wounds, and it is also applied on burns. The infusion is given to relieve gastric and intestinal inflammation. Powder made into lozenges for throat irritation. The bark is also used to treat fever, diarrhoea and respiratory infections.

Umbellularia Nutt.
Family : Lauraceae

Umbellularia californica Nutt. (= Oreodaphne californica Linn.) :

This is a slightly tender evergreen flowering tree upto 30 m tall. Leaves are shiny and leathery upto 12.8 cm long, pungently aromatic. Flowers appear in clusters, yellowish green in colour. Fruits are purple, pear shaped upto 2.8 cm long.

The plant is native to California and Oregon. Commonly known as "California olive" or "Balsam of Heaven".

Uses : Leaves are useful for medicinal "purpose. A tea prepared from leaves is used to treat stomach trouble and headache. Leaves are also put into hot baths for treatment of rheumatism.

Uncaria Schreb.
Family naucleaceae (= Rubiaceae)

Uncaria gambier (Hunter) Roxb. :

This is a climbing shrub which climbs by hooks, usually with angular branchlets. Seeds are minute, usually flattened with endosperm.

The species is native to Malaysia and Indonesia, cultivated in India. Commonly known as "Gambijer" or "Pale Catechu".

Uses : Extract of leaves and twigs have medicinal value which contains 8 to 22% catechins; 25 to 60% catechutannic acid; quercetin, and catechu red.

This has astringent properties and is used to treat diarrhoea and sore throat. It is a valuable source of tan. The resinous substance, obtained from leaves and young branchlets, used as a dye-stuff, musticatory and as a tan.

Urginea Steinh.
Family : Liliaceae

In this genus there are about eighty species and six species are native to India. Which are all bulbous herbns with recemed flowers.

Urginea indica (Roxb.) Kunth :

This is a bulbous plant in which basal leaves are flat. Flowering scapes are upto 50 cm tall and flowers are light brown.

The species is native to India and Burma and distributed in Southern and Eastern India. Commonly known as "Indian Squill" or "Jangli Piaz".

Uses : Bulbs are used in heart ailments, cough, bronchitis and for promoting urination.

Urginea maritima (Linn.) Baker :

The Plants are upto 95, cm tall. Bulbs are large and flowers are whitish in colour.

It is distributed in Mediterranean Europe and Algeria. Commonly known as "Red Squill".

Uses : It is a source of "Squill" of commerce. The dried bulb possess the medicinal properties. It contains glycosides related to digitalis, scillaroside A and scillaroside B; and also a phytosterol and calcium oxalate. It has diuretic, expectorant and emetic properties. It was also regarded as cardiotonic. Primarily, it is employed as rat poison.

Urtica Linn.
Family : Urticaceae

This genus belong to annual and perennial herbs, with stinging hairs, consisting of about fifty two species, and four species are native to India.

Urtica dioica Linn. :

This is a herb with opposite leaves covered with stinging hairs. In this species, there is panicle, but each sex is confined to its own plant. The stamens are bent down inwards in the bud and when ripe spring violently upwards and bend out of the flowers.

Commonly found between 560-2200 m at Arakot of Uttaranchal. Commonly called "Larger Common Nettle" or Khusha.

The paste of root is used as astringent. The extract of root is given tor easy delivery. Young tops can be eaten like spinach. Useful fibre is available from stems. The root extract is dropped in hollow cavity of teeth in teethache.

Different species of Urtica like 17. Pilulifera L. (Roman Nettle)and U. ureus L (Small Nettle) are also used for medicinal purposes.

The plant has astringent properties and has been recommended for haemoptysis, methorrhagia and haematemesis. Roots of this plant is used for treatment of jaundice.

Vaccinium Linn.
Ericaceae

Vaccinium arboreum Marsh. :

This is a tree growing of about 10 m in height. Leaves are leathery, thick and shiny. Flowers are white in colour. Berries are black.

The species is native to America. Commonly known as "Sparkle Berry" or "Farkle Berry".

Uses : Mature leaves, root barks and berries have medicinal properties. The decoctions of leaves and barks are used to treat sore throat and diarrhoea. Drink prepared from fruit is used in case of chronic dysentery.

Vaccinium myrtillus L. :

This is a hardy deciduous flowering and berry bearing shrub upto 45 cm tall. Flowers are pale pink in colour. Fruits are acidic.

The species is native to Britain and North Asia and common in the hilly districts. Commonly known as "Belberry or Wortle Berry".

Uses : The fruits are sedative and astringent and its preparations are used in the treatment for diarrhoea and dysentery.

Valeriana Linn.
Family : Valerianaceae

Under this genus there are about two hundred and fifty species and twelve are native to India. Which are perennial herbs. Rootstocks are strong smelling.

Valeriana officinalis Linn. :

This is a perennial glabrous herb, attain a height upto 1.5 m. Stem solitary, erect, furrowed and hollow. Leaves are opposite, pinate, lower leaves with long petioles and upper small. Flowers appear in cymose panicle, corymb, white or dull white in colour, fragrant, small, calyx many toothed and corolla 5-lobed. Fruit oblong ovate, small, smooth, without hairs, one seeded. Rootstock is thicker, short, with many fasciculate rootlets.

This species is native to Europe and Kashmir in India at 3000 m altitude. The plant is cultivated in Belgium, France, Netherlands and East European countries and it is naturalised in USA.

Commonly known as "Common Vallerian" or "Garden Heliotrope".

Uses : The root contains epoxy-iridoid esters called valepotriates which include valtrate didrovaltrate, acevaltrate, isovaltrate and isovaleroxy-hydroxy didrovaltrate. It also contains chatinine and valerine alkaloids and volatile oil.

The drug is said to be sedative,hypotensine, antispasmodic and stomachic. The juice of fresh rhizomes is used in treatment of hysterical fits and other nervous disorders and flatulence. The root preparation is also know as tonic and stimulant. The fresh juice is administered in case of insomnia and in certain cardiac prepartations. The essential oil of the roots is used as a flavour in food products and beverages..

Valeriana wallichii DC. (= V. jatamansi Jones) :

This is another species of similar properties as that of V. officinalis Linn. It is a perennial herb upto 16 cm tall and stem tuffed. Flowers are white tinged with pink. Rootstock thick and horizontal, aromatic and nodular.

This species is native to Himalayan region of India, Nepal and South WesternChina. It is distributed abundantly in India in Himalayas from Kashmir to Assam. Rarely occurring at Chyangseel of Uttar Kashi between 1500-2800 m elevation.

Uses : Roots of these plants are useful in diseases of eye, blood, liver and spleen enlargement. They are also useful for clearing voice. The crushed leaves are rubbed on the forehead in extreme headache. The root preparations are used as cosmetic and hair oil.

Valeriana hardwichii Wall. :

This plant is a large herb, dichotomously branched. Flowers are white in colour.

The species is found in Uttara Khand Himalaya from 2500 m to 3200 m.

Uses : It has same medicinal properties as that of above two species.

Vanda R. Br.
Family : Orchidaceae

This genus have about sixty species of which ten species are native to India. These are epiphytic orchids with leafy stems.

Vanda cristata Linn. :

This is an epiphytic orchid with erect and stout stem of 16 cm tall. Leaves are narrowly oblong, bilobed at the apex upto 15 cm long and 1.5 cm broad, horizontally arranged and leathery. Inflorescence is erect, shorter than leaves, 4 to 7 flowered. Flowers are 6 cm across, yellowish or olive green with purplish blotches.

It is distributed in Kumaon Hills, Dehradun Mussorie and Sikkim between 1200 to 1500 m altitude.

Uses : Leaves of these plants are used as expectorant.

Vanda tessellata HK. f :

This is an epiphytic herb with scandent stem of 70 cm long. Leaves are keeled, linear-oblong, leathery, recurved, channeled to 22.6 cm long and 2.6 cm broad. Inflorescence suberect to 56 cm long, 6 to 8 flowered. Flowers are 6 cm across, sepals and petals tessellated, yellow with brown lines and white margin lip bluish dotted with purple.

It is distributed in West Bengal, Bihar, Orissa, South India, Uttar Pradesh, Madhya Pradesh, Assam and Gujarat.

Uses : Leaves of these plants are used to treat ear infections, fractures and rheumatism.

Vanda parviflora Rchb. f.

This is an epiphytic orchid. Stem is 16 cm long, densely leafy. Leaves are leathery, upto 22 cm long, unequally bilobed at apex/In florescence with stout peduncle, erect, to 22 cm long, rather loosely 6-14 flowered. Flowers are upto 3 cm in diameter, flesh coloured flushed with greenish tinge and the lip is white.

It is distributed in India, Ceylon and Burma.

Uses : Leaves and roots of these plants are medicinal. These are used to treat nervous disorders, rheumatism ad scorpion stings.

Vanilla Mill.
Family : Orchidaceae

Vanilla fragraus SW. (= V. planifolia Andrews) :

This is tropical climbing orchid. A herbaceous, fleshy, perennial vine which climb up by means of aerial adventitious roots. The leaves are upto 26 cm long and 9 cm broad, fleshy, subsessile, alternate, oblong and elliptic in shape. Flowers are borne in axillary or terminal racemose inflorescences, usually 7 to 32 per inflorescence. Flowers are bisexual, large and greenish yellow with short stalks, sweet smelling. Fruits are capsules, known as beans in trade which are pendulous, cylindrical, upto 26 cm long and 1.6 cm broad. They contain hundreds of minute seed.

This species is distributed in Northern South America, Central America, West Indies and South-East Mexico. It is commercially cultivated in Madagascar, Indonesia, Comoro Islands, Mexico and India. In India, Vanilla can be commercially cultivated in Kerala, Karnataka, Tamilnadu, Lakshadweep, Andaman and Nicobar Islands, and certain parts of North Eastern India, Kerala, Karnataka, Tamil Nadu and Lakshadweep.

Uses : The Vanilla beans contain protein, fatty oil, volatile oil, nitrogen free extract, carbohydrate, fibre, vanillin, resins and minerals.

It is one of the costliest spice crop. This is a stimulant and aphrodisiac. In cases of nervous fevers it is used as stimulant. Different processed products are made out of vanilla beans. These are vanilla extract or essence, vanilla tincture, vanilla vanillin, vanilla oleoresin, vanilla powder and vanilla absolute. The vanilla flavour is primarily used in ice creams, chocolates, bakery products, pudding, liquors, perfumes and pharmaceuticals.

Ventilego Gaertn.
Family : Rhamnaceae

Ventilego calyculata Tulasne. :

This is a large perennial climber, woody in nature. Leaves are oblong-elliptic. Flowers are small greenish, numerous. The bark is grey coloured.

It distributed in the lower valleys of Garhwal. Commonly known as "Lal Beli".

Uses : The root bark of this plant yield dye and is used for dyeing cotton fabrics.

Ventilego denticulata Willd. :

This is a scandent shrub. Flowers are small in axillary and terminal panicles. Fruits are with wing on upper ends, formed from style after fertilisation.

It is native to India, occuring at Chindi of Madhya Pradesh commonly known as Raktokai.

Uses : The decoction of root bark (about 15 ml) is taken 3 times a day for a fortnight in case of debility.

Veratrum Linn.
Family : Liliaceae

Veratrum album Linn. :

A hardy herbaceous perennial plant which is 1.30 m tall. The stem is cylindrical, erect. Leaves are 12 to 14 in numbers, much ribbed, oblong upto 32 cm long and 1& cm wide, yellowish green in colour. Flowers are produced in dense terminal raceme of 65 cm tall, yellowish-white tinged with green. Bulbs are blackish or yellowish externally, whitish within, bitter and acrid in taste, and odour is onion like, having fibrous roots.

It is native to Southern Europe and Japan. Commonly known as white Hellebore.

Uses : For medicinal purpose generally the bulbs are collected from the wild. It contains protoveratrine A and protoveratrine B alkaloids. In the bulbs germerine, veratrosine, veratrarine and jevine are present.

It was used to treat cholera and also applied in the form of ointments externally in gout, skin affletions, herpes and scabies. It is known as emetic, purgative and diaphoretic. This plant is mainly used for preparation of veterinary medicine. A tincture of rhizomes lowers blood pressure promptly, used for hypertension.

Veratrum californicum E. Durand. :

This is a hardy herbaceous perennial upto 2.5 m tall. The stems are leafy, unbranched and arise from thick rootstocks. Leaves are ribbed, alternate, elliptical, with parallel veining. Flowers are greenish wite in colour. Roots are poisonous.

The species is native to California. Commonly known as "False Helleborie".

Uses : A decoction of roots is used for preventing conception. For killing head lice, the water in which the roots are boiled is used.

Veratrum viride Ait. :

The species of this plant is erect of about upte 2.5 tall. Leaves are ovate with parallel venation. Flowers are yellowish green.

The species is native to North America.

Uses : Roots and bulbs are used for treatment of herpes, to reduce blood pressure, heart disease, asthma, rheumatism, gout and indigestion.

Verbascum Linn.
Family : Scrophulariaceae

Under this genus there are about three hundred species and six species are native to India. Which are erect, tomentose or woolly tall herbs.

***Verbascum nigrum* Linn. :**

This is a hardy perennial herb, grows upto 1.5 height. Tap roots are stout. Inflorescence is primarily racemose, but lateral flowers often replaced by condensed dichasia. Flowers are yellow.

The species is native to Britain. Commonly known as "Dark Mullein".

Uses : Leaves have mucilaginous principles and in decoction they soothe stomachache. It is also used in tea to cure cough. The plant has demulcent properties. The flowers contain yellowish volatile oil.

***Verbascum thapsus* Linn. :**

This is a hardy biennial plant, attain a height of 2.5 m. Leaves are hairy, lance-shaped and oblong. Flowers are arranged densely along a club like spike, yellow in colour.

The species is native to Europe and Asia. Sporadically found in H.P. betwen 1500-3500 m and also common at Thunara of Utter Kashi between <800-2gOO m elevations.

Commonly known as "Common Mullein" or "Adam's Flannd".

Uses : Mature leaves, flowers and roots are used as medicine. Leaves are smoked for asthma and sore throat. Leaves and flowers are considered soothing for mucous membrane. Small pills of crushed leaves are given to cure constipation. Infusion of plant is orally given as a cure to snakebite. Decoction of its leaves with onion forms a remedy for stomach complaints. Tea prepared from leaves are used to treat colds and dysentery. The oil extracted from flower is used to treat earache.

Verbena Linn
Family : Verbenaceae

This genus is of pubescent herbs or sub-shrubs, consisting of two hundred and sixty species, out of which only one species is wild in India.

Verbena hastata Linn. :

This is a half hardy perennial plant and attain a height of 130 cm. Leaves are with serrated edges and three lobed, where two sharp lobes arises from the base. Flowers appear in numerous terminal spikes, blue in colour. The plant is native to America. Commonly known as "Blue Vervain".

Uses : To clear up cloudy urine, tea prepared from roots is administered. A ground parts and roots are used as medicine. Verbena is used "Induce vomiting, to clear the respiratory tracts of mucous, to increase perspiration. Tea made from boiled leaves is used as stomachic.

Verbena officinalis Linn. :

It is distributed from Kashmir to Bhutan and also in the plains of W.B. Jharkhand, Maharashtra and Andhra Pradesh and Commonly Known as "Vervain", "Ash Throat" or "Berbine".

Uses : It has been said to increase mother's milk and prevent miscarriage. Formerly this species was of great repute as a remedy in eye disease. It is stomachic and anti-spasmodic. The plant is regarded as an astringent, resolvent, diaphoretic, antispasmodic and vulverary. It is used in the treatment for intermittent fevers, cellular effusions, jaundice, ulcers, ophthalmia and pleurisy.

Verbesina Linn.
Family : Compositae (= Asteraceae)

Verbesina helianthoides Michx. :

This is a perennial plant of 95 cm tall. Stems are hairy. Leaves are rough above and hairy beneath. Flowers are yellow in colour.

The species is native to warm parts of America. Commonly known as "Crown Beard".

Uses : Plants in bloom and roots are useful. A tea prepared from the plant is used to treat stomach trouble, haemorrhoids. An infusion of roots is used for retention of water and bladder inflammation and also to purify blood.

Veronica Linn.
Family : Scrophulariaceae

Veronica officinalis Linn. :

Veronica is a hardy herbaceous perennial.

The species is native to Europe and Asia. Commonly known as "Speedwell".

Uses : Plant plant is used as a light tonic in the convalescence of certain fevers, asthma and bronchitis. It acts as a stimulant, provoking urination and facilitating expectoration. It is also used for the treatment of chest diseases, such as chronic pulmonary catarrh, phthisis and bronchitis.

Veronicastrum Moench
Family : Scrophulariaceae

Veronicastrum virginicum (Linn.) Farw. :

This is a perennial plant of 2.5 rn tall. The plant is unbranched and erect in habit. Leaves are lance-shaped, arranged in whorls on the nodes of the stem. Flowers appear in dense spikes on the top of the plant, white or blue in colour.

The species is native to North America.

Uses : Rhizomes and roots of these plants are useful as a mild laxative. It also induces vomiting.

Viburnum Linn.
Family : Caprifoliaceae

Under this genus there are about two hundred species of shrubs and trees. Seventeen speices are native to India.

Viburnum nudum Linn. :

This is a perennial tall shrub upto 8 m tall. Leaves are 12.6 cm long. Flowers are white in colour.

The species is native to North America. Commonly known as "Passumhaw Virburnum".

Uses : Stem, bark and roots of these plants are medicinal. The bark , prepartion is used as tonic and employed to treat uterine disorder, malaria and promoting urine flow.

Viburnum opulus Linn. :

This is a shrub upto 5 m height. Branches are smooth, light grey in colour. Leaves are three lobed, toothed, hairy beneath upto 10 cm long. Flowers are white in colour. Fruits are 1.6 cm across, scarlet in colour.

Commonly known as "Guelder Rose" or "Cramp Bark".

Uses : Plant bark and ripe berries of this species are medicinal. It has tonic and diuretic properties. Bark preparation is used to treat uterine infection, septic poisoning during child birth and also to promote urine flow. To treat scurvy, berries are used, which are rich in vitamin C.

Vicia Linn.
Family : Leguminasae (= Fabaceae)

Vicia Faba Linn. :

This is an annual leafy tendrilous climber, vigorous and erect in habit upto 1.5 m tall. Leaflets are in 1 to 3 pairs, alternately arranged on the stem. The terminal leaflet is a tendril. Flowers are white with purple tinge. Beans are flat, variously coloured, upto 4 cm long.

The species is native to North Africa and South-West Asia. It is cultivated in gardens for beans which are edible. Commonly known as "Broad Bean".

Uses : Roots and plants are useful. The poultice of the plant is applied to skin abrasions. Plants and roots are laxative and also induce vomiting.

Victoria Schomb (Lindl.)
Family : Euryalaceae (= Nymphaeaceae)

Victoria amazonica (Poepp.) SW. :

This is a tray like circular floating leaves, 2.1 to 2.6 m in diameter. The flower stalks are about 3 to 3.6 m long. Flowers are borne on thick prickly stalks above the surface of the water. They are about 30-46 cm in diameter, fragrant.

In India, Victoria was first introduced in the Indian Botanic Garden, Howrah. It is now widely cultivated in the lakes there. It is native to tropical America. Commonly known as "Queen Victoria water Lily" or "Giant Water Platter".

Uses : Acclaimed as one of the wonders of the plant world V. amazonica, attracts a large number of visitors to the garden where it is grown. Apart from the ornamental and decorative value of the plants, seeds are also edible. Excellent pastry is made from the floury starch of seeds.

Viola Linn.
Family : Violaceae

Under this genus there are nearly five hundred species of which fifteen are distributed in India. These are perennial herbs, rarely shrubby below.

***Viola biflora* Linn. :**

This is a small glabrous, erect annual herb with Twin Flowered yellow flowers. It is found in nature in the marshy slopes at Chandrabadni, Bhilagana valley and Kedarnath of Garhwal hills, India, upto 3,600 m altitude.

Uses : The leaves are used as laxative and the roots as emetic.

***Viola canadensis* Linn. :**

This is a hardy herbaceous plant upto 32 cm tall. The leaves have deep indentation. Flowers are white, yellow or violet in colour.

The species is native to North America. Commonly known as "Canada Violet".

Uses : Entire plants and roots are medicinal. Tops and roots are laxative and used to induce vomiting. The poultice of the plant is applied to skin abrasions.

***Viola canescens* Wall. :**

Uses : It is a perennial herb of temperate zone and found wild in Himalayan Hills, Kashmir, Himachal Pradesh, Kumaon hills and Darjeeling. Plants are valued for their medicinal values and the drugs obtained from it are used for treatment of several ailments.

***Viola pedata* Linn. :**

This is a dwart herbaceous plant with violet flowers.

It is found in nature in Garhwal Himalaya.

Uses : The leaves and flowers contain fragrant essential oil. The plant is used in the treatment of lung trouble and blood disorder.

***Viola tricolour* Linn. :**

This is one of the parents of pansy. A small herbaceous plant, produce yellow, purple or white flowers.

It is native to Europe, Asia and America.

Uses : Infusions of flowers and leaves are prescribed to children in cases of cutaneous eruptions and diarrhoea. The roots are used in preparing depurative teas. Commonly known as "Beady's Eye" or "Heart's Ease".

Viscum Linn.
Family : Loranthaceae (= Viscaceae)

Viscum album Linn. :

This is an evergreen hardy semi-parasitical and sacred plant. The plant grows as parasite on apple, hawthorn, oak etc. and draws nourishment from its host by suckers. Male and female flowers are borne on separate plants. Flowers are green in colour. Flowers secrete honey and are visited by flies. The layers of viscia prevents the birds to eat the berry from swallowing the seed, which it scrapes off to bill on to a branch, where it adheres and germinates.

The species is native to Britain. Commonly known as "Misttetoe".

Uses : Most suitable trees for mistletoe culture are apple, hawthorn, poplar, mountain ash, cedar, oak etc. For growing these plants a notch is cut in the bark of the above plants on underside of branches and the ripe berry of mistletoe is pressed. Young plants grow slowly for two to three years.

All parts of plant has diuretic properties but it should be used cautiously.

Vitex Linn.
Family : Verbenaceae

This is a deciduous shrub of 4 m tall and aromatic. Flowers are lavender blue. There is a variety "incisa" in this species, which is with deeply toothed leaves.

It is native to China and India. Commonly distributed at Ishyalee of Uttar Kashi betwen 600-1600 m. This species is also seen throughout the greater part of India and often gown for reclamation of forest land. Commonly known as "Chinese Chaste Tree" or "Sanboa".

Uses : The leaves are mixed with cowdung and are massaged in backache. Leaves are also used to treat headache.

Withania Pauquy
Family : Solanaceae

Withania somnifera Dunal. :

This is native to Mediterranean region. This is a medium sized undershrub attain an ultimate height of~T7& m. Stems and branches are covered with minute star shaped hairs. Leaves are ovate to ItLcm long, hairy. Flowers are greenish yellow to l-5cm long, borne in axillary clusters. Fruits are smooth, red when ripe, globose, enclosd in the inflated and membranous calyx. Several types of alkaloids are found in this plant, out of which withamine and somniferine are important. It is found growing naturally in forests particularly in drier regions of India. This is considered as an important medicinal cash crop in Mandsaur district of Madhya Pradesh, adjoining areas in Rajasthan villages and Garhwal hills.

Commonly known as "Cluster Winter Cherry" or "Ashwagandha".

Uses : The drug consists of dried root of the plants and are used in Ayurvedic and Unani preparations. The root is used as aphrodisiacs, tonic and diuretic. It is useful in sexual and general weakness and rheumatism. It promotes urination, acts as a narcotic, and removes functional obstruction of body. The root powder is applied locally on ulcers and inflammation and on tubercular glands. Warm leaves are sometimes used for providing comfort during eye diseases. Fruits are diuretic in nature.

Withania coagulans Dunal :

This plant species is found in North-Western India.

Uses : The fruits of these plants are administered in digestive and liver complaints. It is also used to prepare cheese.

Woodfordia Salisb.
Family : Lythraceae

Woodfordia fruticosa Kurz. :

This is shrub with spreading branches. Flowers are red, appear in axillary cymes. Flowers produce red dye and used for dyeing silk.

It is native to Asia and Africa. Found at Rater of Chhindwara District, M.P. It is also commonly found in lower valleys of Garhwal Himalaya and different parts of India. It is a common ornamental shrub, known as "Fulse Flower" or Dhawai or "Dhaula".

Uses : The dried flowers are astringent and used twice daily for one week in dysentery. The dried flower is useful in affections of mucous membrane of bilious complaint. The leaf paste is used in skin diseases. The leaf juice is applied three times a day for one week in conjunctivitis.

Wrightia R. Br.
Family : (Apocynaceae)

This genus have twenty five species of Palaeotropical region. Five species are native to India.

Wrightia tinctoria R. Br. :

These are lacticiferous shrubs or trees, occasionally climbers. Plants are deciduous or evergreen.

It occurs throughout India in dry deciduous forests. Commonly known as "Pala Indigo Plant".

Uses : A decoction of leaves and bark of this plant is taken as stomachic. Seeds possess aphrodisiac properties ; the bark and seeds are used in flatulence and bilious affections. The juice of the fresh unripe fruits contain proteolytic-enzyme and used for coagulating milk. Flowers and fruits of this species yield a blue dye.

Wrightia tomentosa (Roxb.) Rhem. and Schult. :

This species is native to Bihar, Assam, Punjab, Rajasthan and W. Peninsula.

Uses : This plant is used for relief from toothache and fever. The milky juice of this species is used to stop haemorrhage. The leaves and tender flowers are edible. The seeds and roots yield a yellow dye. Stems and root bark jardedas an antidote to snake-bite and sting of scorpions.

Xanthium Linn.
Family : Compositae (= Asteraceal)

Xanthium strumarium Linn. :

This is a herbaceous annual of about 150 m tall and the plant is covered with hair. Flowers are inconspicuous, greenish in colour. Flowers appear in unisexual heads, single or in axillary cymes. Fruits are hard, woody, which is covered with hooks and well suited to animal distribution. It has been so widely distributed by man, that it is hard to discover its native place, a cosmopolitan specis.

It is found throughout the warmer regions of India as weed. Commonly known as "Broad Burweed" or "Common Cocklebur".

Uses : Mature leaves possess the medicinal properties, which is used to treat skin and bladder infections, to stop bleeding from cuts and abrasions, neck gland tuberculosis and herpes.

Xanthorhiza Marshall
Family : Ranunculaceae

Xanthorhiza simplicissima Marsh. (= X. Apiifolia Linn.) :

This is a hardy deciduous flowering shrub of 65 cm tall. The wood of this shrub is bright yellow. Leaves are compound, five lance-shaped to oval leaflets with toothed margin. Flowers appear on spikes, small and purple in colour.

It is native to Eastern America. Commonly known as "Yellow Root".

Uses : Roots of this plant constitute the drug which is considered a tonic. A tea made from root is used to treat stomach and throat disorder. A decoction of root is used to ease child birth of mother. Boiled roots cure colds, sore mouth and jaundice. Root is also used to treat dyspepsia.

Zanthoxylum Linn.
(Xanthoxylum J.F. Gmel.)
Family : Rutaceae

Under this genus there are about two hundred species of scandent or erect shrubs of trees mainly pantropical. About ten species are found in India.

Zanthoxylum armatum Linn. :

This is a hardy deciduous prickly shrub, commonly found at Saran of Uttar Kashi between 900-1600 m elevation. Commonly known as Prickly Ashok Toothache Tree.

Uses : The stem is used as tooth brush and mouth purifiers. The leaves and fruits are chewed in the teeth enamel disease.

Zanthoxylum clava-herculis Linn. :

This is prickly shrub of 2.5 m tall. Leaves are compound with 6 to 12 leaflets, arranged alternately. Flowers are small, inconspicuous and greenish white in colour. Fruit capsules are with thorns, aromatic, round and reddish in colour. Commonly known as "Hercules Club" or "Prickly Yellow Wood".

Uses : Berries, bark, wood and roots are useful. To treat various kinds of itches, a lotion made by boiling inner bark in water is applied. Berries are considered stimulant, tonic, antirheumatic and also to treat colic, gas, muscle spasms and chest ailments. A decoction of bark is useful for treating gonorrhoea. To increase perspiration a decoction of boiled roots is prescribed. The wood of this plant is said to relieve toothache.

Plant other species like Z. fraxineum Willd. is used in America as a remedy of toothache; and Z.piperitum DC. "Japan Pepper" yields fruit used as a condiment.

Zehneria Linn.
Family : Cucurbitaceae

Zehneria umbellata Thwaites :

It is a tendril climber with elliptic fruits, red on ripening. Roots are perennial with tip portion of the root fibrous, swollen, forming spindle shaped tubers.

The species is native to Palaeotropical region and distributed in different parts of H.P. between 1600-3000 m elevation. Commonly known as Mokri, Gawal Kakru.

Uses : The cleaned, dried and powdered roots are given in the dose of 2 gm twice a day with milk in seminal debility and spermatorrhoea. Ripe fruits are considered as stomachic and tonic.

Zingiber Boehm.
Family : Zingiberaceae

Zingiber officinale Rosc. :

This is a herbaceous plant with underground stem. These herbs possess tuberous, horizontal and aromatic rootstocks.

It occurs naturally in Pacific Islands and widely cultivated in different countries.'India is the largest producer of ginger and other countries where it is cultivated are Australia, Taiwan, Japan, China, Nigeria and Jamaica. In India, ginger is cultivated in Kerala, Meghalaya, Orissa, Sikkim, Andhra Pradesh, Madhya Pradesh, Uttar Pradesh, West Bengal and Mizoram. Commonly known as "Ginger" or "East Indian Ginger".

Uses : Dry ginger contains protein, fat, fibre, carbohydrate, minerals, vitamins A, B and C. On steam distillation it yields 1.0 to 3.0% aromic oil. The essential oil contains sesquiterpene, zingiberene, terpene, camphene and phellandrene. It also constitute linalool, geraniol and cineol. Ginger oleoresin is extracted from powdered dried ginger which contains gingerol, zingerone, shogaol, volatile oil, resins and phenols.

It is used in food preparation, soft drinks, candy and alcoholic beverages. It has carminative and stimulative properties and used in mild diarrhoea, colic and dyspepsia. Also, used for-veterinary purposes.

Zizyphus Adams (Ziziphus Mill.)
Family : Rhamnaceae

The genus Zizyphus consists of forty five species of tropical and subtropical regions.

In India, Z. mauritiana Lamk is grown commercially in different parts of the country. Besides India, this fruit is also grown in Burma, Iran, Syria, Afganisthan, Europe, Australia, America and Africa.

Zizyphus juguba Lam. :

This is a medium sized tree. Leaves are variable. Flowers are greenish yellow in colour. Fruits are green and then turn to pale yellow. Fruits are sweet-sour in taste. The roots go deep into the earth. Bark is dark grey in colour.

It is commonly distributed in different parts of India and also in Garhwal Hills. Commonly known as "Beri", "Ber".

Uses : Fruits are edible, rich in Vitamin C, and contain many rare minerals.

Zizyphus mauritiana Lamk. :

This is a very hardy fruit crop. The tree is thorny with spreading crown and drooping branches. The tree bark is rough, dull black or grey coloured. Leaves are elliptic, oblong and prominently three nerved, tomentose on the under surface. Flowers are greenish yellow in colour, appear in axillary cymes. Fruits are globose or ovoid, red, orange or yellowish in colour when ripe. Fruits are sweet-sour in taste. There are innumerable varieties under this species.

Uses : The fruit contains vitamins A, B and C, sugar, protein and minerals.

In sore throat, a leaf decoction is used as gargle. For dysentery and diarrhoea, docoction of root and bark is prescribed. To treat bleeding gums, leaf decoction is also useful. The seed kernels are considered aphrodisiac. Various products like jam, squash, murabba and chutney are prepared from ripe fruits. Fruits are edible, rich in vitamin C and contain many rare minerals.

Zizyphus nummularia (Burnf.) W. & A. :

This plant species is generally found at Karrapani of Madhya Pradesh. Commonly known as "Ber".

It grows in tropical and subtropical areas ideal.

Uses : The stem bark decoction (about 16 ml) is taken thrice a day for six days in diarrhoea.

Zizyphus xylopyrus (Retz.) Willd. :

This is a large erect shrub. Occur at Rater of areas Madhya Pradesh.

Uses : The leaves (two and half) are chewed thrice a day for 20 days in urinary troubles.

Advice for Improvement of Medicinal plant Industry in Scientific Method.

The Human beings has used plants, Since ancient times to cure diseases and relieve physical sufferings. Understanding of herbal drugs was very innovative and critical from ancient times and this was restricted to the usually medicinal men of the tribal people. Regarding source and use of medicinal plants knowledge was inherited through generations. Several developing countries have at present well organised and deeprooted traditional systems of medicine. Vast natural resources of medicinal plants are available in India. In Ayurvedic, Siddha and Unani Systems of medicine, more than one thousand plants are regularly used which are generally collected in their wild form from forests. A vast untapped heritage of knowledge exists of medicinal plants in this country which can be adopted with advantage in traditional system as well as modern medicine.

It has been observed that a umber of products obtained from plant sources finds a place in the list of two hundred fifty drugs recommended by World Health Organisation. More than four hundred different types of medicinal herbs are marketed commercially in Western Europe and Hamburg. Several hundred genera of plants are used medicinally mostly in herbal products, where either some plant parts or whole plant is used for its extraction. The active constituents isolated from a limited number of plant species are used in some medicines. Certain plant ingredients are also used as a precursor for drug synthesis. Since synthetic products with highly specific actions have become popular in many markets, the medicinal plant markets, in the industrialised countries are shrinking. However, in China, Hongkong, Indonesia, Singapore and Thailand, the trade for these plants is expanding.

The products Identification, which are in good demand in the markets is very much relevant. In medicinal plants the business of trading depends on trust between the suppliers and traders. The traders and users complain Very often about decline in active ingredients and the use of illegal pesticides and other residues.

There is an urgent need to undertake intensive research for tKeproduction and development of these products both in the developing and under-developed countries not only for ensuring general health care of the citizens but also to exploit the export potential of these products.

This is a very good goal to provide each and every human being in this world with minimum standard of health care. This noble goal can be achieved only by developing medicinal products and drugs based on medicinal plants. There is no denying of the fact that the phytopharmaca represent the best single source of therapeutic agents for men.

Identification of plant species and authentication of plant maTerial are very important in this regard. Collection" of germplasm from all over the world and their maintenance and assessment will give further impetus to this programme.

First top priority should be given to the conservation of existing germplasm. On the basis of ethnomedical information, location of indigenous source of germplasm must be determined and protected. Selection of plants which are highly effective or with more and rare active ingredients are to be made.

It is a necessity for preservation of plant germplasm available in different parts of the world for use as genetic resources, the basic raw material for successful drug improvement. For these group of plants Gene Bank will be helpful for characterisation of genotypes pharmacologically and pharmacognostically. Further detailed studies are required on the topics given below :

i. Seed propagated plants. Floral biology.

ii. Suitable species and types for commercial cultivation.

iii. Application of tissue culture technique for large scale multiplication of plants.

iv. Fixation of Standardization of optimum conditions required for germination of seeds.

v. Ecological adaptations and Morphological variation of plants from seedling to harvest stage in order to find out the stage when the active principle content is maximum.

vi. Influencing Ecological factors on growth and active principles content of plants under different altitudes and climates.

vii. Plant protection chemicals, organic. Effect of trace elements and inorganic manures on the variation of active principle content of plants.

viii. For controlling diseases of plants Pathological investigations.

ix. To find out the optimum conditions under which the highest active principles may be obtained by Collection, during, drying and storage of plant parts so as.

x. By sexual method Improvement of plants of hybridization or by mutation through physical mutagens and chemical.

xi. Biotechnology and Genetic manipulation.

xii. Isolation of active ingredients, Chemical studies of plant parts synthetic studies and structural determination.

xiii. Studies of bioactivity, toxicology Pharmacological and teratology.

xiv. Clinical evaluation Drug information new dose forms, quality control, stabilisation and standardization.

xv. Feasiblity studies and Market research.

BIBLIOGRAPHY

Anonymous (1948-1976) : Wealth of India (Raw Materials), vols, 1-11, C.S.I.R., New Delhi.

Anonymous (1970) : Medicinal plants of Nepal, Bull. No.3, Department of Medicinal Plants, Kathmandu, Nepal.

— (1976) : Arunachal, Report of Medico-botanical Survey Team, Cent. Coun. Res. Ind. Med. & Homeo., New Delhi.

Aiyappan, A. (1948) : Report on the Socio-economic studies on the aboriginal tribes of the Province of Madras, Govt. Press, Madras.

Akita, S. (1976) : Photorespiration and its effect on dry matter production, Jap. Agri. Res. Q-10(1) : 1-6.

Anonymous (1958) : List of medicinal plants cultivated in National Botanic Gardens, Bulletin No. 25, Lucknow.

Ahluwalia, K.S. (1952) : Medicinal plants of Kangra Valley, Ind. For. 78(4) : 188-194.

Arora, R.K. (1977) : Jobs-tears (Coix lacryma-Jobi) - a minor food and fodder crop of north-eastern India, Econ. Bot., 31(3) : 358-366.

Agarwal V.S. and S. Saha (1968) : Unreported medicinal plants of India, J. Andhra Pradesh Acad. Sci., 2 : 21-33.

Altschul Siri von Reis (1968) : Useful food plants in herbarium record, Econ. Bot., 22 : 293-296.

Bodding, P.O. (1925) : Studies in Santhal medicine and connected folklore - I. Santhals and disease, Mem. Asiat. Soc. Bengal, 10(1) : 1-132.

— (1927) : Studies in Santhal medicine and connected folklore - II. Santhal medicine, Ibid. 10(2) : 133-426.

Biswas, K. (1956) : Common medicinal plants of Darjeeling and the Sikkim Himalayas, Govt. Printing Press, West Bengal.

Banerjee, M.L. 91955) : Some edible and medicinal plants from East Nepal, J. Bombay nat. Hist. Soc., 53 : 153-155.

Bressers, J. 91951) : The Botany of Ranchi District, Bihar, Govt. of Bihar.

Birdwood, H.M. (1887-1897) : Catalogue of the flora of Mahabaleshwar and Matheran, J. Bombay nat. Hist. Soc., 2 : 107-132, 1887; 10 : 385-430, 899.

Blatter, E. (1926-1935) : Revision of the flora of Bombay Presidency, J. Bombay nat. Hist. Soc., Parts 3-11, 15-19, 21-27, published jointly with C. McCann.

Chopra, R.N., I.C. Chopra and B.S. Verma 91969) : Supplement to Glossary of Indian Medicinal Plants, C.S.I.R., New Delhi.

Chari, R.B. (1975) : Census of India, 1971; Series 1, Paper 1 of 1975.

Chopra, R.N., S.L. Nayar and I.C. Chopra (1956) : Glossary of Indian Medicinal Plants, C.S.I.R., New Delhi.

Chakravarty, H.L. (1975) : Herbal heritage of India, Bull. bot. Soc. Bengal, 29 : 97-103.

Cooke, T. (1901-1908) : The flora of the Presidency of Bombay, I-II, London.

Dastur, J.F. (1975) : Medicinal Plants of India and Pakistan, Bombay.

Deb, D.B. 91968) : Medicinal Plants of Tripura State. Ind. For., 94 : 753-765.

Dixit, R.D. (1975) : Ferns — a much neglected group of medicinal plants, J. Res. Ind. Med., 74-90.

Dhar, M.L., M.M. Dhar, B.N. Dhawan, B.N. Mehrotra and C. Roy (1968) : Screening of Indian plants for biological activity, Pt. I, Indian J. exp. Biol., 6 : 232.

Das, B.L. (1972) : Ranchi District Census Handbook, Census 1971, Series 4, Bihar, Part X, Indian Press (P) Ltd., Allahabad.

Dalton, E.T. (1872) : Descriptive Ethnology of Bengal, Calcutta.

De, J.N. (1968) : Ethnobotany — a new science in India, Sci. Cult., 34(8) : 326-328.

Dastur, J.F. (1964) : Useful plants of India and Pakistan, Taraporevala & Sons, Bombay.

Drury, Col. Heber (1873) : The Useful Plants of India, London.

Effron, D.H. (ed.) (1967) : Ethnopharmacologic search for psychoactive drugs, U.S. Public Health Publ. No. 1645, Washington.

Elwin, V. (1947) : The Maria and their Ghotul, Bombay.

— (1950) : Maria murder and suicide, Bombay.

Francis, W (1908) : Madras District Gazetters, The Nilgiris, Government Press, Madras.

Faulks, P.J. (958) : An Introduction of Ethnobotany, Moredale Publication London.

Gaur, D.S. and L.P. Gupta (1970) : A study on drug evaluation in ancient India, in Advances in Res. in Indian Med., Udupa et. al., (Ed.), Varanasi.

Gunther, Erna (1945) : Ethnobotany of Western Washington, Seattle.

Gupta, S.P. (1963) : An appraisal of Chotanagpur tribal pharmacopoeia, Bull. Bihar trib. Res. Inst., 5(2) : 1-18.

Gupta, S.M. (1971) : Plant Myths and Traditions in India, Leiden, Netherlands.

Griffith Ralph, T.H. (1917) : The Hymns of Atharvaveda, 2, E.J. Lazarus & Co., Varanasi.

Gupta, Rajendra (1961) : Flora of Lam Dal, Ind. For., 87(5) 316-24.

— (1964) : Survey Record of Medicinal and Aromatic plants of Chamba Forest Division, Himachal Pradesh, Ibid., 90(7) : 454-68.

Gupta, Rajkumar (1960) : Some useful and Medicinmal Plants of Nainital in Kumaon Himalayas, Ibid., 59(2) : 309-29.

— (1962) : Medicinal Plants of Western Himalayas, J. agric. Trop. Bot., Applied : 9(1 & 2) : 1-54.

Gowda, M. (1950) : The story of pan chewing in India, Bull. Mus. Leaflet, 14 : 181-214, Harvard University.

Hartwell, J.L. (1967-71) : Plants used against cancer — a survey. Lloydia, 30(4) : 379-436; 31(2) : 71-170, 32(1) : 79-107; 32(2) : 153-205; 32(3) : 247-296; 33(1) : 97-194; 33(3) : 288-392; 34(1) : 103-160; 34(2) : 204-255; 34(3); 310-360; 34(4); 386-425.

Hoffer, A. and H. Osmund (1967) : Hallucinogens, New York.

Harshburger, J.W. (1896) : The purpose of ethnobotany, Bot. Gazette, 31 : 146-154.

Hoffman, J. (1950) : Encyclopaedia Mundarica, Vols. 1-13.

Haines, H.H. (1921-25) : The Botany of Bihar and Orissa, Adlard & Sons and West Newman Ltd., London.

Jain S.K. (1963a) : Studies in Indian ethnobotany — Plants used in medicine by the tribals of Madhya Pradesh, Bull. reg. Res. Lab., Jammu, 1 : 126-128.

Jain, S.K. (1963b) : Studies in Indian ethnobotany ¡ Less known uses of fifty common plants from the tribal areas of Madhya Pradesh, Bull.bot. Surv. India, 5 : 223-226.

Jain, S.K. (1965a) : Medicinal plant lore of the tribals of Bastar, Econ. Bot., 19 : 236-250.

Jain, S.K. (1966) : Observations on ethnobotany of Purulia, West Bengal, Bull. bot. Surv. India, 8 : 237-251.

Jain, S.K. (1967) : Plants in Indian medicine and folklore associated with healing of bones, Indian J. Orthop., 1 : 95-104.

Jain, S.K. (1975) : Medicinal Plants National Book Trust, New Delhi.

Jain, S.K., D.K. Banerjee and D.C. Pal (1973) : Medicinal Plants among certain Adibasis in India, Bull. bot. Surv. India, 15 : 85-91.

Jain, S.K. and C.R. Tarafder (1973) : Native plant remedies for snake-bite among the Adivasis of Central India, Indian med. J., 57 : 307-309.

Jain, S.K. and R.R. Rao (1977) : Handbook of field and herbarium methods, Today & Tomorrow Publishers, New Delhi.

Kortchak, H.P. and L.G. Nickell (1970) : Calvin type carbon dioxide fixation in sugarcane. Stalk parenchyma tissue, Plant Physiol., 45 : 515-516.

Krenzer, E.G. (Jr.), D.N. Moss and R.K. Crookston (1975) : Carbon dioxide compensation points of flowering plants, Plant Physiol., 56 : 194-206.

Krishnaswamy, N. and P. Chanmdrasekharan (1958) : A new species of Oryza L., Madras Agril. J., 45(12) : 471-472.

Kosambi, D.D. (1962) : Myth and Reality, Popular Press, Bombay.

Kirtikar, K.R. and B.D. Basu (1935) : Indian Medicinal plants, Vols. 1-3, Allahabad.

Kanjilal, U.N. et. al. (1934-40) : Flora of Assam, 4 vols., Govt. of Assam, Shillong.

Kazmi, M.A. (1953) : On the drug plants of Kalat State, Pak. J. For., 3(4) : 217-33.

Krishnamurthy, S. (1954) Horticultural and Economic Plants of the Nilgiris, Coimbatore.

Lewin, L. (1964) : Phantastica : Narcotics and Stimulating Drugs, Their Use and Abuse, New Delhi.

Laszlo, Henry G. de (1951) : Library of Medicinal Plants, Heffer & Sons Ltd., London.

Luiz, A.A.D. (1962) : Tribes of Kerala, Bharatiya Adimjati Sevak Sangh, New Delhi.

Madhihasan, S. (1963) : Identifying Soma as ephedra, Pak. J. For., 13(4) : 370-371.

Minnis, E. Paul (1976) : Directory of Ethnobotanists, Univ. of Michigan, USA.

Macdonell, A.A. (1958) : Vedic Index of Names and Subjects, 1 : 44, John Murray & Co., London.

Nairne, E.K. (1894) : The Flowering Plants of Western India, London.

Om Prakash (1961) : Food Habits (Food and Drinks in ancient India from earliest times to about 1200 A.D.), Delhi.

Pattnaik, H. (1956) : Some useful weeds in and around Cuttack, J. Bombay nat. Hist. Soc., 54 : 141-152.

Prasad, N. (1961) : Land and People of Tribal Bihar, Ranchi.

Prain, D. (1903) : Bengal Plants, Vols. 1-2, Calcutta.

Randhawa, M.S. (1964) : The Cult of Trees and Worship in Buddhist and Hindu Sculpture. All India Fine Arts and Crafts Society, New Delhi.

Roy, J.K. and R.K. Rao (1957) : Investigations on the diet of the Muria of Bastar District, Bull. Dep. Anthrop. Govt. of India, 6 : 33-45.

Rao, R.R. (1977) : Medicobotany of Mysore plants, J. Res. Indian med. Hom., 12 : 53-58.

Singh, H.B. and R.K. Arora (1978) : Wild Edible Plants of India, Ind. Coun. Agr. Res., New Delhi.

Singh, M.P. (1986) : Flora of Patna, International Publisher, New Delhi.

Shah, N.C. and M.C. Joshi (1971) : An ethnobotanical study of Kumaon region of India, Econ. Bot., 25(4) : 414-422.

Sithole, R.V. (1976) : Plants represented in ancient Indian sculpture, Geophysics, 6(1) : 15-26.

Sukumardev, V.P. (1976) : Rice in Ayurveda, Kalpadhenu, 3(4) : 157-158.

Santapau, H. (1953) : The flora of Khandala on the Western Ghats of India, Rec. bot. Surv. India, 16(1).

Todd, James (1829) : Annals and Antiquities of Rajasthan, Routledge & Kegan Paul, London.

Tripathi, R.N. Vidyarthi and U.S. Tripathi (1978) : Ayurvedyeea-Itihaas, Sachitra Ayurveda, May : 280-284.

Vidyalankar, Jai Dev (1959) : Carak Samhita, Moti Lal Banarsi Das, Varanasi.

Watts, G. (1889-99) : A Dictionary of the Economic Products of India, Vols. 1-6, London.

Wasson, R.G. (1971) : The Soma of the Rgveda; What was it ? J. Amer. orient. Soc., 91(2) : 169-187.

Watt, J.M. (1968) In Tony Swan (Ed.) : Magic and Witchcraft in relation to Plants and Folk Medicine, Plants in the Development of Modern Medicine, Harv. Univ. Press, Massachusetts, pp. 75, 79.

Wilson, H.H. (1961) : Vishnu Purana, Punthi Pustak, Calcutta.

List of Figures & Plates

Plate-11

Plate-12

Plate -13

Plate -14

Plate-15

Plate-16

INDEX